Innovative and Applied Research on Platinum-Group and Rare Earth Elements

Innovative and Applied Research on Platinum-Group and Rare Earth Elements

Dedicated to the Work and Memory of Dr. Demetrios G. Eliopoulos, IGME (Greece)

Special Issue Editors

Maria Economou-Eliopoulos
Federica Zaccarini
Giorgio Garuti

MDPI • Basel • Beijing • Wuhan • Barcelona • Belgrade • Manchester • Tokyo • Cluj • Tianjin

Special Issue Editors

Maria Economou-Eliopoulos
University of Athens
Greece

Federica Zaccarini
University of Leoben
Austria

Giorgio Garuti
University of Leoben
Austria

Editorial Office
MDPI
St. Alban-Anlage 66
4052 Basel, Switzerland

This is a reprint of articles from the Special Issue published online in the open access journal *Minerals* (ISSN 2075-163X) (available at: https://www.mdpi.com/journal/minerals/special_issues/ PGE_REE).

For citation purposes, cite each article independently as indicated on the article page online and as indicated below:

LastName, A.A.; LastName, B.B.; LastName, C.C. Article Title. *Journal Name* **Year**, *Article Number*, Page Range.

ISBN 978-3-03936-597-5 (Hbk)
ISBN 978-3-03936-598-2 (PDF)

Cover image courtesy of Federica Zaccarini.

Contents

About the Special Issue Editors

Maria Economou-Eliopoulos Geologist and Chemist, was a Professor (1993–2014) of Economic Geology and Geochemistry and Director of the Postgraduate programme "Applied Environmental Geology" (2000–2014), and is currently Professor Emeritus at the National and Kapodistrian University of Athens. Her scientific interests include geochemistry of ore deposits, from the ore-forming processes to the environmental problems. Her post-doctoral research at the University of Toronto, Imperial College, U.S. Geological Survey (Reston) and Universities of Europe was mainly focused on the platinum-group elements in mafic-ultramafic rocks and associated deposits (chromite, sulfides, magnetite, and laterites), and porphyry Cu-Au±Pd±Pt systems, which are her main research topic currently. Maria is the author of more than 150 publications in scientific journals, thematic issues, chapters in books, monographs, a fellow of the SEG (Society-Economic-Geologists), evaluator of research proposals (national and international), a reviewer of scientific papers in more than 30 leading geological journals, has over 2500 citations, and an h-index of 29. Her distinctions include awards from the National Academy of Athens (2008, 2001, 1991, 1986, and 1981) for the discovery of a new mineral (Theophrastite) and palladium in a porphyry-Cu deposit of Greece and constraints on the genesis of ore deposits.

Federica Zaccarini from Modena, Italy obtained an Italian-European habilitation as a full Professor in Mineralogy, Petrology, Geochemistry, Volcanology, and Ore Deposits. She is a Senior Scientist at the University of Leoben (Austria) and Head of the E.F. Stumpfl Electron Laboratory. Federica has received a large number of international awards including the following:

- 1998 Ph.D. Student grant, Society of Economic Geology (SEG), USA
- 2001 Johndino Nogara, Italian Society of Mineralogy and Petrology, ITALY
- 2004–2006 FWF Lise Meitner fellowship, University of Leoben, AUSTRIA
- 2005 Johndino Nogara, Italian Society of Mineralogy and Petrology, ITALY
- 2006 Juan de la Cierva Fellowship, University of Granada, SPAIN
- 2007 Ramon y Cajal fellowship, SPAIN.

Her main research interests include the characterization of opaque mineral species, particularly using optical and electron microscopes, and electron microprobe analysis, investigation and characterization of accessory minerals, analytical techniques for the noble metals and their geochemistry, mineralogy, and origin of ore deposits related with mafic-ultramafic rocks, platinum-group minerals in chromitites from ophiolite, stratiform complexes and Ural-Alaskan type intrusion, and sulfides deposits and sulfide deposits in ophiolitic complexes. Her work has been recognized by the international scientific community and the International Mineralogical Association has adopted a new mineral name in her honor (Zaccarinite). Federica is the author of more than 100 papers and several book chapters. She has over 2000 citations and an *h*-index of 24.

Giorgio Garuti from Modena, Italy was a Professor of Ore Deposits at the University of Modena from 1982 to 2009. Currently, he is a lecturer at the University of Leoben (Austria). Giorgio is an internationally recognized expert in petrology, geochemistry, and mineralogy of mafic-ultramafic rocks and associated ore deposits. His main research interests include the mineralogy and geochemistry of noble metals, especially platinum group elements and their deposits, such as magmatic sulfides and chromitites. Due to his experience, Giorgio worked intensively in several areas worldwide (North–Central–South America, Africa, Russia, Europe, and Asia). His current research includes the investigation of ore deposits related to ophiolites, with a special focus on the presence of gold and uranium minerals. He was the Associate Editor of the *Canadian Mineralogist* and presently he is a member of the Editorial Board of the journals *Geologica Acta* and *Ofioliti*. His work has been recognized by the international scientific community and the International Mineralogical Association has adopted a new mineral name in his honor (Garutite). Giorgio is the author of more than 100 papers and several book chapters. He has over 2000 citations and an *h*-index of 29.

Editorial

Editorial for the Special Issue "Innovative and Applied Research on Platinum-Group and Rare Earth Elements"

Maria Economou-Eliopoulos [1],*, Federica Zaccarini [2] and Giorgio Garuti [2]

[1] Department of Geology and Geoenvironment, University of Athens, 15784 Athens, Greece
[2] Department of Applied Geological Sciences and Geophysics, University of Leoben, A-8700 Leoben, Austria; federica.zaccarini@unileoben.ac.at (F.Z.); giorgio.garuti1945@gmail.com (G.G.)
* Correspondence: econom@geol.uoa.gr

Received: 19 May 2020; Accepted: 27 May 2020; Published: 28 May 2020

This Special Issue "Innovative and Applied Research on Platinum-group and Rare Earth Elements" is dedicated to the work and memory of Demetrios Eliopoulos, IGME (Institute of Geology and Mineral Exploration), Greece who passed away on 19 April 2019.

Demetrios Eliopoulos (1947–2019) started his research work (nearly 40 years ago) on base and precious metals with his Master's thesis (Western University, Canada) and Ph.D thesis (Southampton, UK). Over the course of his career at IGME (Greek Institute of Geology and Mineral Exploration), Demetrios was the IGME representative of R&D, European Commission Projects, Head of Economic Geology, and a contributor to numerous fundamental research reports and publications. His main distinctions include: his membership on the Editorial Board of Mineralium Deposita, being a Fellow of the SEG (Society-Economic-Geologists) and SGA (Society for Geology Applied to Mineral Deposits), and the Award from the Academy of Athens (discovery of palladium in a porphyry-Cu deposit of Greece). Demetrios was the chairman of the Organizing Committee of the 7th Biennial SGA Meeting, held in Athens, Greece, August 24–28, 2003. He was involved in fruitful scientific collaboration over the years with his wife and colleague Maria Economou-Eliopoulos. Their work was recognized by the international community, and the IMA has adopted a new mineral name in honor of Demetrios and Maria (Eliopoulosite, IMA 2019-96, this volume). This Special Issue contains a series of papers, including two reporting on the finding of new minerals—Grammatikopoulosite, NiVP, a new phosphide, and Eliopoulosite, V_7S_8, a new pyrrhotite like mineral—both from chromitites of the Othrys ophiolite complex, approved by the Commission of New Minerals, Nomenclature, and Classification of the International Mineralogical Association (IMA 2019-090 and 2019-96) [1,2]. The geological environment, mode of the occurrence, physical and chemical properties, and crystallographic data are provided. Problems concerning the distribution, mineralogy and field relationships of PGE-enriched ores, which are important for our understanding of the metallogenic controls on the concentration of PGE in ophiolite complexes and their exploration, are addressed in three papers. Kiseleva et al. [3] present the first detailed data on the chromitites and PGE mineralization in the Ulan-Sar'dag ophiolite located in the Central Asian Fold Belt (East Sayan, Russia), and propose a scheme for the sequence of the formation and transformation of the PGM at various stages of the evolution of that ophiolite complex. Kravtsova et al. [4] present the peculiarities of the distribution and binding forms of PGE in the arsenopyrites and pyrites of the Natalkinskoe gold ore deposit in NE Russia. They conclude that the possibility of their extraction from that type of ore may significantly increase the quality and value of the extracted materials at the Natalkinskoe deposit. Zaccarini and Garuti [5] describe the occurrence of zoned laurite found in the Merensky Reef of the Bushveld layered intrusion, South Africa. Specifically, zoned laurites from the Merensky Reef are characterized by textural position, composition, and zoning—suggesting that they are "hydrothermal" in origin, having crystallized in the presence of a Cl- and As-rich hydrous solution, at temperatures much lower than those typical of the precipitation of magmatic laurite. The paper by Sadeghi et al. [6] deals with the study of the rare earth elements (REE) in till and bedrock, as well as in mineral deposits associated with apatite-iron oxide mineralizations, various skarn deposits, hydrothermal deposits, and alkaline-carbonatite intrusions in Sweden. In this study, analytical data of samples collected from REE mineralizations during the EURARE (www.eurare.org) project are compared with bedrock and till REE geochemistry—both sourced from databases available at the Geological Survey of Sweden (www.sgu.se). These results are useful in the assessment of REE mineral potential in areas where REE mineralizations are poorly explored or even undiscovered. The paper by Stouraiti et al. [7] presents the effect of mineralogy on the beneficiation of REE from heavy mineral sands from Nea Peramos, Kavala, northern Greece. They provide the characterization of the fractions separated by a magnetic separator using X-ray powder diffraction (XRD) and bulk Inductively Coupled Plasma and Mass Spectrometry (ICP-MS) chemical analyses, and show that at high intensities of magnetic separation, they were strongly enriched in Light Rare Earth Elements (LREE) compared to the non-magnetic fraction. This new information can contribute to the optimization of the beneficiation process, that can be applied for REE recovery from Heavy Mineral (HM) black sands. A group of papers deal with the genetic significance of a spectrum of trace elements and stable chromium isotopes in chromitites. The paper by Economou-Eliopoulos et al. [8] on the factors controlling the chromium isotope compositions in podiform chromitites is an application of Cr stable isotope (δ^{53}Cr values) compositions to the investigation of magmatic and post-magmatic effects on chromitites associated with ophiolite complexes. Chromitites from the Balkan Peninsula depict a wide range of δ^{53}Cr values. Signatures range from positively to slightly negatively fractionated δ^{53}Cr values—even in individual relatively large deposits. Positively fractionated δ^{53}Cr values of all chromitite samples from Othrys and of high-Al chromitites from Skyros, coupled with a negatively correlated trend between δ^{53}Cr and Cr/(Cr + Al), may reflect the control of δ^{53}Cr by degree of partial melting and by magma fractionation. This is best exemplified by high-Al chromitites from the cumulate sequence of the Vourinos complex. The paper by Eliopoulos and Eliopoulos [9] on the factors controlling the gallium preference in high-Al chromitites presents geochemical—including Ga—and mineral chemistry data on chromitites associated with the ophiolite complexes of Greece and elsewhere. Potential factors controlling the lower Ga contents in high-Cr chromitites compared to high-Al ones are the composition of the parent magma, temperature, redox conditions, the disorder degree of spinels, and the ability of Al^{3+} to occupy both octahedral and tetrahedral sites, in contrast to the competing Cr^{3+} that can occupy only octahedral sites (due to its electronic configuration) while the Ga^{3+} shows a strong preference on tetrahedral sites. The paper by Eliopoulos and Economou-Eliopoulos [10] on the trace element distribution in magnetite separates of varying origin and genetic/exploration significance provides SEM/EDS and ICP-MS data on representative magnetite samples covering various geotectonic settings and rock-types, such as calc-alkaline and ophiolitic rocks, porphyry-Cu deposits, skarn-type, ultramafic lavas, black coastal sands, and metamorphosed Fe–Ni-laterites deposits. They concluded that, despite the potential re-distribution of trace elements—including Rare Earth Elements (REE) and Platinum Group Elements (PGE) in magnetite-bearing deposits—the data may provide valuable evidence for their origin and exploration. In particular, the trace element content and the presence of abundant magnetite provide valuable evidence for discrimination between Cu–Au–Pd-Pt- porphyry systems and those lacking precious metals. The paper by Gray and Van Rythoven [11] deals with a comparative study between porphyry-Cu deposits of the Western Cordillera, using portable X-ray fluorescence analysis (pXRF) and optical microscopy. Although the proposed research methodology cannot reach the levels of precision and accuracy required to meet standards for mineral resources, the combined use of pXRF in the field with optical microscopy provides a fast and cost-effective method for the exploration of large unexplored porphyry intrusions characterized by extensive haloes of alteration. The application of the proposed methodology to porphyry intrusions can influence decisions regarding geological mapping, cross-sections, and the selection of samples for a more detailed petrographic and geochemical investigation of porphyry-type deposits, as well as influence exploration for potential sources of Cu, Mo and precious metals (Au and platinum-group elements). The paper by Eliopoulos et al. [12] is focused on sulphides of both Cyprus-type and Fe-Cu-Co-Zn sulphides associated with magnetite within gabbro, close to its tectonic contact with serpentinized harzburgite of the Pindos ophiolite complex. The higher Zn, Se, Mo, Au, Ag, Hg, and Sb contents and lower Ni contents in the Pindos compared to the Othrys sulphides may reflect inheritance of a primary magmatic signature. Relatively high Cu/(Cu+Ni) and Pt/(Pt+Pd), and low Ni/Co ratios for sulphides from the deeper part of the complex suggest either no magmatic origin, or a complete transformation of preexisting magmatic assemblages. However, the recorded textural and mineralogical features of both sulphide types from Pindos resemble the Fe-Cu-Zn-Co-Ni mineralization reported in other mafic–ultramafic-hosted VMS deposits, which have shown mineralogical, compositional and textural analogies with present-day counterparts on ultramafic-rich substrate seafloor VMS deposits formed by hydrothermal processes.

Author Contributions: M.E.-E., F.Z. and G.G. wrote the editorial. All authors have read and agreed to the published version of the manuscript.

Conflicts of Interest: The authors declare no conflict of interest.

References

1. Bindi, L.; Zaccarini, F.; Ifandi, E.; Tsikouras, B.; Stanley, G.; Garuti, G.; Mauro, D. Grammatikopoulosite, NiVP, a New Phosphide from the Chromitite of the Othrys Ophiolite, Greece. *Minerals* **2020**, *10*, 131.
2. Bindi, L.; Zaccarini, F.; Bonazzi, P.; Grammatikopoulos, T.; Tsikouras, B.; Stanley, C.; Garuti, G. Eliopoulosite, V₇S₈, A New Sulfide from the Podiform Chromitite of the Othrys Ophiolite, Greece. *Minerals* **2020**, *10*, 245. [CrossRef]
3. Kiseleva, O.N.; Airiyants, E.V.; Belyanin, D.K.; Zhmodik, S.M. Podiform Chromitites and PGE Mineralization in the Ulan-Sar'dag Ophiolite (East Sayan, Russia). *Minerals* **2020**, *10*, 141. [CrossRef]
4. Kravtsova, R.G.; Tauson, V.L.; Makshakov, A.S.; Bryansky, N.V.; Smagunov, N.V. Platinum Group Elements in Arsenopyrites and Pyrites of the Natalkinskoe Gold Deposit (Northeastern Russia). *Minerals* **2020**, *10*, 318.
5. Zaccarini, F.; Garuti, G. Zoned Laurite from the Merensky Reef, Bushveld Complex, South Africa: "Hydrothermal" in Origin? *Minerals* **2020**, *10*, 373. [CrossRef]
6. Sadeghi, M.; Arvanitidis, N.; Ladenberger, A. Geochemistry of Rare Earth Elements in Bedrock and Till, Applied in the Context of Mineral Potential in Sweden. *Minerals* **2020**, *10*, 365. [CrossRef]
7. Stouraiti, C.; Angelatou, V.; Petushok, S.; Soukis, K.; Eliopoulos, D. Effect of mineralogy on the beneficiation of REEs from heavy mineral sands: The case of Nea Peramos, Kavala, northern Greece. *Minerals* **2020**, *10*, 387. [CrossRef]
8. Economou-Eliopoulos, M.; Frei, R.; Ioannis, I. Factors Controlling the Chromium Isotope Compositions in Podiform Chromitites. *Minerals* **2020**, *10*, 10. [CrossRef]
9. Eliopoulos, I.P.D.; Eliopoulos, G.D. Factors Controlling the Gallium Preference in High-Al Chromitites. *Minerals* **2020**, *9*, 623. [CrossRef]
10. Eliopoulos, D.G.; Economou-Eliopoulos, M. Trace Element Distribution in Magnetite Separates of Varying Origin: Genetic and Exploration Significance. *Minerals* **2020**, *9*, 759. [CrossRef]
11. Gray, C.A.; Van Rythoven, A.D. A Comparative Study of Porphyry-Type Copper Deposit Mineralogies by Portable X-ray Fluorescence and Optical Petrography. *Minerals* **2020**, *10*, 431. [CrossRef]
12. Eliopoulos, D.G.; Economou-Eliopoulos, M.; Economou, G.; Skounakis, V. Mineralogical and Geochemical Constraints on the Origin of Mafic–Ultramafic-Hosted Sulphides: The Pindos Ophiolite Complex. *Minerals* **2020**, *10*, 454.

 *minerals*

Article

Eliopoulosite, V₇S₈, A New Sulfide from the Podiform Chromitite of the Othrys Ophiolite, Greece

Luca Bindi [1,*], Federica Zaccarini [2], Paola Bonazzi [1], Tassos Grammatikopoulos [3], Basilios Tsikouras [4], Chris Stanley [5] and Giorgio Garuti [2]

[1] Dipartimento di Scienze della Terra, Università degli Studi di Firenze, I-50121 Florence, Italy; paola.bonazzi@unifi.it

[2] Department of Applied Geological Sciences and Geophysics, University of Leoben, A-8700 Leoben, Austria; federica.zaccarini@unileoben.ac.at (F.Z.); giorgio.garuti1945@gmail.com (G.G.)

[3] SGS Canada Inc., 185 Concession Street, PO 4300, Lakefield, ON K0L 2H0, Canada; Tassos.Grammatikopoulos@sgs.com

[4] Faculty of Science, Physical and Geological Sciences, Universiti Brunei Darussalam, BE 1410 Gadong, Brunei Darussalam; basilios.tsikouras@ubd.edu.bn

[5] Department of Earth Sciences, Natural History Museum, London SW7 5BD, UK; c.stanley@nhm.ac.uk

* Correspondence: luca.bindi@unifi.it; Tel.: +39-055-275-7532

Received: 22 February 2020; Accepted: 6 March 2020; Published: 8 March 2020

Abstract: The new mineral species, eliopoulosite, V₇S₈, was discovered in the abandoned chromium mine of Agios Stefanos of the Othrys ophiolite, located in central Greece. The investigated samples consist of massive chromitite hosted in a strongly altered mantle tectonite, and are associated with nickelphosphide, awaruite, tsikourasite, and grammatikopoulosite. Eliopoulosite is brittle and has a metallic luster. In plane-reflected polarized light, it is grayish-brown and shows no internal reflections, bireflectance, and pleochroism. It is weakly anisotropic, with colors varying from light to dark greenish. Reflectance values of mineral in air (R_o, $R_{e'}$ in %) are: 34.8–35.7 at 470 nm, 38–39 at 546 nm, 40–41.3 at 589 nm, and 42.5–44.2 at 650 nm. Electron-microprobe analyses yielded a mean composition (wt.%) of: S 41.78, V 54.11, Ni 1.71, Fe 1.1, Co 0.67, and Mo 0.66, totali 100.03. On the basis of $\Sigma_{atoms} = 15$ apfu and taking into account the structural data, the empirical formula of eliopoulosite is $(V_{6.55}Ni_{0.19}Fe_{0.12}Co_{0.07}Mo_{0.04})_{\Sigma = 6.97}S_{8.03}$. The simplified formula is $(V, Ni, Fe)_7S_8$ and the ideal formula is V₇S₈, which corresponds to V 58.16%, S 41.84%, total 100 wt.%. The density, based on the empirical formula and unit-cell volume refined form single-crystal structure XRD data, is 4.545 g·cm⁻³. The mineral is trigonal, space group $P3_221$, with $a = 6.689(3)$ Å, $c = 17.403(6)$ Å, $V = 674.4(5)$ Å³, $Z = 3$, and exhibits a twelve-fold superstructure ($2a \times 2a \times 3c$) of the NiAs-type subcell with V-atoms octahedrally coordinated by S atoms. The distribution of vacancies is discussed in relation to other pyrrhotite-like compounds. The mineral name is for Dr. Demetrios Eliopoulos (1947–2019), a geoscientist at the Institute of Geology and Mineral Exploration (IGME) of Greece and his widow, Prof. Maria Eliopoulos (nee Economou, 1947), University of Athens, Greece, for their contributions to the knowledge of ore deposits of Greece and to the mineralogical, petrographic, and geochemical studies of ophiolites, including the Othrys complex. The mineral and its name have been approved by the Commission of New Minerals, Nomenclature, and Classification of the International Mineralogical Association (No. 2019-96).

Keywords: eliopoulosite; sulfide; chromitite; Agios Stefanos mine; Othrys; ophiolite; Greece

1. Introduction

Only eight minerals containing V and S are in the list of valid species approved by the International Mineralogical Association (IMA). They include, in alphabetic order: colimaite, K₃VS₄ [1]; colusite,

$Cu_{13}VAs_3S_{16}$ [2]; germanocolusite, $Cu_{13}VGe_3S_{16}$ [3]; merelaniite, $Mo_4Pb_4VSbS_{15}$ [4]; nekrasovite, $Cu_{13}VSn_3S_{16}$ [5]; patronite, VS_4 [6]; stibiocolusite, $Cu_{13}V(Sb,Sn,As)_3S_{16}$ [7]; and sulvanite, Cu_3VS_4 [8]. All of them, with the exception of patronite, a mineral discovered in 1906 in the Minas Ragra of Peru [6], are sulfides or sulfosalts characterized by a complex composition including other metals besides vanadium.

During a recent investigation of the heavy mineral concentrates from a chromitite collected in the Othrys ophiolite (central Greece), three new minerals were discovered. Two of them are phosphides, namely tsikourasite, $Mo_3Ni_2P_{1+x}$ [9] and grammatikopoulosite, NiVP [10]. A chemical and structural study revealed the third mineral to be a new sulfide, trigonal in symmetry and having the ideal formula V_7S_8. The mineral and its name were approved by the Commission of New Minerals, Nomenclature and Classification of the International Mineralogical Association (No. 2019-096). The new mineral has been named after Dr. Demetrios Eliopoulos (1947–2019), a geoscientist at the Institute of Geology and Mineral Exploration (IGME) of Greece and his widow, Prof. Maria Eliopoulos (nee Economou, 1947), University of Athens, Greece, for their contributions to the knowledge of ore deposits of Greece and to the mineralogical, petrographic, and geochemical studies of ophiolites, including the Othrys complex. Holotype material is deposited in the Mineralogical Collection of the Museo di Storia Naturale, Università di Pisa, Via Roma 79, Calci (Pisa, Italy), under catalogue number 19911 (same type specimen of grammatikopoulosite).

2. Geological Background and Occurrence of Eliopoulosite

Eliopoulosite was discovered in a heavy-mineral concentrate obtained from massive chromitite collected in the mantle sequence of the Mesozoic Othrys ophiolite, located in central Greece (Figure 1A). The geology, petrography, and geodynamic setting of the Otrhys ophiolite have been discussed by several authors [11–24]. The studied sample was collected in the abandoned chromium mine of Agios Stefanos, located a few km southwest of the Domokos village (Figure 1B). In the studied area, a mantle tectonite composed of plagioclase lherzolite, harzburgite, and minor harzburgite-dunite occurs in contact with rocks of the crustal sequence (Figure 1C). The investigated chromitite was collected from the harzburgite-dunite that is cut across by several dykes of gabbro (Figure 1C).

The ophiolite of Othrys is a complete but dismembered suite (Mirna Group) and consists of three structural units: the uppermost succession with variably serpentinized peridotites, which is structurally bounded by an ophiolite mélange; the intermediate Kournovon dolerite, including cumulate gabbro and local rhyolite; and the lower Sipetorrema Pillow Lava unit also including basaltic flows, siltstones, and chert. The Mirna Group constitutes multiple inverted thrust sheets, which were eventually obducted onto the Pelagonian Zone, during the Late Jurassic-Early Cretaceous [11–14]. Three types of basalts with different geochemical signatures have been described: i) alkaline within-plate (WPB), occurring in the mélange, which is related to oceanic seamounts or to ocean-continent transition zones; ii) normal-type mid-ocean ridge (N-MORB); and iii) low-K tholeiitic (L-KT) rocks, which are erupted close to the rifted margin of an ocean-continent transition zone. Radiolarian data from interbedded cherts suggest Middle to Late Triassic ages [15]. The Othrys ophiolite is structurally divided into the west and east Othrys suites, which are thought to derive from different geotectonic environments. The former is related to an extension regime (back-arc basin or MORB [16–18]), while the latter formed in a supra-subduction zone (SSZ) setting [17–19]. This difference is reflected in the contrasting geochemical compositions of their ultramafic rocks, as well as in the diverse platinum group minerals (PGM) assemblages of their host chromitites [20,21], which suggest large mantle heterogeneities in this area. Conflicting views with regards to the evolution of the Othrys ophiolite suggest the involvement of Pindos [11,22,23] or Vardar (Axios) [15] oceanic domains.

The studied spinel-supergroup minerals that host eliopoulosite can been classified as magnesiochromite, although their composition is rather heterogeneous [9,10,25] with the amounts of Cr_2O_3 (44.96–51.64 wt.%), Al_2O_3 (14.18–20.78 wt.%), MgO (13.34–16.84 wt.%), and FeO (8.3–13.31 wt.%) varying significantly through the sampled rock. The calculated Fe_2O_3 is relatively high, and ranges from 6.72 to 9.26 wt.%. The amounts of MnO (0.33–0.60 wt.%), V_2O_3 (0.04–0.30 wt.%), ZnO (up to 0.07 wt.%), and NiO (0.03–0.24 wt.%) are low. The maximum TiO_2 content is 0.23 wt.%, which is typical for the podiform chromitites hosted in the mantle sequence of ophiolite complexes.

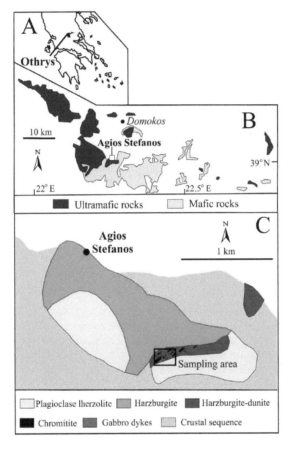

Figure 1. Location of the Othrys complex in Greece (**A**); general geological map of the Othrys ophiolite showing the location of the Agios Stefanos chromium mine (**B**) and (**C**) detailed geological setting of the Agios Stefanos area (modified after [10] and [5]).

3. Analytical Methods

The heavy minerals were concentrated at SGS Mineral Services, Canada, following the methodology described by several authors [9,10,21,25,26]. After the concentration process, the heavy minerals were embedded in epoxy blocks and then polished for mineralogical examination.

Quantitative chemical analyses and acquisition of back-scattered electron images of eliopoulosite were performed with a JEOL JXA-8200 electron-microprobe, installed in the E. F. Stumpfl laboratory, Leoben University, Austria, operating in wavelength dispersive spectrometry (WDS) mode. Major and minor elements were determined at 20 kV accelerating voltage and 10 nA beam current, with 20 s as the counting time for the peak and 10 s for the backgrounds. The beam diameter was about 1 μm in size. For the WDS analyses, the following lines and diffracting crystals were used: S = ($K\alpha$, PETJ), V, Ni, Fe, Co = ($K\alpha$, LIFH), and Mo = ($L\alpha$, PETJ). The following standards were selected: metallic vanadium for V, pyrite for S and Fe, millerite for Ni, molybdenite for Mo, and skutterudite for Co.

The ZAF correction method was applied. Automatic correction was performed for interference Mo-S. Representative analyses of eliopoulosite are listed in Table 1.

Table 1. Electron microprobe analyses (wt.%) of eliopoulosite.

Analysis	S	V	Ni	Fe	Co	Mo	Total
1	41.34	53.08	1.24	0.87	0.48	0.55	99.01
2	41.35	53.45	1.33	0.92	0.52	0.57	99.29
3	41.36	53.55	1.39	0.97	0.55	0.57	99.37
4	41.39	53.71	1.41	1.00	0.55	0.57	99.43
5	41.42	53.90	1.51	1.01	0.61	0.58	99.72
6	41.44	53.91	1.54	1.09	0.63	0.59	99.75
7	41.57	53.97	1.55	1.09	0.63	0.59	99.77
8	41.65	54.03	1.58	1.09	0.63	0.60	99.82
9	41.67	54.03	1.65	1.10	0.64	0.63	99.86
10	41.73	54.06	1.67	1.11	0.64	0.64	99.92
11	41.76	54.09	1.69	1.11	0.65	0.65	100.18
12	41.82	54.16	1.70	1.12	0.68	0.67	100.20
13	41.90	54.21	1.72	1.12	0.69	0.68	100.26
14	41.94	54.25	1.72	1.13	0.72	0.68	100.27
15	41.99	54.40	1.94	1.16	0.72	0.69	100.28
16	42.03	54.40	1.99	1.16	0.73	0.69	100.31
17	42.12	54.56	2.02	1.18	0.76	0.70	100.78
18	42.28	54.64	2.13	1.19	0.78	0.75	100.78
19	42.33	54.80	2.22	1.21	0.79	0.81	100.84
20	42.62	55.02	2.27	1.47	0.91	0.90	100.84
average	41.78	54.11	1.71	1.1	0.67	0.66	100.03

A small grain of eliopoulosite was hand-picked from the polished section under a reflected light microscope. The crystal (about 80 µm in size) was carefully and repeatedly washed in acetone and mounted on a 5 µm-diameter carbon fiber, which was, in turn, attached to a glass rod in preparation of the single-crystal X-ray diffraction measurements.

Single-crystal X-ray diffraction data were collected at the University of Florence (Italy) using a Bruker D8 Venture equipped with a Photon II CCD detector, with graphite-monochromatized MoKα radiation (λ = 0.71073 Å). Intensity data were integrated and corrected for standard Lorentz-polarization factors with the software package *Apex*3 [27,28]. A total of 1289 unique reflections was collected up to 2θ = 62.24°.

The reflectance measurements on eliopoulosite were carried out using a WTiC standard and a J&M TIDAS diode array spectrophotometer at the Natural History Museum of London, UK.

4. Physical and Optical Properties

In the polished section, eliopoulosite occurs as tiny crystals (from 5 µm up to about 80 µm) and is anhedral to subhedral in habit. It consists of polyphase grains associated with other minerals, such as tsikourasite, nickelphosphide, awaruite, and grammatikopoulosite (Figure 2).

In plane-reflected polarized light, eliopoulosite is grayish-brown and shows no internal reflections, bireflectance, and pleochroism. It is weakly anisotropic, with colors varying from light to dark greenish. Reflectance values of the mineral in air (R in %) are listed in Table 2 and shown in Figure 3.

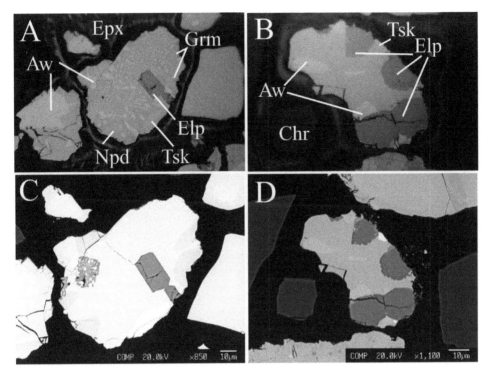

Figure 2. Digital image in reflected plane polarized light (**A,B**) and back-scattered electron image (**C,D**) showing eliopoulosite from the chromitite of Agios Stefanos. Abbreviations: Elp = eliopoulosite, Grm = grammatikopoulosite, Tsk = tsikourasite, Aw = awaruite, Npd = nickelphosphide, Chr = chromite, Epx = epoxy.

Table 2. Reflectance values of eliopoulosite, those required by the Commission on Ore Mineralogy (COM) are given in bold.

λ (nm)	R_o (%)	$R_{e'}$ (%)
400	32.0	33.1
420	32.7	33.7
440	33.5	34.5
460	34.3	35.3
470	**34.8**	**35.7**
480	35.2	36.1
500	36.1	37.0
520	36.9	37.8
540	37.7	38.7
546	**38.0**	**39.0**
560	38.6	39.7
580	39.5	40.8
589	**40.0**	**41.3**
600	40.5	41.9
620	41.2	42.9
640	42.1	43.8
650	**42.5**	**44.2**
660	42.8	44.6
680	43.5	45.5
700	44.3	46.2

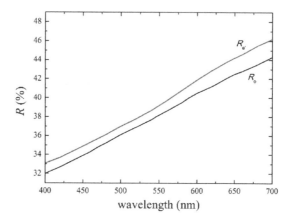

Figure 3. Reflectance data for eliopoulosite.

Due to the small amount of available material, density was not measured. The calculated density was = 4.545 g·cm^{-3}, based on the empirical formula and the unit-cell volume refined from single-crystal XRD data.

5. X-Ray Crystallography and Chemical Composition

The mineral is trigonal, with a = 6.689(3) Å, c = 17.403(6) Å, V = 674.4(5) Å^3, and Z = 3. The reflection conditions (00l: l = 3n), together with the observed R_{int} in the different Laue classes, point unequivocally to the choice of the space group $P3_221$. The structure solution was then carried out in this space group. The positions of most of the atoms (all the V positions and most of the S atoms) were determined by means of direct methods [29]. A least-squares refinement on F^2 using these heavy-atom positions and isotropic temperature factors produced an R factor of 0.123. Three-dimensional difference Fourier synthesis yielded the position of the remaining sulfur atoms. The program Shelxl-97 [30] was used for the refinement of the structure. The site occupancy factor (s.o.f.) at the cation sites was allowed to vary (V vs. structural vacancy) using scattering curves for neutral atoms taken from the International Tables for Crystallography [30]. Four V sites (i.e., V1, V2, V3, and V4) were found to be partially occupied by vanadium (75%), while V5 and V6 were found to be fully occupied by V and fixed accordingly. Sulfur atoms were found on four fully occupied general positions leading to an ideal formula V$_7$S$_8$. Given the almost identical partial occupancy of four V sites (i.e., 75%), we carefully checked either the possible presence of twinning or the acentricity of the model. The lack of the inversion center was confirmed using the Flack parameter in Shelxl (0.07(1)) and the trigonal model was double-checked with the Platon *addsymm* routine. At the last stage, with anisotropic atomic displacement parameters for all atoms and no constraints, the residual value settled at $R1$ = 0.0363 for 398 observed reflections ($Fo > 4\sigma(Fo)$) and 79 parameters and at $R1$ = 0.0537 for all 1289 independent reflections. Refined atomic coordinates and isotropic displacement parameters are given in Table 3, whereas selected bond distances are reported in Table 4. Crystallographic Information File (CIF) is deposited.

The calculated X-ray powder diffraction pattern (Table 5) was computed on the basis of the unit-cell data above and with the atom coordinates and occupancies reported in Table 3.

Table 3. Atoms, site occupancy, fractional atom coordinates (Å), and isotropic atomic displacement parameters (Å^2) for eliopoulosite.

Atom	Site Occupancy	x/a	y/b	z/c	U_{iso}
V1	V$_{0.749(6)}$	0.0001(6)	0	2/3	0.0088(6)
V2	V$_{0.762(6)}$	0.5005(6)	0	2/3	0.0095(6)
V3	V$_{0.755(6)}$	0.0006(7)	0	1/6	0.0113(6)
V4	V$_{0.749(7)}$	0.5010(7)	0	1/6	0.0187(8)
V5	V$_{1.00}$	0.4996(4)	0.4997(5)	0.16664(5)	0.0060(3)
V6	V$_{1.00}$	0.4997(5)	0.4996(5)	0.33327(5)	0.0120(4)
S1	S$_{1.00}$	0.1663(7)	0.3343(6)	0.08309(11)	0.0227(4)
S2	S$_{1.00}$	0.1659(7)	0.3328(7)	0.41639(11)	0.0226(4)
S3	S$_{1.00}$	0.1661(7)	0.3331(6)	0.74976(11)	0.0226(4)
S4	S$_{1.00}$	0.3333(6)	0.6673(7)	0.25009(11)	0.0228(4)

Table 4. Bond distances (Å) in the structure of eliopoulosite.

V1-S2	2.409(5) (×2)		V5-S1	2.413(4)
V1-S3	2.411(3) (×2)		V5-S4	2.416(3)
V1-S1	2.418(4) (×2)		V5-S1	2.417(4)
mean	2.413		V5-S2	2.418(4)
			V5-S4	2.419(3)
V2-S1	2.412(4) (×2)		V5-S3	2.419(4)
V2-S2	2.413(5) (×2)		mean	2.417
V2-S4	2.414(3) (×2)			
mean	2.413		V6-S1	2.407(4)
			V6-S4	2.411(4)
V3-S3	2.414(4) (×2)		V6-S3	2.413(4)
V3-S2	2.418(3) (×2)		V6-S3	2.413(4)
V3-S1	2.422(3) (×2)		V6-S2	2.415(4)
mean	2.418		V6-S4	2.417(4)
			mean	2.413
V4-S4	2.413(3) (×2)			
V4-S2	2.417(4) (×2)			
V4-S3	2.421(5) (×2)			
mean	2.417			
V1-V3	2.9006(10) (×2)			
V2-V5	2.9001(13) (×2)			
V4-V6	2.9018(14) (×2)			
V5-V6	2.8998(12) (×2)			

Table 5. Calculated X-ray powder diffraction data (d in Å) for eliopoulosite. The strongest observed reflections are given in bold.

hkl	d_{calc}	I_{calc}
200	**2.8964**	**29**
023	**2.5914**	**45**
026	**2.0495**	**100**
220	**1.6723**	**40**
029	1.6082	10
012	1.4503	8
400	1.4482	3
046	1.2957	20
22$\underline{12}$	1.0956	15
246	1.0242	12
600	0.9655	4

The chemical data (Table 1), yielding to a mean composition (wt.%) of S 41.78, V 54.11, Ni 1.71, Fe 1.1, Co 0.67, Mo 0.66, and total 100.03, were then normalized taking into account the structural results. The empirical formula of eliopoulosite on the basis of Σatoms = 15 apfu is $(V_{6.55}Ni_{0.19}Fe_{0.12}Co_{0.07}Mo_{0.04})_{\Sigma = 6.97}S_{8.03}$. The simplified formula is $(V,Ni,Fe)_7S_8$ and the ideal formula is V_7S_8, which corresponds to V 58.16, S 41.84, and total 100 (wt.%).

The structure can be considered as a twelve-fold superstructure ($2a \times 2a \times 3c$) of the NiAs-type subcell with V atoms octahedrally coordinated by S atoms and sulfur located in a trigonal prism. There are six octahedral layers in the structure with rods of fully occupied V sites alternate to rods of partially occupied sites, so that every layer contains 3.5 vanadium atoms. In successive layers rods are directed along [110], [100], and [010] accordingly to the threefold screw axis. As shown in Figure 5, the distribution of fully (blue in color) and partially (pale blue) occupied sites determine the superstructure along the a_1 and a_2 axes. Differences in electron density between fully and partially occupied sites are indeed rather modest (22 vs. 16.5): Accordingly, *hkl* reflections with $h,k = 2n + 1$ are rather weak ($<I/\sigma(I)> = 27.1$ vs. 4890.3).

Since the octahedral vacancies' distribution within the layers at $z = 0$, $z = 1/6$, and $z = 1/3$ is perfectly replicated at layers at $z = 1/2$, $z = 2/3$, and $z = 5/6$ (Figure 5), no contribution to *hkl* reflections with $l = 2n + 1$ is given by metal atoms. However, their intensities ($<I/\sigma(I)> = 564.3$ vs. 3349.2) are even higher than those leading to doubling of **a**-axis, due to the different position of sulfur atoms related to the different octahedral orientation in the layer at $z = 0$ with respect to $z = 1/2$ (Figure 4).

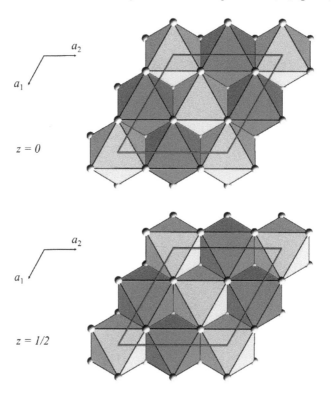

Figure 4. Octahedral layers in eliopoulosite projected down [001]. Layer at $z = 0$ (top of picture) repeats at $z = 1/3$ and $z = 2/3$; layer at $z = 1/2$ (bottom of picture) repeats at $z = 1/6$ and $z = 5/6$. Blue and pale blue colors represent full and partial occupancy. Sulfur atoms are depicted in yellow. The unit cell and the orientation of the structure are indicated.

The eliopoulosite structure shows strong analogies with the superstructures observed in the pyrrhotite-group of minerals. It is identical to the structure inferred for synthetic $VS_{1.125}$ [31]. The mean <V-S> bond distances in eliopoulosite (in the range 2.413–2.418 Å) are in excellent agreement with those found for synthetic vanadium sulfides. To achieve the charge balance, one should hypothesize the presence of both divalent and trivalent vanadium in eliopoulosite (i.e., $V^{2+}{}_5V^{3+}{}_2S_8$). The octahedral sites, however, do not show any significant difference symptomatic of a V^{2+}-V^{3+} ordering. Likewise, we did not find any evidence of ordering of the minor substituents (i.e., Ni and Fe) at a particular structural site. Furthermore, eliopoulosite shows, for the analogy of stoichiometry, unit cell and space group, close relationships with the metastable, trigonal form of 3C-Fe_7S_8 ($a = 6.852(6)$, $c = 17.046(2)$, $P3_121$) [32]. However, the distribution of vacancies is quite different (Figure 5b): As in other types of pyrrhotites [33], in 3C-Fe_7S_8 octahedral vacancies are completely ordered on one position alone so that the layers containing a void (i.e., three metal atoms) are alternating with fully occupied layers (i.e., four metal atoms).

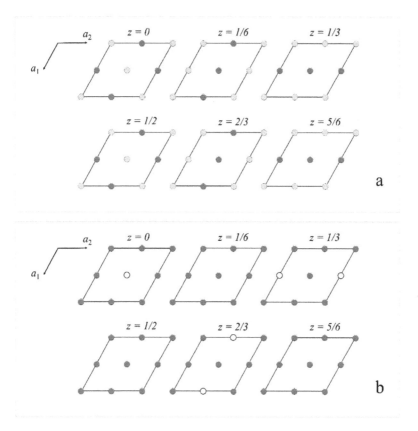

Figure 5. Metal distribution in eliopoulosite (**a**) and in synthetic 3C-Fe_7S_8 pyrrhotite (**b**). Blue and pale blue circles refer to metal positions with full and partial (75%) occupancy, respectively. White circles represent completely empty sites in synthetic 3C-Fe_7S_8 pyrrhotite [32]. The unit cell and the orientation of the structure are indicated.

Eliopoulosite does not correspond to any valid or invalid unnamed mineral [34]. According to literature data [1–8] eliopoulosite is the ninth mineral containing V and S that has been accepted by IMA. Noteworthy, eliopoulosite is the second V^{2+}-bearing mineral, the first being dellagiustaite, Al_2VO_4 [35]. The presence of divalent V requires extremely reducing conditions (well below the iron-wüstite buffer). Furthermore, based on its chemical composition, eliopoulosite is the second sulfide that contains only V as a major component discovered so far. Recently, Ivanova et al. [36] reported on the presence of a V,Cr,Fe-sulfide in the extremely reduced assemblage of a rare CB chondrite. Due to its small size (one micron) the grain was studied only by EBSD (Electron BackScattered Diffraction) and no crystallographic or optical data were provided. However, its ideal composition $(V,Cr,Fe)_4S_5$ seems different compared to that of eliopoulosite in terms of S content and V/Fe ratio and the presence of abundant Cr.

Interestingly, Selezneva et al. [37] recently showed that synthetic V_7S_8 shows differences in the magnetic behavior with respect to trigonal Fe_7S_8. Indeed, V_7S_8 is observed to exhibit a Pauli-paramagnetic behavior opposite to the classic ferrimagnetic ordering usually observed in pyrrhotite. Unfortunately, the small amount of the natural material precludes any possible measurement of both the magnetic susceptibility and magnetization. Such an experiment would have allowed us to verify if the amount of structural vacancies observed for eliopoulosite are enough to compensate the magnetic moments to avoid the ferromagnetic order.

6. Remarks on the Origin of Eliopoulosite

Eliopoulosite was found in the same sample in which tsikourasite and grammatikopoulosite were discovered, and the three minerals occur in the same mineralogical assemblage [9,10]. Therefore, we can argue that they formed under the same chemical-physical condition (i.e., in a strongly reducing environment). This assumption is fully supported by the chemical composition of eliopoulosite, that points to a low valence state for V. However, as already proposed for the origin of tsikourasite and grammatikopoulosite, it is still not possible to provide a conclusive model to explain exhaustively the genesis of eliopoulosite. The following hypothesis can be formulated, since all of them imply the presence of a reducing environment [9,10,38–43]: i) low-temperature alteration of chromitite during serpentinization process; ii) high-temperature reaction of the chromitites with reducing fluids, at mantle depth; iii) post-orogenic surface lightning strike; or iv) meteorite impact.

However, the probability of intercepting a fragment of a meteorite or a fulgurite in the Otrhys ophiolite during the sampling of the studied chromitite seems very unlikely, although a V-bearing sulfide has been reported in a meteorite [36].

Supplementary Materials: The following are available online at http://www.mdpi.com/2075-163X/10/3/245/s1, CIF: eliopoulosite.

Author Contributions: L.B., F.Z., and P.B. wrote the manuscript; F.Z. performed the chemical analyses; L.B. and P.B. performed the diffraction experiments; T.G. and B.T. provided the concentrate sample and information on the sample provenance and petrography of Othrys chromitite; G.G. discussed the chemical data and C.S. obtained the optical data. All the authors provided support in the data interpretation and revised the manuscript. All authors have read and agreed to the published version of the manuscript.

Funding: The authors are grateful to the University Centrum for Applied Geosciences (UCAG) for the access to the E. F. Stumpfl electron microprobe laboratory. SGS Mineral Services, Canada, is thanked to have performed the concentrate sample. S. Karipi is thanked for sample collection and participating in the field work. L.B. thanks MIUR, project "TEOREM deciphering geological processes using Terrestrial and Extraterrestrial ORE Minerals", prot. 2017AK8C32 (PI: Luca Bindi).

Acknowledgments: The authors acknowledge Ritsuro Miyawaki, Chairman of the CNMNC and its members for helpful comments on the submitted new mineral proposal. Many thanks are due to the editorial staff of Minerals.

Conflicts of Interest: The authors declare no conflicts of interest.

References

1. Ostrooumov, M.; Taran, Y.; Arellano-Jimenez, M.; Ponse, A.; Reyes-Gasga, J. Colimaite, K_3VS_4—A new potassium-vanadium sulfide mineral from the Colima volcano, State of Colima (Mexico). *Rev. Mex. Cienc. Geol.* **2009**, *26*, 600–608.
2. Zachariasen, W.H. X-Ray examination of colusite, $(Cu,Fe,Mo,Sn)_4(S,As,Te)_{3-4}$. *Am. Mineral.* **1933**, *18*, 534–537.
3. Spiridonov, E.M.; Kachalovskaya, V.M.; Kovachev, V.V.; Krapiva, L.Y. Germanocolusite $Cu_{26}V_2(Ge,As)_6S_{32}$—a new mineral. *Vest. Moskov. Univers., Ser. 4, Geologiya* **1992**, *1992*, 50–54. (In Russian)
4. Jaszczak, J.A.; Rumsey, M.S.; Bindi, L.; Hackney, S.A.; Wise, M.A.; Stanley, C.J.; Spratt, J. Merelaniite, $Mo_4Pb_4VSbS_{15}$, a new molybdenum-essential member of the cylindrite group, from the Merelani Tanzanite Deposit, Lelatema Mountains, Manyara Region, Tanzania. *Minerals* **2016**, *6*, 115. [CrossRef]
5. Kovalenker, V.A.; Evstigneeva, T.L.; Malov, V.S.; Trubkin, N.V.; Gorshkov, A.I.; Geinke, V.R. Nekrasovite $Cu_{26}V_2Sn_6S_{32}$—A new mineral of the colusite group. *Mineral. Zh.* **1984**, *6*, 88–97.
6. Hillibrand, W.F. Vanadium sulphide, patronite, and its mineral associates from Minasragra, Peru. *Am. J. Sci.* **1907**, *24*, 141–151. [CrossRef]
7. Spiridonov, E.M.; Badalov, A.S.; Kovachev, V.V. Stibiocolusite $Cu_{26}V_2(Sb,Sn,As)_6S_{32}$: A new mineral. *Dokl. Akad. Nauk* **1992**, *324*, 411–414. (In Russian)
8. Trojer, F.J. Refinement of the structure of sulvanite. *Am. Mineral.* **1996**, *51*, 890–894.
9. Zaccarini, F.; Bindi, L.; Ifandi, E.; Grammatikopoulos, T.; Stanley, C.; Garuti, G.; Mauro, D. Tsikourasite, $Mo_3Ni_2P_{1+x}$ (x < 0.25), a new phosphide from the chromitite of the Othrys Ophiolite, Greece. *Minerals* **2019**, *9*, 248.
10. Bindi, L.; Zaccarini, F.; Ifandi, E.; Tsikouras, B.; Stanley, C.; Garuti, G.; Mauro, D. Grammatikopoulosite, NiVP, a new phosphide from the chromitite of the Othrys Ophiolite, Greece. *Minerals* **2020**, *10*, 131. [CrossRef]
11. Smith, A.G.; Rassios, A. The evolution of ideas for the origin and emplacement of the western Hellenic ophiolites. *Geol. Soc. Am. Spec. Pap.* **2003**, *373*, 337–350.
12. Hynes, A.J.; Nisbet, E.G.; Smith, G.A.; Welland, M.J.P.; Rex, D.C. Spreading and emplacement ages of some ophiolites in the Othris region (Eastern Central Greece). *Z. Deutsch Geol. Ges.* **1972**, *123*, 455–468.
13. Smith, A.G.; Hynes, A.J.; Menzies, M.; Nisbet, E.G.; Price, I.; Welland, M.J.; Ferrière, J. The stratigraphy of the Othris Mountains, Eastern Central Greece: A deformed Mesozoic continental margin sequence. *Eclogue Geol. Helv.* **1975**, *68*, 463–481.
14. Rassios, A.; Smith, A.G. Constraints on the formation and emplacement age of western Greek ophiolites (Vourinos, Pindos, and Othris) inferred from deformation structures in peridotites. In *Ophiolites and Oceanic Crust: New Insights from Field Studies and the Ocean Drilling Program*; Dilek, Y., Moores, E., Eds.; Geological Society of America Special Paper: Boulder, CO, USA, 2001; pp. 473–484.
15. Bortolotti, V.; Chiari, M.; Marcucci, M.; Photiades, A.; Principi, G.; Saccani, E. New geochemical and age data on the ophiolites from the Othrys area (Greece): Implication for the Triassic evolution of the Vardar ocean. *Ofioliti* **2008**, *33*, 135–151.
16. Barth, M.G.; Mason, P.R.D.; Davies, G.R.; Drury, M.R. The Othris Ophiolite, Greece: A snapshot of subduction initiation at a mid-ocean ridge. *Lithos* **2008**, *100*, 234–254. [CrossRef]
17. Barth, M.; Gluhak, T. Geochemistry and tectonic setting of mafic rocks from the Othris Ophiolite, Greece. *Contr. Mineral. Petrol.* **2009**, *157*, 23–40. [CrossRef]
18. Dijkstra, A.H.; Barth, M.G.; Drury, M.R.; Mason, P.R.D.; Vissers, R.L.M. Diffuse porous melt flow and melt-rock reaction in the mantle lithosphere at a slow-spreading ridge: A structural petrology and LA-ICP-MS study of the Othris Peridotite Massif (Greece). *Geochem. Geophys. Geosyst.* **2003**, *4*, 278. [CrossRef]
19. Magganas, A.; Koutsovitis, P. Composition, melting and evolution of the upper mantle beneath the Jurassic Pindos ocean inferred by ophiolitic ultramafic rocks in East Othris, Greece. *Int. J. Earth Sci.* **2015**, *104*, 1185–1207. [CrossRef]
20. Garuti, G.; Zaccarini, F.; Economou-Eliopoulos, M. Paragenesis and composition of laurite from chromitites of Othrys (Greece): Implications for Os-Ru fractionation in ophiolite upper mantle of the Balkan Peninsula. *Mineral. Dep.* **1999**, *34*, 312–319. [CrossRef]

21. Tsikouras, B.; Ifandi, E.; Karipi, S.; Grammatikopoulos, T.A.; Hatzipanagiotou, K. Investigation of platinum-group minerals (PGM) from Othrys chromitites (Greece) using superpanning concentrates. *Minerals* **2016**, *6*, 94. [CrossRef]
22. Robertson, A.H.F. Overview of the genesis and emplacement of Mesozoic ophiolites in the Eastern Mediterranean Tethyan region. *Lithos* **2002**, *65*, 1–67. [CrossRef]
23. Robertson, A.H.F.; Clift, P.D.; Degnan, P.; Jones, G. Palaeogeographic and palaeotectonic evolution of the Eastern Mediterranean Neotethys. *Palaeogeogr. Palaeoclim. Palaeoecol.* **1991**, *87*, 289–343. [CrossRef]
24. Economou, M.; Dimou, E.; Economou, G.; Migiros, G.; Vacondios, I.; Grivas, E.; Rassios, A.; Dabitzias, S. Chromite deposits of Greece. In *Chromites, UNESCO's IGCP197 Project Metallogeny of Ophiolites*; Petrascheck, W., Karamata, S., Eds.; Theophrastus Publ. S.A.: Athens, Greece, 1986; pp. 129–159.
25. Ifandi, E.; Zaccarini, F.; Tsikouras, B.; Grammatikopoulos, T.; Garuti, G.; Karipi, S.; Hatzipanagiotou, K. First occurrences of Ni-V-Co phosphides in chromitite of Agios Stefanos mine, Othrys ophiolite, Greece. *Ofioliti* **2018**, *43*, 131–145.
26. Zaccarini, F.; Ifandi, E.; Tsikouras, B.; Grammatikopoulos, T.; Garuti, G.; Mauro, D.; Bindi, L.; Stanley, C. Occurrences of new phosphides and sulfide of Ni, Co, V, and Mo from chromitite of the Othrys ophiolite complex (Central Greece). *Per. Mineral.* **2019**, *88*. [CrossRef]
27. Bruker. *APEX3*; Bruker AXS Inc.: Madison, WI, USA, 2016. Available online: https://www.bruker.com/products/x-ray-diffraction-and-elemental-analysis/single-crystal-x-ray-diffraction/sc-xrd-software/apex3.html (accessed on 6 March 2020).
28. Bruker. *SAINT and SADABS*; Bruker AXS Inc.: Madison, WI, USA, 2016. Available online: https://www.bruker.com/products/x-ray-diffraction-and-elemental-analysis/single-crystal-x-ray-diffraction/sc-xrd-software/apex3.html (accessed on 6 March 2020).
29. Sheldrick, G.M. A short history of SHELX. *Acta Crystallogr.* **2008**, *A64*, 112–122. [CrossRef]
30. Wilson, A.J.C. *International Tables for Crystallography: Mathematical, Physical, and Chemical Tables*; International Union of Crystallography: Chester, UK, 1992; Volume 3.
31. Grønvold, F.; Haraldsen, H.; Pedersen, B.; Tufte, T. X-ray and magnetic study of vanadium sulfides in the range V_5S_4 to V_5S_8. *Rev. Chim. Minéral.* **1969**, *6*, 215.
32. Nakano, A.; Tokonami, M.; Morimoto, N. Refinement of 3C pyrrhotite, Fe_7S_8. *Acta Crystallogr.* **1979**, *B35*, 722–724. [CrossRef]
33. Morimoto, N. Crystal structure of a monoclinic pyrrhotite. *Rec. Progr. Nat. Sci. Japan* **1978**, *3*, 183–206.
34. Smith, D.G.W.; Nickel, E.H. A system for codification for unnamed minerals: report of the Subcommittee for Unnamed Minerals of the IMA Commission on New Minerals, Nomenclature and Classification. *Can. Mineral.* **2007**, *45*, 983–1055. [CrossRef]
35. Cámara, F.; Bindi, L.; Pagano, A.; Pagano, R.; Gain, S.E.M.; Griffin, W.L. Dellagiustaite: A novel natural spinel containing V^{2+}. *Minerals* **2019**, *9*, 4. [CrossRef]
36. Ivanova, M.A.; Ma, C.; Lorenz, C.A.; Franchi, I.A.; Kononkova, N.N. A new unusual bencubbinite (cba), Sierra Gorda 013 with unique V-rich sulfides. *Met. Plan. Sci.* **2019**, *54*, 6149.
37. Selezneva, N.V.; Ibrahim, P.N.G.; Toporova, N.M.; Sherokalova, E.M.; Baranov, N.V. Crystal structure and magnetic properties of pyrrhotite-type compounds $Fe_{7-y}V_yS_8$. *Acta Phys. Polon.* **2018**, *A133*, 450–452. [CrossRef]
38. Malvoisin, B.; Chopin, C.; Brunet, F.; Matthieu, E.; Galvez, M.E. Low-temperature wollastonite formed by carbonate reduction: a marker of serpentinite redox conditions. *J. Petrol.* **2012**, *53*, 159–176. [CrossRef]
39. Etiope, G.; Tsikouras, B.; Kordella, S.; Ifandi, E.; Christodoulou, D.; Papatheodorou, G. Methane flux and origin in the Othrys ophiolite hyperalkaline springs, Greece. *Chem. Geol.* **2013**, *347*, 161–174. [CrossRef]
40. Etiope, G.; Ifandi, E.; Nazzari, M.; Procesi, M.; Tsikouras, B.; Ventura, G.; Steele, A.; Tardini, R.; Szatmari, P. Widespread abiotic methane in chromitites. *Sci. Rep.* **2018**, *8*, 8728. [CrossRef]
41. Xiong, Q.; Griffin, W.L.; Huang, J.X.; Gain, S.E.M.; Toledo, V.; Pearson, N.J.; O'Reilly, S.Y. Super-reduced mineral assemblages in "ophiolitic" chromitites and peridotites: The view from Mount Carmel. *Eur. J. Mineral.* **2017**, *29*, 557–570. [CrossRef]

42. Pasek, M.A.; Hammeijer, J.P.; Buick, R.; Gull, M.; Atlas, Z. Evidence for reactive reduced phosphorus species in the early Archean ocean. *Proc. Nat. Acad. Sci. U.S.A.* **2013**, *110*, 100089–100094. [CrossRef]
43. Ballhaus, C.; Wirth, R.; Fonseca, R.O.C.; Blanchard, H.; Pröll, W.; Bragagni, A.; Nagel, T.; Schreiber, A.; Dittrich, S.; Thome, V.; et al. Ultra-high pressure and ultra-reduced minerals in ophiolites may form by lightning strikes. *Geochem. Perspec. Lett.* **2017**, *5*, 42–46. [CrossRef]

 *minerals*

Article

Grammatikopoulosite, NiVP, a New Phosphide from the Chromitite of the Othrys Ophiolite, Greece

Luca Bindi [1], Federica Zaccarini [2,*], Elena Ifandi [3,4], Basilios Tsikouras [3], Chris Stanley [5], Giorgio Garuti [2] and Daniela Mauro [6]

1   Dipartimento di Scienze della Terra, Università degli Studi di Firenze, I-50121 Florence, Italy; luca.bindi@unifi.it
2   Department of Applied Geological Sciences and Geophysics, University of Leoben, A-8700 Leoben, Austria; giorgio.garuti1945@gmail.com
3   Faculty of Science, Physical and Geological Sciences, Universiti Brunei Darussalam, BE 1410 Gadong, Brunei Darussalam; selena.21@windowslive.com (E.I.); basilios.tsikouras@ubd.edu.bn (B.T.)
4   Department of Geology, Section of Earth Materials, University of Patras, 265 00 Patras, Greece
5   Department of Earth Sciences, Natural History Museum, London SW7 5BD, UK; c.stanley@nhm.ac.uk
6   Dipartimento di Scienze della Terra, Università degli Studi di Pisa, I-56126 Pisa, Italy; daniela.mauro@dst.unipi.it
*   Correspondence: federica.zaccarini@unileoben.ac.at; Tel.: +43-(0)3842-402-6218

Received: 22 December 2019; Accepted: 27 January 2020; Published: 31 January 2020

Abstract: Grammatikopoulosite, NiVP, is a new phosphide discovered in the podiform chromitite and hosted in the mantle sequence of the Othrys ophiolite complex, central Greece. The studied samples were collected from the abandoned chromium mine of Agios Stefanos. Grammatikopoulosite forms small crystals (from 5 µm up to about 80 µm) and occurs as isolated grains. It is associated with nickelphosphide, awaruite, tsikourasite, and an undetermined V-sulphide. It is brittle and has a metallic luster. In plane-polarized light, it is creamy-yellow, weakly bireflectant, with measurable but not discernible pleochroism and slight anisotropy with indeterminate rotation tints. Internal reflections were not observed. Reflectance values of mineral in air (R_1, R_2 in %) are: 48.8–50.30 at 470 nm, 50.5–53.5 at 546 nm, 51.7–55.2 at 589 nm, and 53.2–57.1 at 650 nm. Five spot analyses of grammatikopoulosite give the average composition: P 19.90, S 0.41, Ni 21.81, V 20.85, Co 16.46, Mo 16.39, Fe 3.83, and Si 0.14, total 99.79 wt %. The empirical formula of grammatikopoulosite—based on $\Sigma(V + Ni + Co + Mo + Fe + Si) = 2$ apfu, and taking into account the structural results—is $(Ni_{0.57}Co_{0.32}Fe_{0.11})_{\Sigma1.00}(V_{0.63}Mo_{0.26}Co_{0.11})_{\Sigma1.00}(P_{0.98}S_{0.02})_{\Sigma1.00}$. The simplified formula is (Ni,Co)(V,Mo)P and the ideal formula is NiVP, which corresponds to Ni 41.74%, V 36.23%, P 22.03%, total 100 wt %. The density, calculated on the basis of the empirical formula and single-crystal data, is 7.085 g/cm^3. The mineral is orthorhombic, space group *Pnma*, with $a = 5.8893(8)$, $b = 3.5723(4)$, $c = 6.8146(9)$ Å, $V = 143.37(3)$ Å^3, and $Z = 4$. The mineral and its name have been approved by the Commission of New Minerals, Nomenclature and Classification of the International Mineralogical Association (IMA 2019-090). The mineral honors Tassos Grammatikopoulos, geoscientist at the SGS Canada Inc., for his contribution to the economic mineralogy and mineral deposits of Greece.

Keywords: grammatikopoulosite; phosphide; chromitite; Agios Stefanos mine; Othrys; ophiolite; Greece

1. Introduction

Natural phosphides are very rare phases, representing only 3% of the minerals approved by the International Mineralogical Association (IMA) ([1] and references therein). Most of these natural phosphides have been discovered and described in meteorites. Only recently have several new phosphides been discovered in terrestrial rocks, most of which are associated with the pyrometamorphic rocks of the Hatrurim Formation, Southern Levant [2]. Phosphides are documented in podiform chromitites hosted in the mantle sequence of ophiolite complexes of Greece and Russia [3–6] but are very rare. The phosphide found in the Russian chromitite was found in situ, in contact with serpentine of the altered interstitial silicates. The phosphides associated with the Greek chromitite were documented in heavy concentrates. Recently, the new mineral tsikourasite ($Mo_3Ni_2P_{1+x}$ ($x < 0.25$) was discovered in heavy mineral concentrates from a chromitite from the Othrys ophiolite (Greece)) [1]. Further investigation of the same mineral concentrates led to the discovery of a second new mineral. Quantitative chemical analysis and crystal structure proved that the studied phase is a new phosphide, characterized by the simplified formula $(Ni,Co)(V,Mo)P$ and by the ideal formula NiVP, which corresponds to Ni 41.74%, V 36.23%, P 22.03%, total 100 wt %. The mineral is orthorhombic, space group *Pnma*, with $a = 5.8893(8)$, $b = 3.5723(4)$, $c = 6.8146(9)$ Å, and $Z = 4$. The mineral and its name have been approved by the Commission of New Minerals, Nomenclature and Classification of the International Mineralogical Association (IMA 2019-090). The mineral honors Tassos Grammatikopoulos (b. 1966), a geoscientist at the SGS Canada Inc., for his contribution to the investigation of economic mineralogy and mineral deposits of Greece. Holotype material is deposited in the Mineralogical Collection of the Museo di Storia Naturale, Università di Pisa, Via Roma 79, Calci (Pisa, Italy), under catalogue number 19,911.

2. Geological Background and Occurrence of Grammatikopoulosite

Grammatikopoulosite was discovered in a heavy mineral concentrate, which was prepared from chromitite specimens hosted in the mantle sequence of the Mesozoic Othrys ophiolite, central Greece (Figure 1A–C).

Othrys ophiolite is a complete but dismembered suite (Mirna Group) and consists of three structural units: the uppermost succession with variably serpentinized peridotites, which is structurally bounded by an ophiolite mélange; the intermediate Kournovon dolerite, including cumulate gabbro and local rhyolite; and the lower Sipetorrema Pillow Lava unit including also basaltic flows, siltstones, and chert. The Mirna Group constituted multiple inverted thrust sheets [7], which were eventually obducted onto the Pelagonian Zone during the Late Jurassic–Early Cretaceous [8–14]. Three types of basalts with different geochemical signatures have been described: (i) alkaline within-plate (WPB), (ii) normal-type mid-ocean ridge (N-MORB), and (iii) low-K tholeiite (L-KT). A biostratigraphic investigation indicated that radiolarites associated with N-MORB were deposited in the Middle and Late Triassic. Radiolarites deposited over the L-KT basalts are Early Carnian–Middle Norian–Late Norian in age [15]. N-MORB erupted during the Middle–Late Triassic period. The L-KT basalts erupted during the Middle–Late Triassic period, in part simultaneously with N-MORB, in the ocean–continent transition zone, close to the rifted continental margin. Finally, the alkaline WPB are interpreted to have formed in oceanic seamounts or in the ocean–continent transition zone adjacent to the rifted continental margin [15].

Electron microprobe analyses of the assemblages of these chromitites revealed that the studied spinel-supergroup minerals are magnesiochromites [1,5,6]. However, spinel mineral chemistry is rather heterogeneous with Cr_2O_3 (44.96–51.64 wt %), Al_2O_3 (14.18–20.78 wt %), MgO (13.34–16.84 wt %), and FeO (8.3–13.31 wt %). The calculated Fe_2O_3 ranges from 6.72 to 9.26 wt %. The amounts of MnO (0.33–0.60 wt %), V_2O_3 (0.04–0.30 wt %), ZnO (up to 0.07 wt %), and NiO (0.03–0.24 wt %) exhibit minor variations. The TiO_2 content is low (0.03–0.23 wt %), in agreement with the typical value reported for the mantle-hosted podiform chromites. The $Cr/(Cr + Al)$ ratios of the investigated magnesiochromite are lower than those of chromites formed in the supra-subduction zone (SSZ) and those related to boninitic melt [1,5,6].

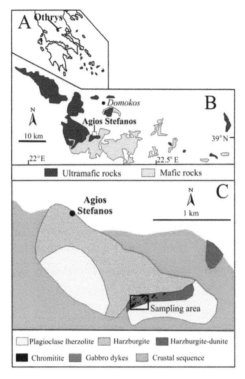

Figure 1. Location of the Othrys complex in Greece: (**A**) general geological map of the Othrys ophiolite showing the location of the Agios Stefanos chromium mine; (**B**) and (**C**) detailed geological setting of the Agios Stefanos area (modified after [5,10,16,17]).

3. Analytical Methods

The processing and recovery of the heavy minerals were carried out by treating about 10 kg of massive chromitite at SGS Mineral Services, Canada, following the procedure described by several authors [1,5,6,18]. Approximately 10 kg of chromitite were collected in the abandoned mine of Agios Stefanos. Subsequently, they were crushed and blended to generate a homogenous composite sample. About 500 g of the composite sample were riffled and stage crushed to a P80 (80% passing) of 75 μm. After, the sample was treated with heavy liquid (the density of the heavy liquids was 3.1 g/cm^3) to separate the heavy minerals. This procedure produced a heavy and light mineral concentrate. The heavy fraction was selected and processed with a superpanner. This method is designed for small amounts of sample and is closely controlled, leading to very effective separation. It consists of a tapering triangular deck with a "V" shape cross section. The table reproduces the concentrating action of a gold pan. Initially, the sample is swirled to stratify the minerals. Then, the heaviest minerals settle to the bottom and are deposited on the deck surface. The less dense material moves towards the top, overlying the heavy minerals. The operation of the deck is then changed to a rapid reciprocal motion, with an appropriate "end-knock" at the up-slope end of the board, and a steady flow of wash water is introduced. The "end-knock" forces the heavy minerals to migrate to the up-slope end of the deck. The wash water carries the light minerals to move to the narrower, down-slope end of the deck. The heaviest fractions were split into the heaviest fraction (tip) followed by a less dense fraction (middling). The "tip" and the "middlings" of the superpanner are the densest fractions and included liberated grains of chromite, sulphides, alloys, and phosphides, while the lighter tail consisted of a particle mixture of chromite and silicates. No source of contamination is likely during sample collection and subsequent treatment. The heavy minerals were prepared in epoxy blocks, and then polished for mineralogical examination. Quantitative chemical analyses and acquisition of back-scattered electron images of grammatikopoulosite were performed with a JEOL JXA-8200 electron microprobe, installed in the E. F. Stumpfl laboratory, Leoben University, Austria, operating in Wavelength Dispersive Spectrometry (WDS) mode. Major and minor elements were determined at a 20 kV accelerating voltage and a 10 nA beam current, with 20 s as the counting time for the peak and 10 s for the backgrounds. The beam diameter was about 1 μm in size. For the WDS analyses, the following lines and diffracting crystals were used: P, S, Si = ($K\alpha$, PETJ), V, Fe, Co, Ni = ($K\alpha$, LIFH), and Mo = ($L\alpha$, PETJ). The following standards were selected: synthetic Ni_3P for Ni and P, molybdenite for S and Mo, synthetic Fe_3P for Fe, synthetic metallic vanadium for V, skutterudite for Co, quartz for Si, and chromite for Cr. The ZAF correction method was applied. Automatic corrections were performed for interferences P-Mo, Cr-V, and Mo-S. Representative analyses of grammatikopoulosite are listed in Table 1.

Table 1. Chemical data (wt % of elements) for grammatikopoulosite.

Sample	P	S	Ni	V	Co	Mo	Fe	Si	Total
VP40-1	20.38	0.41	21.98	21.02	16.33	16.72	3.82	0.14	100.79
VP40-2	19.83	0.42	21.70	20.48	16.66	16.36	3.83	0.14	99.41
VP40-3	19.65	0.39	21.72	20.73	16.51	16.35	3.86	0.13	99.33
VP40-4	19.65	0.40	21.95	21.05	16.37	16.31	3.85	0.13	99.71
VP40-5	20.01	0.41	21.69	20.98	16.45	16.20	3.78	0.16	99.67
average	19.90	0.41	21.81	20.85	16.46	16.39	3.83	0.14	99.79

Single-crystal and powder X-ray diffraction data were collected at the University of Florence using a Bruker D8 Venture equipped with a Photon II CCD detector, with graphite-monochromatized MoKα radiation (λ = 0.71073 Å) using a crystal fragment hand-picked from the polished section under a reflected light microscope (cif file, see Supplementary Materials). The crystal (about 80 μm in size) was carefully and repeatedly washed in acetone. It did not show any other visible phase attached to the surface. Single-crystal X-ray diffraction intensity data were integrated and corrected for standard Lorentz polarization factors with the software package *Apex3* [19,20].

The reflectance measurements on grammatikopoulosite were carried out using a WTiC standard and a J&M TIDAS diode array spectrophotometer at the Natural History Museum of London, UK.

4. Physical and Optical Properties

More than 30 grains of grammatikopoulosite were found in the studied polished sections. Grammatikopoulosite occurs as anhedral to subhedral grains. Most of them were less than 10 μm in size and only two of them were up to about 80 μm. Grammatikopoulosite consists of single (Figure 2A) or poly-phase grains associated with other minerals, such as tsikourasite, nickelphosphide, awaruite, and a V-sulphide, which likely represents another new mineral (Figure 2B,C). Compositions of these minerals have been reported in previous papers [5,6]. In plane-polarized light, grammatikopoulosite is creamy-yellow, weakly bireflectant, with measurable but not discernible pleochroism and slight anisotropy with indeterminate rotation tints. Internal reflections were not observed. Reflectance values of the mineral in air (R in %) are reported in Table 2 and in Figure 3. Density was not measured because of the small amount of available material and the presence of fine intergrowths with awaruite. The calculated density is equal to 7.085 g·cm^{-3}, based on the empirical composition and unit–cell volume refined from single-crystal XRD data.

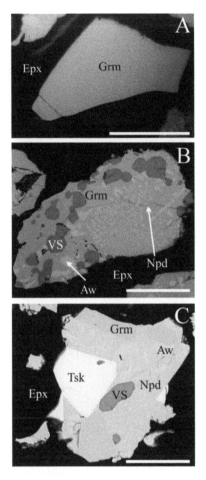

Figure 2. Digital image in reflected plane-polarized light (**A,B**) and a back-scattered electron image (**C**) showing grammatikopoulosite from the chromitite of Agios Stefanos. Abbreviations: Grm = grammatikopoulosite, Tsk = tsikourasite, VS = V-sulphide, Aw = awaruite, Npd = nickelphosphide, Epx = epoxy. Scale bar = 50 microns.

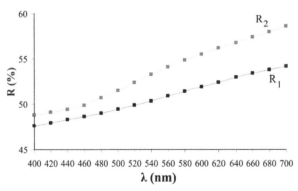

Figure 3. Reflectance data for grammatikopoulosite.

Table 2. Reflectance values for grammatikopoulosite. The values required by the Commission on Ore Mineralogy are given in bold.

λ nm	R_1	λ nm	R_2
400	47.6	400	48.8
420	47.9	420	49.1
440	48.3	440	49.4
460	48.6	470	49.9
470	**48.8**	**470**	**50.3**
480	49.0	480	50.7
500	49.4	500	51.5
520	49.9	520	52.4
540	50.3	540	53.3
546	**50.5**	**546**	**53.5**
560	50.9	560	54.1
580	51.4	580	54.9
589	**51.7**	**589**	**55.2**
600	51.9	600	55.5
620	52.4	620	56.2
640	53.0	640	56.8
650	**53.2**	**650**	**57.1**
680	53.8	680	58.0
700	54.2	700	58.6

5. Chemical Composition and X-Ray Crystallography

Chemical composition and X-ray data reveal that the empirical formula of grammatikopoulosite, based on $\Sigma(V + Ni + Co + Mo + Fe + Si) = 2$ apfu and taking into account the structural results (see below), is $(Ni_{0.57}Co_{0.32}Fe_{0.11})_{\Sigma1.00}(V_{0.63}Mo_{0.26}Co_{0.11})_{\Sigma1.00}(P_{0.98}S_{0.02})_{\Sigma1.00}$. The simplified formula is $(Ni,Co)(V,Mo)P$ and the ideal formula is $NiVP$, which corresponds to the composition of Ni = 41.74%, V = 36.23%, and P 22.03% (total 100 wt %).

A small grammatikopoulosite grain (about 80 μm in size) was handpicked from the polished section under a reflected light microscope and mounted on a 5 μm diameter carbon fiber, which was, in turn, attached to a glass rod in preparation for the single-crystal X-ray diffraction study. The fragment consists of crystalline grammatikopoulosite associated with minor, fine-grained polycrystalline awaruite. A total of 493 unique reflections were collected up to $2\theta = 80.33°$. The mineral is orthorhombic, space group *Pnma*, with $a = 5.8893(8)$, $b = 3.5723(4)$, $c = 6.8146(9)$ Å, and $Z = 4$. Given the similarity in unit–cell values and space groups, the structure was refined starting from the atomic coordinates reported for allabogdanite [21] using the software Shelxl-97 [22]. The site occupancy factor (s.o.f.) at the two cation sites was allowed to vary (Ni vs. Mo) using scattering curves for neutral atoms taken from the International Tables for Crystallography [23], leading to 27.5 and 28.4 e^- for M1 and M2 sites, respectively. The P site was found to be fully occupied (P vs. structural vacancy) by phosphorous (site scattering = $15.0 \ e^-$) and fixed accordingly. Taking into account the site distribution observed in florenskyite [24], allabogdanite [21], and andreyivanovite [25], and in most Co_2Si-structure-type synthetic compounds [26], we assigned Ni, Fe, and Co to fill the M1 site (i.e., $Ni_{0.57}Co_{0.32}Fe_{0.11}$) and all the other elements (i.e., $V_{0.63}Mo_{0.26}Co_{0.11}$) to the M2 site. The mean electron numbers calculated with such a site distribution were identical (27.36 and 28.38 e^-) to those obtained in the refinement. For this reason, the subsequent cycles of refinement were run with the above constrained site populations, yielding an $R_1 = 0.0276$ for 465 reflections with Fo > 4σ(Fo) and $R_1 = 0.0291$ for all the 493 independent reflections and 19 parameters. Refined atomic coordinates and isotropic displacement parameters are given in Table 3, whereas selected bond distances are reported in Table 4.

X-ray powder diffraction data for grammatikopoulosite (Table 5) were obtained with a Bruker D8 Venture equipped with a Photon II CCD detector, with graphite-monochromatized CuKα radiation (λ = 1.54138 Å). The least squares refinement gave the following values: $a = 5.8088(2)$, $b = 3.5993(2)$, $c = 6.8221(3)$ Å, and $V = 142.634(8)$ Å^3. The calculated powder diffraction pattern obtained using the site occupancies and atomic coordinates is reported in Table 3.

Table 3. Atoms, site occupancies, atom coordinates, and isotropic displacement parameters (Å^2) for grammatikopoulosite.

Atom	Site Occupancy	x/a	y/b	z/c	U_{iso}
M1	$Ni_{0.57}Co_{0.32}Fe_{0.11}$	0.35709(6)	$\frac{1}{4}$	0.93703(5)	0.00578(10)
M2	$V_{0.63}Mo_{0.26}Co_{0.11}$	0.47087(6)	$\frac{1}{4}$	0.33109(5)	0.00595(9)
P	$P_{1.00}$	0.23639(12)	$\frac{1}{4}$	0.62449(10)	0.00547(13)

Table 4. Selected bond distances (Å) for grammatikopoulosite.

Atoms	Bond Distance
M1–P	2.2453(8)
M1–P (×2)	2.2639(5)
M1–P	2.2728(8)
M1–M1 (×2)	2.6000(6)
M1–M2 (×2)	2.7280(5)
M1–M2 (×2)	2.7487(5)
M1–M2	2.7677(5)
M1–M2	2.7696(5)
M2–P	2.4299(8)
M2–P (×2)	2.5008(6)
M2–P (×2)	2.5811(6)
M2–M1 (×2)	2.7280(5)
M2–M1 (×2)	2.7487(5)
M2–M1	2.7677(5)
M2–M1	2.7696(5)
M2–M2 (×2)	2.9339(6)

Table 5. Observed and calculated X-ray powder diffraction data (d in Å) for grammatikopoulosite. The strongest observed reflections are given in bold.

Indices	1		2	
hkl	d_{obs}	I_{obs}	d_{calc}	I_{calc}
101	4.43	10	4.4559	14
002	-	-	3.4073	5
102	**2.950**	**20**	**2.9493**	**19**
111	**2.785**	**25**	**2.7872**	**24**
201	2.699	5	2.7031	5
112	**2.273**	**60**	**2.2743**	**65**
210	2.269	10	2.2722	8
201	2.230	10	2.2279	9
211	**2.157**	**100**	**2.1555**	**100**
103	**2.118**	**25**	**2.1194**	**27**
013	1.915	15	1.9168	14
301	1.888	10	1.8864	12
113	1.824	15	1.8227	20
020	**1.784**	**20**	**1.7861**	**21**
004	1.702	10	1.7036	10
302	1.700	15	1.7010	22
213	1.608	10	1.6065	7
114	1.489	5	1.4878	6
303	1.482	5	1.4853	6
400	1.470	5	1.4723	6
123	1.367	10	1.3658	7
322	1.233	10	1.2318	9
314	1.211	5	1.2106	6
215	1.170	10	1.1688	9
511	1.102	5	1.1038	6
513	1.005	5	1.0035	5

1 = observed diffraction pattern; 2 = calculated diffraction pattern obtained with the atomic coordinates reported in Table 3 (only reflections with $I_{rel} \geq 4$ are listed).

6. Description of the Structure and Relations to Other Species

In the structure of grammatikopoulosite (Figure 4), $M1$ links four P atoms and eight $M2$ (Figure 5—left), whereas $M2$ links five P, six $M1$, and two $M2$ (Figure 5—right). The metal–phosphorous distances are much shorter in the $M1$ coordination sphere than in that of $M2$ (Table 4). Interestingly, if only the M–P distances are considered in the coordination polyhedra of the M atoms, $M1P_4$ tetrahedra (Figure 6) forming corner-sharing chains along the **b**-axis or $M2P_5$ square pyramids forming zig-zag chains along the **a**-axis (Figure 7) can be observed.

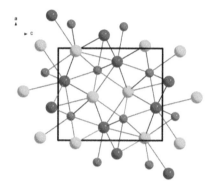

Figure 4. The crystal structure of grammatikopoulosite projected down [10]. Blue, green, and red circles refer to $M1$, $M2$, and P, respectively.

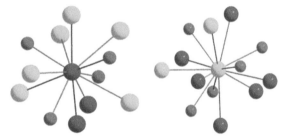

Figure 5. Coordination environment of $M1$ (left) and $M2$ (right) sites in the crystal structure of grammatikopoulosite. Colors and symbols as in Figure 4.

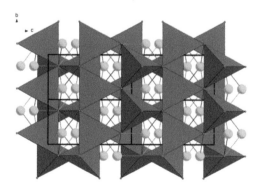

Figure 6. Polyhedral representation of the $M1P_4$ tetrahedra in the crystal structure of grammatikopoulosite. Colors and symbols as in Figure 4.

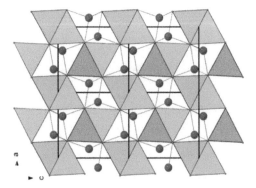

Figure 7. Polyhedral representation of the $M2P_5$ distorted pyramids in the crystal structure of grammatikopoulosite. Colors and symbols as in Figure 4.

It was reported that the departure from the hexagonal, high-temperature P-$62m$ structure (barringerite structure) was linked to some ordering between the metals at the M sites. This was not immediately evident in both allabogdanite [21] and andreyivanovite [25] because of the close site scattering of Fe, Ni, and Cr. On the contrary, the ordering is evident if we take into consideration florenskyite [24] and grammatikopoulosite with ideal formulae FeTiP and NiVP, respectively. Ti and V preferentially enter the larger $M2$ site in agreement with their larger metallic radius (1.47 and 1.35 Å, respectively [27]).

Grammatikopoulosite does not correspond to any valid or invalid unnamed mineral [28] and it belongs to the group of natural phosphides with the Co_2Si orthorhombic *Pnma* structure, i.e., florenskyite [24], allabogdanite [21], and andreyivanovite [25].

7. Discussion and Genetical Implications

Grammatikopoulosite, similar to tsikourasite, is natural in origin despite the fact that both phases were found in heavy mineral concentrates and not in situ [1]. This assumption is supported by the following observations: during the sampling, the only tool used to collect the chromitite was a steel hammer; the concentrates were obtained in a laboratory that uses a grinder manufactured by TM Engineering, with media Alloy 1 composed of Cr and Mo, and does not contain Ni, V, Co, Fe, or P; the used material is very hard and is recognized to be best for applications where abrasion and impact is the norm; and the chemical composition of grammatikopoulosite indicates crystallization under reducing conditions, similar to what has been previously proposed for other phosphides found in ophiolitic chromitites [1,3–6]. According to literature data, the crystallization temperature of most of the terrestrial phosphides is generally comprised between 700 °C and 1150 °C [4]. However, a conclusive model to explain the origin of grammatikopoulosite cannot be provided yet, since it may have a terrestrial origin or represent fragments of a meteorite. The following models, which imply the crystallization in a terrestrial environment from low to high temperatures, can be suggested for the origin of grammatikopoulosite: (i) alteration of chromitite during sub-oceanic or on-land serpentinization at temperatures between 400 °C and 150 °C, which is a typical reducing process [29]; (ii) reaction of the chromitite with high-temperature reducing fluids in the mantle; and (iii) interaction of the chromitite and their host serpentinized peridotites with a surface lightning strike. Phosphide minerals occurring in ophiolitic chromitites in Othrys and Gerakini–Ormylia (Greece), as well as in Alapaevsk (Russia), are closely associated with highly reducing phases such as awaruite, typically forming during serpentinization [1,3–6]. Likewise, grammatikopoulosite is also associated with awaruite and other reducing phases, likely suggesting a low-temperature origin during serpentinization. This hypothesis is strongly supported by the discovery of a Ni-phosphide in situ in the serpentine-rich matrix of the Alapaevsk chromitite of Russia [3]. Recently, Etiope et al. [30,31] have shown that the Agios Stefanos chromitite is the source rock of abiotic CH_4 measured in the west Othrys springs. These authors argued that CH_4 formation took place at temperatures below 150 °C via Sabatier reaction during oceanic and on-land serpentinization. This observation provides further evidence that reducing conditions can be achieved at a low temperature during the alteration of mantle-derived rocks. Alternatively, Xiong et al. [32] suggested that reducing conditions can be achieved in mantle rocks, including chromitites, in the shallow lithosphere due to the interaction of mantle-derived fluids enriched in CH_4 and H_2 with basaltic melt. However, no phosphides are reported in the ultra-reduced, high-temperature mineralogical assemblage described by Xiong et al. [32]. Natural phosphides and other ultra-reduced minerals have been rarely described in few fulgurites [33,34] but they are very common accessory phases in meteorites [21,24,25,35–39] or. The rare and localized occurrence of the Othrys phosphides may provide some support to the two last models. However, the probability to have collected a fragment of a meteorite or a fulgurite in the Othrys ophiolite during the sampling of the studied chromitite seems improbable, and for the discrimination of the meteoritic materials, a more detailed isotopic study may be needed. To conclude, an indisputable genetic model to explain how grammatikopoulosite is crystallized still needs further investigation.

Supplementary Materials: The following are available online at http://www.mdpi.com/2075-163X/10/2/131/s1, CIF: grammatikopoulosite.

Author Contributions: L.B. and F.Z. wrote the manuscript; F.Z. and L.B. performed the chemical analyses and the diffraction experiments, respectively; E.I. and B.T. provided the concentrate sample and information on the sample provenance and petrography of Othrys chromitite; G.G. and D.M. discussed the chemical data; and C.S. obtained the optical data. All authors provided support in the data interpretation and revised the manuscript. All authors have read and agreed to the published version of the manuscript.

Funding: The authors are grateful to the University Centrum for Applied Geosciences (UCAG) for the access to the E.F. Stumpfl electron microprobe laboratory. The authors also thank SGS Mineral Services, Canada, for performing the concentrate sample. S. Karipi is thanked for collecting the sample and participating in the field work. L. Bindi thanks MIUR, project "TEOREM deciphering geological processes using Terrestrial and Extraterrestrial ORE Minerals", prot. 2017AK8C32 (PI: Luca Bindi).

Acknowledgments: The authors acknowledge Ritsuro Miyawaki, Chairman of the CNMNC and its members for helpful comments on the submitted new mineral proposal. We are very grateful to the four anonymous referees for constructive comments that improved the manuscript. Many thanks are due to the editorial staff of *Minerals* and to the consulted referees. This paper is dedicated to the memory of our friend and colleague Demetrios.

Conflicts of Interest: The authors declare no conflicts of interest.

References

1. Zaccarini, F.; Bindi, L.; Ifandi, E.; Grammatikopoulos, T.; Stanley, C.; Garuti, G.; Mauro, D. Tsikourasite, $Mo_3Ni_2P_{1+x}$ ($x < 0.25$), a new phosphide from the chromitite of the othrys ophiolite, Greece. *Minerals* **2019**, *9*.
2. Britvin, S.N.; Murashko, M.N.; Vapnik, Y.; Polekhovsky, Y.S.; Krivovichev, S.V. Earth's phosphides in levant and insights into the source of Archean prebiotic phosphorus. *Sci. Rep.* **2015**, *5*. [CrossRef]
3. Zaccarini, F.; Pushkarev, E.; Garuti, G.; Kazakov, I. Platinum-group minerals and other accessory phases in chromite deposits of the Alapaevsk ophiolite, Central Urals, Russia. *Minerals* **2016**, *6*, 108. [CrossRef]
4. Sideridis, A.; Zaccarini, F.; Grammatikopoulos, T.; Tsitsanis, P.; Tsikouras, B.; Pushkarev, E.; Garuti, G.; Hatzipanagiotou, K. First occurrences of Ni-phosphides in chromitites from the ophiolite complexes of Alapaevsk, Russia and GerakiniOrmylia, Greece. *Ofioliti* **2018**, *43*, 75–84.
5. Ifandi, E.; Zaccarini, F.; Tsikouras, B.; Grammatikopoulos, T.; Garuti, G.; Karipi, S.; Hatzipanagiotou, K. First occurrences of Ni-V-Co phosphides in chromitite of Agios Stefanos mine, Othrys ophiolite, Greece. *Ofioliti* **2018**, *43*, 131–145.
6. Zaccarini, F.; Ifandi, E.; Tsikouras, B.; Grammatikopoulos, T.; Garuti, G.; Mauro, D.; Bindi, L.; Stanley, C. Occurrences of of new phosphides and sulfide of Ni, Co, V, and Mo from chromitite of the Othrys ophiolite complex (Central Greece). *Per. Ital. Mineral.* **2019**, *88*. [CrossRef]

7. Smith, A.G.; Rassios, A. The evolution of ideas for the origin and emplacement of the western Hellenic ophiolites. *Geol. Soc. Am. Spec. Pap.* **2003**, *373*, 337–350.
8. Hynes, A.J.; Nisbet, E.G.; Smith, G.A.; Welland, M.J.P.; Rex, D.C. Spreading and emplacement ages of some ophiolites in the Othris region (eastern central Greece). *Z. Deutsch Geol. Ges.* **1972**, *123*, 455–468.
9. Smith, A.G.; Hynes, A.J.; Menzies, M.; Nisbet, E.G.; Price, I.; Welland, M.J.; Ferrière, J. The stratigraphy of the Othris Mountains, eastern central Greece: A deformed Mesozoic continental margin sequence. *Eclogue Geol. Helv.* **1975**, *68*, 463–481.
10. Rassios, A.; Smith, A.G. Constraints on the formation and emplacement age of western Greek ophiolites (Vourinos, Pindos, and Othris) inferred from deformation structures in peridotites. In *Ophiolites and Oceanic Crust: New Insights from Field Studies and the Ocean Drilling Program*; Dilek, Y., Moores, E., Eds.; Geological Society of America: Boulder, CO, USA, 2001; pp. 473–484.
11. Barth, M.G.; Mason, P.R.D.; Davies, G.R.; Drury, M.R. The Othris Ophiolite, Greece: A snapshot of subduction initiation at a mid-ocean ridge. *Lithos* **2008**, *100*, 234–254. [CrossRef]
12. Barth, M.; Gluhak, T. Geochemistry and tectonic setting of mafic rocks from the Othris Ophiolite, Greece. *Contrib. Mineral. Petrol.* **2009**, *157*, 23–40. [CrossRef]
13. Dijkstra, A.H.; Barth, M.G.; Drury, M.R.; Mason, P.R.D.; Vissers, R.L.M. Diffuse porous melt flow and melt-rock reaction in the mantle lithosphere at a slow-spreading ridge: A structural petrology and LA-ICP-MS study of the Othris Peridotite Massif (Greece). *Geochem. Geophys. Geosyst.* **2003**, *4*. [CrossRef]
14. Magganas, A.; Koutsovitis, P. Composition, melting and evolution of the upper mantle beneath the Jurassic Pindos ocean inferred by ophiolitic ultramafic rocks in East Othris, Greece. *Int. J. Earth Sci.* **2015**, *104*, 1185–1207. [CrossRef]
15. Bortolotti, V.; Chiari, M.; Marcucci, M.; Photiades, A.; Principi, G.; Saccani, E. New geochemical and age data on the ophiolites from the Othrys area (Greece): Implication for the Triassic evolution of the Vardar ocean. *Ofioliti* **2008**, *33*, 135–151.
16. Economou, M.; Dimou, E.; Economou, G.; Migiros, G.; Vacondios, I.; Grivas, E.; Rassios, A.; Dabitzias, S. Chromite deposits of Greece. In *Chromites, UNESCO's IGCP197 Project Metallogeny of Ophiolites*; Petrascheck, W., Karamata, S., Eds.; Theophrastus Publ. S.A.: Athens, Greece, 1986; pp. 129–159.
17. Garuti, G.; Zaccarini, F.; Economou-Eliopoulos, M. Paragenesis and composition of laurite from chromitites of Othrys (Greece): Implications for Os-Ru fractionation in ophiolite upper mantle of the Balkan Peninsula. *Mineral. Depos.* **1999**, *34*, 312–319. [CrossRef]
18. Tsikouras, B.; Ifandi, E.; Karipi, S.; Grammatikopoulos, T.A.; Hatzipanagiotou, K. Investigation of platinum-group minerals (PGM) from Othrys chromitites (Greece) using superpanning concentrates. *Minerals* **2016**, *6*, 94. [CrossRef]
19. Bruker. *APEX3*; Bruker AXS Inc.: Madison, WI, USA, 2016; Available online: https://www.bruker.com/products/x-ray-diffraction-and-elemental-analysis/single-crystal-x-ray-diffraction/sc-xrd-software/apex3.html (accessed on 31 January 2020).
20. Bruker. *SAINT and SADABS*; Bruker AXS Inc.: Madison, WI, USA, 2016. Available online: https://www.bruker.com/products/x-ray-diffraction-and-elemental-analysis/single-crystal-x-ray-diffraction/sc-xrd-software/apex3.html (accessed on 31 January 2020).
21. Britvin, S.N.; Rudashevskii, N.S.; Krivovichev, S.V.; Burns, P.C.; Polekhovsky, Y.S. Allabogdanite, $(Fe,Ni)_2P$, a new mineral from the Onello meteorite: The occurrence and crystal structure. *Am. Mineral.* **2002**, *87*, 1245–1249. [CrossRef]
22. Sheldrick, G.M. A short history of SHELX. *Acta Crystallogr.* **2008**, *A64*, 112–122. [CrossRef]
23. Wilson, A.J.C. *International Tables for Crystallography: Mathematical, Physical, and Chemical Tables*; International Union of Crystallography: Chester, UK, 1992; Volume 3.
24. Ivanov, A.V.; Zolensky, M.E.; Saito, A.; Ohsumi, K.; Yang, S.V.; Kononkova, N.N.; Mikouchi, T. Florenskyite, FeTiP, a new phosphide from the Kaidun meteorite, Locality: Kaidun chondritic meteorite, South Yemen. *Am. Mineral.* **2000**, *85*, 1082–1086. [CrossRef]
25. Zolensky, M.; Gounelle, M.; Mikouchi, T.; Ohsumi, K.; Le, L.; Hagiya, K.; Tachikawa, O. Andreyivanovite: A second new phosphide from the Kaidun meteorite. *Am. Mineral.* **2008**, *93*, 1295–1299. [CrossRef]
26. Fruchart, R.; Roger, A.; Sénateur, J.P. Crystallographic and magnetic properties of solid solutions of the phosphides M2P, M = Cr, Mn, Fe, Co, and N. *J. Appl. Phys.* **1969**, *40*, 1250–1257. [CrossRef]
27. Wells, A.F. *Structural Inorganic Chemistry*, 5th ed.; Clarendon Press: Oxford, UK, 1984; p. 1288.

28. Smith, D.G.W.; Nickel, E.H. A system for codification for unnamed minerals: Report of the subcommittee for unnamed minerals of the IMA commission on new minerals, nomenclature and classification. *Can. Mineral.* **2007**, *45*, 983–1055. [CrossRef]

29. Malvoisin, B.; Chopin, C.; Brunet, F.; Matthieu, E.; Galvez, M.E. Low-temperature Wollastonite formed by carbonate reduction: A marker of serpentinite redox conditions. *J. Petrol.* **2012**, *53*, 159–176. [CrossRef]

30. Etiope, G.; Tsikouras, B.; Kordella, S.; Ifandi, E.; Christodoulou, D.; Papatheodorou, G. Methane flux and origin in the Othrys ophiolite hyperalkaline springs, Greece. *Chem. Geol.* **2013**, *347*, 161–174. [CrossRef]

31. Etiope, G.; Ifandi, E.; Nazzari, M.; Procesi, M.; Tsikouras, B.; Ventura, G.; Steele, A.; Tardini, R.; Szatmari, P. Widespread abiotic methane in chromitites. *Sci. Rep.* **2018**, *8*, 8728. [CrossRef]

32. Xiong, Q.; Griffin, W.L.; Huang, J.X.; Gain, S.E.M.; Toledo, V.; Pearson, N.J.; O'Reilly, S.Y. Super-reduced mineral assemblages in "ophiolitic" chromitites and peridotites: The view from Mount Carmel. *Eur. J. Mineral.* **2017**, *29*, 557–570. [CrossRef]

33. Pasek, M.A.; Hammeijer, J.P.; Buick, R.; Gull, M.; Atlas, Z. Evidence for reactive reduced phosphorus species in the early Archean ocean. *Proc. Natural Acad. Sci. USA* **2013**, *110*, 100089–100094. [CrossRef]

34. Ballhaus, C.; Wirth, R.; Fonseca, R.O.C.; Blanchard, H.; Pröll, W.; Bragagni, A.; Nagel, T.; Schreiber, A.; Dittrich, S.; Thome, V.; et al. Ultra-high pressure and ultra-reduced minerals in ophiolites may form by lightning strikes. *Geochem. Perspec. Lett.* **2017**, *5*, 42–46. [CrossRef]

35. Buseck, P.R. Phosphide from meteorites: Barringerite, a new iron-nickel mineral. *Science* **1969**, *165*, 169–171. [CrossRef]

36. Britvin, S.N.; Kolomensky, V.D.; Boldyreva, M.M.; Bogdanova, A.N.; Krester, Y.L.; Boldyreva, O.N.; Rudashevsky, N.S. Nickelphosphide (Ni,Fe)$_3$P—The nickel analogue of schreibersite. *Zap. Vserossi. Mineral. Obschch.* **1999**, *128*, 64–72.

37. Ma, C.; Beckett, J.R.; Rossman, G.R. Monipite, MoNiP, a new phosphide mineral in a Ca-Al-rich inclusion from the Allende meteorite. *Am. Mineral.* **2014**, *99*, 198–205. [CrossRef]

38. Pratesi, G.; Bindi, L.; Moggi-Cecchi, V. Icosahedral coordination of phosphorus in the crystal structure of melliniite, a new phosphide mineral from the Northwest Africa 1054 acapulcoite. *Am. Mineral.* **2006**, *91*, 451–454. [CrossRef]

39. Skala, R.; Cisarova, I. Crystal structure of meteoritic schreibersites: Determination of absolute structure. *Phys. Chem. Mineral.* **2005**, *31*, 721–732. [CrossRef]

Article

Podiform Chromitites and PGE Mineralization in the Ulan-Sar'dag Ophiolite (East Sayan, Russia)

Olga N. Kiseleva [1,*], Evgeniya V. Airiyants [1], Dmitriy K. Belyanin [1,2] and Sergey M. Zhmodik [1,2]

[1] Sobolev Institute of Geology and Mineralogy, Siberian Branch Russian Academy of Science, pr. Academika Koptyuga 3, Novosibirsk 630090, Russia; jenny@igm.nsc.ru (E.V.A.); bel@igm.nsc.ru (D.K.B.); zhmodik@igm.nsc.ru (S.M.Z.)

[2] Faculty of Geology and Geophysics, Novosibirsk State University, Novosibirsk 630090, Russia

* Correspondence: kiseleva_on@igm.nsc.ru; Tel.: +7-983-126-70-72

Received: 26 December 2019; Accepted: 2 February 2020; Published: 7 February 2020

Abstract: In this paper, we present the first detailed study on the chromitites and platinum-group element mineralization (PGM) of the Ulan-Sar'dag ophiolite (USO), located in the Central Asian Fold Belt (East Sayan). Three groups of chrome spinels, differing in their chemical features and physical–chemical parameters, under equilibrium conditions of the mantle mineral association, have been distinguished. The temperature and log oxygen fugacity values are, for the chrome spinels I, from 820 to 920 °C and from (−0.7) to (−1.5); for chrome spinels II, 891 to 1003 °C and (−1.1) to (−4.4); and for chrome spinels III, 738 to 846 °C and (−1.1) to (−4.4), respectively. Chrome spinels I were formed through the interaction of peridotites with mid-ocean ridge basalt (MORB)-type melts, and chrome spinels II were formed through the interaction of peridotites with boninite melts. Chrome spinels III were probably formed through the interaction of andesitic melts with rocks of an overlying mantle wedge. Chromitites demonstrate the fractionated form of the distribution of the platinum-group elements (PGE), which indicates a high degree of partial melting at 20–24% of the mantle source. Two assemblages of PGM have been distinguished: The primary PGE assemblage of Os-Ir-Ru alloys-I, $(Os,Ru)S_2$, and IrAsS, and the secondary PGM assemblage of Os-Ir-Ru alloys-II, Os^0, Ru^0, RuS_2, OsS_2, IrAsS, RhNiAs with Ni, Fe, and Cu sulfides. The formation of the secondary phases of PGE occurred upon exposure to a reduced fluid, with a temperature range of 300–700 °C, log sulfur fugacity of (−20), and pressure of 0.5 kbar. We have proposed a scheme for the sequence of the formation and transformation of the PGMs at various stages of the evolution of the Ulan-Sar'dag ophiolite.

Keywords: ophiolite; chromitites; PGE mineralogy; geodynamic setting

1. Introduction

Podiform chromitites are associated with restite peridotites in ophiolite complexes [1–5]. The concentration, platinum-group element distribution, and mineral form and composition of chrome spinels in podiform chromitites are sensitive indicators of mantle processes: the partial melting degree, initial melt sources, and melt saturation with volatile components (S, H_2O, etc.) [6–14]. Thus, the geochemical features of chromitites, their distribution and the mineral composition of PGE can be used to evaluate the physical–chemical parameters, mineral equilibria parameters, initial melt composition and geodynamic setting of ophiolite formations [14–22]. The use of an olivine–chrome spinel and olivine geothermometer, as well as experimental data on the temperature equilibria of the primary PGMs, provides important information on the formation temperatures of the mantle olivine spinel PGM assemblage and deformation processes [23–25]. The chemical composition of chrome spinels and PGMs, the microstructural features of this assemblage, and accessory sulfide mineralization give important information for the reconstruction of magmatic, hydrothermal, and metamorphic processes. In this paper, we describe the chemistry of chrome spinels, the geochemical features of chromitites and

PGMs from chromitites, the physical–chemical parameters of mineral equilibria, and the alteration of PGE–chromite mineralization in the mantle peridotites of the Ulan-Sar'dag ophiolite.

2. Materials and Methods

Twenty-one samples of chromitites from the western part of Ulan-Sar'dag were studied. The chemical composition of chrome spinels and olivines and the Al_2O_3 impurities in olivine were determined by wavelength-dispersive analysis using a Camebax-micro electron microprobe from the Sobolev Institute of Geology and Mineralogy, Russian Academy of Science, Novosibirsk, Russia (Analytical Center for multi-elemental and isotope research SB RAS). A JEOL JXA-8100 analysis methodology specific to olivines to obtain their trace elements has been elaborated: The accelerating voltage is 20 kV, the probe current is 400 nA, and the counting time per line and background measurement are both 10 s. The number of measurements per analysis is 25 [26]. For an error of 10 relative %, the quantitation limit of Al_2O_3 is within 45–100 ppm. The analytical conditions for olivines and chrome spinels are as follows: The accelerating voltage is 20 kV, the probe current is 30 nA, the beam size is 2 μm, and the signal accumulation time is 10 s. The standards used were natural and synthetic silicates and oxides. The detection limit for oxides was 0.01–0.03 wt. %.

The content of PGE in chromitites and dunites was determined in the rocks using the ICP-MS microprobe and kinetic methods, along with pre-concentration (nickel matte), at the TsNIGRI analytical laboratory (Moscow, Russia). The ICP-MS microprobe method with pre-concentration (nickel matte) involves melting the probe (50 g) at 1100 °C, during which two phases are formed: A sulfide (nickel matte) and an oxide (skimming) phase. These phases are segregated according to the concentration of all noble elements in nickel matte. The collector is an alloy of sulfides and a metal alloy, in which the bulk of the noble metals is collected [27]. The received matte is cleaned through skimming and crushed for 10–15 s. A portion of the matte is dissolved in hydrochloric acid, while the Ni and Cu of the matte are almost completely dissolved. Platinum metals in the nonsolute residue are dissolved in a mixture of hydrochloric and nitric acids. The mass concentration of Pt, Pd, Rh, Ru, and Ir in the solution is determined by ICP-MS. Osmium is separated from other noble metals by distillation. The method is based on the ability of osmium to form volatile tetraoxides under oxidizing conditions [28]. The determination of the osmium concentration was carried out using a kinetic method. The method is based on the catalytic effect of osmium in indicator oxidation-reduction reactions. The reaction rate depends on the concentration of the catalyst [28–30]. The detection limits were 2 ppb for Os, Ir, Ru, and Rh; and 5 ppb for Pt and Pd.

The studied platinum-group minerals were polished plates (21 pieces) and heavy fractions, extracted from chromite ores. The selected PGM grains were mounted in epoxy blocks and polished with a diamond paste for further analysis. Microtextural observations of PGM were performed by means of reflected-light microscopy. The chemical composition of PGMs was determined using a MIRA 3 LMU scanning electron microscope, with an attached INCA Energy 450 XMax 80 microanalysis energy dispersive system, at the Sobolev Institute of Geology and Mineralogy, Russian Academy of Science, Novosibirsk, Russia (Analytical Center, IGM SB RAS). Pure metals were used as the standards to determine the chemical composition of the PGE, Ni, and Cu; arsenopyrite was used for As; and pyrite was used for Fe. The minimum detection limits of the elements (wt. %) were found to be 0.1–0.2 for S, Fe, Co, Ni, and Cu; 0.2–0.4 for As, Ru, Rh, Pd, Sb, and Te; and 0.4–0.7 for Os, Ir, and Pt.

3. Geological Setting

The ophiolite complexes are widely distributed in the south-eastern part of Eastern Sayan (Siberia, Russia) (Figure 1).

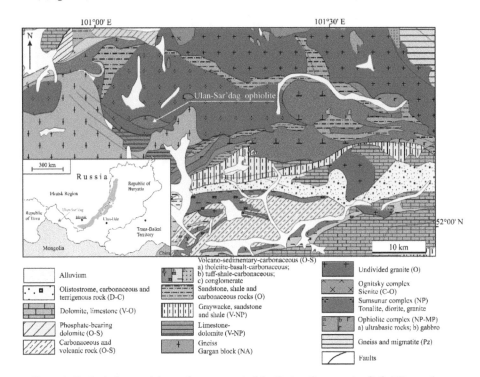

Figure 1. Geological map of the southeastern part of the Eastern Sayan region [31]. NB—northern branch, SB—southern branch.

The ophiolites are localized in the form of extended branches in the south (Il'chir) and north (Holbin-Hairhan). The features of these ophiolites are as follows: rock associations, geochemical and mineralogical characteristics, a geological structure resembling individual "massifs", geodynamic conditions and a formation time, which are actively studied for a considerable time. Dobretsov and Zhmodik et al. obtained data on the heterogeneity of ophiolites of the south-eastern part of Eastern Sayan (SEPES) [32,33]. They indicate that the ophiolites of the southern branch were formed in a mid-ocean ridges setting, and ophiolites of the northern branch were formed in an island arc setting [32–38]. The Ulan-Sar'dag ophiolite (USO) occupies a special structural position in the Eastern Sayan and Central Asian Fold Belt (Dunzhugur island arc). The USO is a tectonic plate, which is located between ophiolites of the southern and northern branches of the Dunzhugur island arc, near the contact zone of the Gargan "block" gneisses, with granites of the Sumsunur complex. The USO is underlaid by volcanogenic and sedimentary rocks of the Ilchir suite and limestones of the Irkut suite. Ophiolites include mantle restites (dunites, harzburgites), podiform chromitites, cumulates (metawehrlite, metapyroxenite, metagabbro), a basic dike and a volcanic complex (Ilchir suite) (Figures 2 and 3). USO has a lenticular body that is elongated in the east–west direction and is 1.5×5 km^2. It is composed of dunites and harzburgites. Harzburgites predominate in the central part, dunites and serpentinites prevail in the margin, and the latter lie in contact with the volcanogenic-sedimentary sequence of the Ilchir suite. The rocks of the cumulative series are metamorphosed under the conditions of epidote-amphibolite and amphibolite facies. The volcanogenic-sedimentary sequence is composed of volcanic, volcanogenic-sedimentary, and metasedimentary rocks (black schists and marbles). Volcanogenic-sedimentary rocks are sulfurized, and sulfide mineralization is confined to the schistosity zones. High-Ti basalts (MORB tholeiites), boninites, and island-arc andesites are distinguished in the volcanic complex. All rocks in the ophiolite nappe bottom are intensively deformed and mark the thrust zone. They include numerous crushing zones, signs of shear displacement and gliding planes. In the contact zone of serpentinites and rocks of the Ilchir suite, the talcum powder zones and zones of actinolite-tremolite composition are widely manifested.

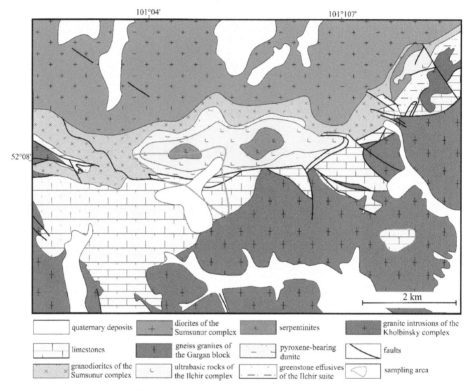

Figure 2. Geological scheme of the Ulan-Sar'dag ophiolite [39] with author additions.

The types of schlieren, lenticular, and vein-like chromite bodies are localized in dunites and serpentinites (Figure 4). The length of schlieren and massive chromitite bodies ranges from a few centimeters to several meters, and their width varies from 5 cm to 0.5 m. The predominant structural-textural type is a massive chromitite. In some schlieren chromitites, the tectonic flow structures (Figure 4e) and the "snowball" type (Figure 4f) are observed.

Figure 3. Photographs of the field relationship between mantle peridotites and volcanic-sedimentary rocks (Ilchir suite), limestone thickness (Irkut suit), Gargan gneisses, and Susunur tonalities: (**a**) view of the south side of the Ulan-Sar'dag ophiolite, (**b**) view of the north side of the Ulan-Sar'dag ophiolite.

Figure 4. Structural features of chromitite pods: (**a**) chromitite seams in dunites; (**b**) schlieren; (**c**) massive pods; (**d**) transformation of the schlieren type into the massive type due to deformation processes; (**e**) folded schlieren-type chromitites; (**f**) structure of the "snowball".

4. Results

4.1. Podiform Chromitites

Massive chromitites are the predominant petrographic variety of rocks in the mantle peridotites of the USO; disseminated, schlieren, and rhythmically-banded ores are less common. Chrome spinels make up 80–95 vol. % of massive chromitites. The intergranular space is filled with olivine and secondary silicates: serpentine, chlorite, and, rarely, talc. The structure of massive chromitites varies from fine-grained to coarse-grained. The crystal bodies vary from subidiomorphic grains (0.1–0.5 cm) to allotrimorphic aggregates of grains of up to 1.5 cm in size. Chrome spinels have a cataclastic texture, and they are partially fragmented. The grains are dissected by cracks filled with secondary silicates. In the central part, chrome spinels are unchanged, and in cracks and on the grain rims, they are often replaced by chrome magnetite. Some grains contain olivine inclusions. A wide range of minerals show accessory mineralization in chromite ores: PGMs, sulfides, sulfarsenides, and sulfosalts of base metals (Figure 5).

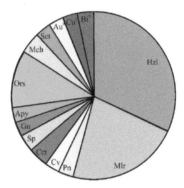

Figure 5. Accessory mineral association in chromitites. Hzl—heazlewoodite Ni_3S_2; Mlr—millerite NiS; Pn—pentlandite $(Fe,Ni)_9S_8$; Cv—covellite CuS; Cct—chalcocite Cu_2S; Orc—orcelite Ni_5As_2; Mh—maucherite $Ni11As_8$; Sct—scutterudite $Co_4(As_4)_3$; Apy—arsenopyrite FeAsS; Sp—sphalerit ZnS; Gn—galena PbS; Au^0, Cu^0, Bi^0—native gold, copper and bismuth.

4.1.1. Chrome Spinels

According to the chemical composition, ore chrome spinels are divided into three groups and are represented by chrome picotite, alumo-chromite, chromite, chrome-magnetite and magnetite (Table 1, Figure 6). In one sample, chrome spinels of all three groups may be present. Chrome spinels of groups I, II, and III have the following compositions (wt. %): Al_2O_3 = (17–43), (7–16), and (9–18); Cr_2O_3 = (26–54), (46–66), and (47–57); MgO = (10–20), (9–13), and (8–10); and FeO = (6–14), (1–18), and (18–20), respectively. In chromitites of the USO, chrome picotites are found, which is typical only of this massif, in contrast to chrome spinels from other ophiolite complexes of the southeastern part of East Sayan [38]. Chrome spinels have a homogeneous composition. Altered chrome spinels demonstrate increased contents of FeO, MnO, NiO, ZnO, and Fe_2O_3 and decreased contents of Cr_2O_3 and Al_2O_3. Chrome spinels are often replaced by chrome magnetite and magnetite along the cracks.

Table 1. Representative chemical composition of Cr spinels and the calculated parental melt composition (wt. %) of Cr spinels. Thermometric and oxybarometric data for Cr spinels.

No.	1	2	3	4	5	6	7	8	9	10	11	12	13	14	15	16	17	18	19	20	21	22
	I Group					II Group													III Group			
N Sample	37	41	74	77	101	130	132	120	10	48	52	47	53	5	112	88	9	60	71	11	22	18
TiO_2	0.1	0.1	0.1	0.1	0.1	0.10	0.1	0.1	0.1	0.05	0.07	0.1	0.1	0.1	0.2	0.2	0.1	0.05	0.04	0.07	0.05	0.07
Al_2O_3	40.3	41.0	24.6	24.1	19.2	12.19	7.97	15.3	14.6	13.52	16.53	14.5	14.8	13.9	12.5	12.4	14.6	14.6	13.0	14.0	16.5	19.6
Cr_2O_3	29.7	28.5	44.2	43.7	49.2	59.57	63.94	56.4	54.2	56.51	53.70	55.6	55.6	56.5	56.4	57.0	57.3	54.0	55.4	49.0	48.0	48.3
MnO	0.5	0.5	0.2	0.3	0.3			0.3	0.3	0.32	0.56	0.3	0.4	0.3	0.5	0.4	0.4	0.3	0.4	0.4	0.7	0.3
FeO	7.9	7.9	14.7	14.8	14.4	15.4	19.1	13.3	14.8	15.8	15.3	15.4	16.4	15.6	16.1	16.3	17.7	19.0	18.3	18.6	20.9	19.8
Fe_2O_3	2.0	2.3	2.1	2.6	3.3	2.4	0.7	1.2	2.1	1.7	2.1	1.6	2.3	1.7	2.7	2.7	1.1	1.8	2.3	5.6	5.8	2.5
MgO	19.7	19.8	13.6	13.3	12.9	12.6	10.8	13.7	12.4	11.7	12.4	12.0	11.7	11.9	11.4	11.3	10.9	9.6	9.8	9.6	8.3	9.6
V_2O_5	0.8	0.8	0.1	0.1	0.1			0.1	0.1	0.13		0.1		0.1	0.1	0.1	0.1	0.1	0.07	0.11	0.12	0.09
NiO	0.3	0.3	0.2	0.1	0.0			0.0	0.1	0.02		0.0		0.0	0.0	0.0	0.1	0.0	0.03	0.06	0.07	0.02
ZnO	0.0	0.0	0.0	0.2	0.7			0.0	0.0	0.14		0.1		0.2	0.3	0.2	0.0	0.3	0.19	0.10	0.46	0.46
Total	101.3	101.1	99.8	99.2	100.2	102.3	234.6	100.4	98.6	99.8	100.7	99.7	101.2	100.2	100.1	100.5	102.2	99.8	99.5	97.6	100.9	100.7
Al'	56	57	35	34	27	23	26	20	20	19	23	20	20	19	20	20	20	20	18	20	23	28
Cr'	41	40	62	62	69	74	84	80	80	79	74	80	80	80	80	80	80	80	78	71	68	69
Fe'	2.7	3	3	3.7	4	2.9	0.9	1.7	3.0	2.3	2.9	2.2	3.2	2.3	3.7	3.7	1.6	2.6	3	8	8	4
Mg'	72	71	48	47	47	59	50	49	54	57	55	56	58	57	59	59	62	66	35	34	29	33
f'	28	29	52	53	53	41	50	49	54	57	55	56	58	57	59	59	62	66	65	66	71	67
TiO_2 melt	0.1	0.1	0.1	0.1	0.1	0.18	0.11	0.1	0.2	0.10	0.13	0.1	0.1	0.1	0.4	0.3	0.2	0.10	0.09	0.14	0.10	0.14
Al_2O_3 melt	18.3	18.5	15.0	14.9	13.5	11.1	9.3	12.3	12.2	11.7	12.7	12.1	12.1	11.8	11.3	11.3	12.0	12.1	11.5	12.0	12.7	13.6
(Fe/Mg)m	0.3	0.3	0.5	0.6	0.5	0.5	0.7	0.4	0.5	0.6	0.5	0.6	0.6	0.6	0.6	0.6	0.7	0.9	0.78	0.80	1.07	0.96
T°C (Ol-Sp)	919.1	893.9	823.4	820.7	885.7	938.62	991.6	1003.9	948.9	931.8	891.6	931.5	921.2	933.2	942.4	913.6	861.2	812.6	846.3	821	742.7	738.1
fO_2	−0.988	−0.765	−1.498	−1.246	−1.1	−2.48	−1.3	−2.1	−1.3	−1.8	−2.5	−1.9	−1.6	−1.9	−1.1	−2.8	−4.4	−3.3	−3.01	−1.49	−1.73	−2.78
T°C (Al in Ol)												893–1332				1073–1225						

Notes: $Al' = Al/(Al + Cr + Fe^{3+})$; $Cr' = Cr/(Cr + Al + Fe^{3+})$; $Fe' = Fe^{3+}/(Cr + Al + Fe^{3+})$; $Mg' = Mg/(Mg + Fe^{2+})$; $f' = Fe^{2+}/(Mg + Fe^{2+})$; Sample No. 1–5—Cr-spinels I; 6–18—Cr-spinels II; 19–22—Cr-spinels III; T°C (Ol-Sp)—olivine-spinel geothermometer; $f(O_2)$—oxygen fugacity; T°C (Al in Ol)—olivine geothermometer.

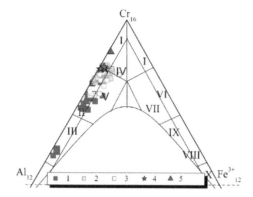

Figure 6. Classification diagram of chrome spinels from the chromitites of Ulan-Sar'dag, based on the structural formula of spinels. Composition fields: I—chromite; II—alumo-chromite; III—chrome picotite; IV—subferrichromite; V—subferrialumochromite; VI—ferrichromite; VII—subferrialumoferrichromite; VIII—chrome magnetite; IX—subalumochrome magnetite; X—magnetite [40]. Notes: 1—chrome spinels-I; 2—chrome spinels-II; 3—chrome spinels-III; 4—chrome spinels-II, with inclusions $(Os,Ru)S_2$; 5—chrome spinels-II, with olivine inclusion.

4.1.2. Olivine

Olivine in chromitites occupies the intergranular space of chrome spinels; it is often replaced by serpentine or chlorite. The data, which correspond to chrysotile and forsterite, are shown in Table 2. The MgO content is 49–54 wt. %, and the NiO variations are insignificant and amount to 0.37–0.42 wt. %, which corresponds to olivines from depleted peridotites. The fraction of the Fo component is 92–97.

Table 2. Chemical composition of the olivine from chromitites (wt. %).

N-Sample	SiO_2	MgO	FeO	NiO	MnO	CaO	Al_2O_3	Cr_2O_3	Total	Mg#
298-1	41.28	52.59	4.98	0.41	0.11	0.001		0.002	99.38	95
298-2	41.83	52.25	4.50	0.42	0.09	0.001		0.002	99.08	95
298-3	41.26	52.20	5.68	0.42	0.12	0.002		0.004	99.68	94
298-4	41.31	52.10	5.71	0.42	0.13	0.001	0.0118	0.002	99.68	94
298-5	41.41	52.68	5.34	0.44	0.11	0.002	0.0052	0.002	99.98	95
3-6	41.08	51.19	6.95	0.37	0.11	0.001	0.0191	0.002	99.73	93
3-7	41.10	51.04	6.98	0.37	0.11	0.003	0	0.001	99.60	93
3-8	41.15	51.03	6.85	0.38	0.12	0.004	0.0119	0	99.54	93
3-9	40.83	50.85	7.13	0.38	0.13	0.015	0.0104	0.002	99.34	93
3-10	41.13	51.17	6.92	0.38	0.10	0.003	0.0015	0	99.69	93
3-11	41.22	51.02	6.89	0.37	0.11	0.004	0.0117	0.001	99.62	93
3-12	41.09	51.16	6.90	0.38	0.13	0.006	0.0005	0.001	99.66	93
3-13	41.07	51.19	6.89	0.37	0.13	0.005	0.0006		99.65	93
3-14	41.10	51.08	6.93	0.37	0.10	0.006	0.0009		99.58	93
3-15	41.24	51.26	6.79	0.38	0.13	0.009	0.003	0.002	99.80	93
305-16	41.33	53.61	3.28					0.61	98.83	97
305-17	41.38	53.86	3.45					0.69	99.38	97
2-18	41.05	51.71	6.47						99.23	93
973-19	40.25	50.24	8.14					0.01	98.64	92
976-20	41.43	51.26	6.90					0.02	99.61	93
977-21	40.09	50.39	6.85			0.02		0.00	97.35	93
939-22	40.94	49.59	7.28			0.01		0.03	97.85	92
939-23	40.74	49.06	9.61			0.04		0.36	99.81	90
970-24	40.85	49.93	7.87			0.00		0.02	98.67	92
980-25	42.39	53.59	5.50			0.01		0.02	101.51	95

Notes: (Samples 16–18)—inclusions in chromite.

4.2. Geochemistry of Platinum-Group Elements

The content of PGE in mantle peridotites and podiform chromitites was determined. The content of PGE in dunites and serpentinites is 94–180 ppb; in chromitites, it ranges from 242 to 992 ppb (Table 3). The PGE content increases with an increase in the volume percentage of chromite in the rock. In addition, depending on the proportion of the chromite and silicate components in the rock, the IPGE/PPGE ratio changes (Figure 7a,b) (IPGE: Os, Ir, and Ru; PPGE: Rh, Pt, and Pd). For example, in massive ores (85–95 vol. % of chromite), IPGE > PPGE, and IPGE/PPGE = 0.86–2.15. In schlieren lenticular chromitites and chromitites with deformational textures, the PPGE proportion increases and amounts to IPGE/PPGE = 0.03–0.67. In chromitites I, the IPGE/PPGE is higher than in chromitites II (Table 3).

Table 3. PGE abundance for the dunites and chromitites of the Ulan-Sar'dag ophiolite (ppb).

No.	1	2	3	4	5	6	7	8	9	10	11
N Sample	6	305	6 mas	7 mas	294 mas	307 mas	6 mas	3 mas	17 mas	20 mas	2 shl
Os	6	8	45	51	58	117	49	37	81	46	7
Ir	21	6	26	58	43	82	57	20	35	20	6
Ru	7	14	68	53	116	221	121	59	44	46	20
Rh	7	3	12	10	8	19	24	16	9	11	21
Pt	35	15	39	49	31	54	49	64	46	41	35
Pd	104	48	177	182	104	122	478	97	87	78	903
Total	180	94	367	403	360	615	778	293	302	242	992
Pt/Ir	1.67	2.5	1.5	0.84	0.72	0.66	0.86	3.20	1.31	2.05	5.83
∑IPGE	34	28	139	162	217	420	227	116	160	112	33
∑PPGE	146	66	228	241	143	195	551	177	142	130	959
IPGE/PPGE	0.23	0.42	0.61	0.67	1.52	2.15	0.41	0.66	1.13	0.86	0.03
∑PGE	180	94	367	403	360	615	778	293	302	242	992
Cr# (Crt)			68	67	65	57	70	69	84	83	75
Al#(Crt)			30	30	42	34	26	27	14	15	22

Notes: 1,2—dunites; 3–11—chromitites. mas—massive pods, shl—shlieren pods; 3–6—group I chromitites; 7–10—group II chromitites; 11—group III chromitite.

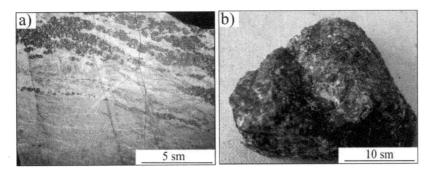

Figure 7. Photographs of chromitites with different values of IPGE/PPGE: (**a**) schlieren densely disseminates in serpentinized dunite (∑PGE = 903ppb, and IPGE/PPGE = 0.03); (**b**) massive (∑PGE = 615 ppb, and IPGE/PPGE = 2.15).

4.3. Mineralogy of Platinum-Group Elements

The first data on PGE mineralization (PGM) in chromitites of USO have been obtained. The chemical composition of the minerals is presented in Table 4, and classification diagrams are presented in Figure 8a,b. Primary and secondary PGMs have been distinguished. The most common platinum-group minerals in the chromitites of the USO are PGE sulfides: laurite-erlichmanite $(Ru,Os)S_2$, with different Ru/(Ru + Os) ratios. Other PGE phases are represented by high-temperature Os-Ir-Ru alloys-I, phases of variable-composition Os-Ir-Ru alloys-II, native Os, Ru, sulfarsenides of (Os, Ir, Ru), and zaccarinite RhNiAs. The primary and secondary PGMs will be described separately.

Table 4. Representative composition of platinum-group elements in the chromitite from the Ulan-Sar'dag ophiolitic massif.

No. an	wt. %											apfu										Ru/(Ru + Os)
	Os	Ir	Ru	Rh	Fe	Ni	S	As	Sb	O	Total	Os	Ir	Ru	Rh	Fe	Ni	S	As	Sb	O	
1	75.01	21.95	3.38		0.72						101.06	0.39	0.11	0.03		0.01						
2	74.84	20.6	5.16		0.33						100.93	0.39	0.11	0.05		0.01						
3	79.75	20.74	2.35		0.35			0			103.19	0.41	0.1	0.02		0.01						
4	23.87	5.23	34.23		0.59		33.36				97.28	0.24	0.05	0.65		0.02		2				0.59
5	23.21	5.18	39.55				34.33				102.87	0.23	0.05	0.73		0.00		2				0.63
6	33.29	5.5	27.26	0.6	0.58		31.27				97.9	0.36	0.06	0.55	0.01	0.02		2				0.45
7	22.92	4.3	39.85				35.23				102.3	0.22	0.05	0.73				2				0.63
8	20.65	6.86	39.48				33.15				100.14	0.21	0.07	0.75				2				0.66
9	8.94	5.76	50.87				37.82				103.39	0.08	0.05	0.85				2				0.85
10	33.62	2.93	34.3				32	0.66			103.51	0.35	0.03	0.67				1.98	0.02			0.51
11	49.97	2.18	20.8				30.12				103.07	0.56	0.02	0.44				2.00				0.29
12		3.5	58.59				38.19	1.67			101.95		0.03	0.95				1.96	0.04			1
13			62.81				39.43				102.24			1.01				2				1
14	1.91	4.45	57.45		0.32	0.49	36.31	0.92			101.85	0.02	0.04	0.99		0.01	0.01	1.98	0.02			0.97
15	92.73	0	8.57			0.47	1.2	0.19			103.16	0.79		0.14		0.00	0.01	0.06				
16	37.98	31.17	29.55		0.88	1.61					101.19	0.29	0.23	0.42		0.02	0.04					
17	28.79	27.29	37.06	4.4	0.71		1.15				99.4	0.21	0.2	0.51		0.02						
18	33.33	11.19	53.61							2.12	100.25	0.19	0.06	0.6	0.06						0.14	
19	34.25	9.51	54.3				1.57			3.19	102.82	0.17	0.05	0.54				0.05			0.19	
20	88.01	11.34			1.09	2.01				0.66	103.31	0.8	0.1			0.03	0.06					
21	59.46	16.78	21.62		0.58	3.06					101.5	0.46	0.13	0.32		0.01	0.07					
22	2.52	2.11	93.14			2.55					100.32	0.01	0.01	0.93			0.04					
23	3.01	3.67	52.84		0.32	0.99	36.78	3.3			100.59	0.03	0.03	0.88		0.01	0.03	1.93	0.07			0.95
24			60.54		0.37	0.46	38.33				99.65			1		0.02	0.01	2				1
25	47.44	4.97	22.55				27.28				102.61	0.59	0.06	0.52				2				0.32
26			60.93				38.64				99.57			1				2				1
27	9.25	3.47	54.02		0.37		34.92				101.66	0.09	0.03	0.98		0.03		2				0.85
28	60.28	6.17	6.88		0.46	1.04	23.79				98.53	0.85	0.09	0.18		0.01	0.05	2				0.1
29			59.89				37.67				98.02			1.01				2				1
30		58.69	2.1	2.31			12.51	23.25			98.86		0.29	0.02	0.02			0.37	0.30			
31	1.18	59.23	1.44	0.35		0.81	11.38	24.82			99.21	0.03	0.41	0.1				0.63	0.37			
32	1.8	56.58	5.84	0.6			12.85	19.07	1		97.74	0.01	0.29	0.06	0.01			0.39	0.25	0.01		
33		48.23	11.47	3.91			16.81	20.93			101.35		0.21	0.09	0.03			0.44	0.23			
34		61.04	2.53	0.87			11.81	23.78	1.16				0.30	0.02	0.01			0.35	0.30	0.01		
35	4.21	3.15	10	37.77		22.35		25.66			103.14	0.02	0.01	0.08	0.3		0.31		0.28			
36	21.34	10.58	26.92	0.57		28.95	6.5	14.23			109.09	0.17	0.08	0.40	0.01		0.74	0.31	0.29			

Notes: Primary high temperature PGM: Os-Ir-Ru alloys-I—(1-3); laurite I—(4-7); altered laurite II—(8-14); laurite III—(23-29); irarsite—(30-34); RhNiAs—35; (Ru,Ni,Os,Ir,Rh)AsS—36. Secondary PGM: Os-Ir-Ru alloys-II—(15-22).

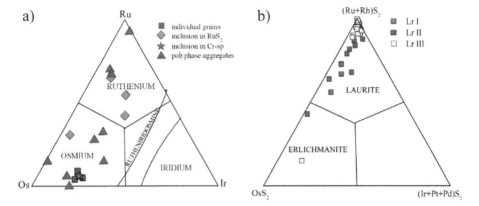

Figure 8. Diagram of the compositions of PGE minerals: (**a**) Os-Ir-Ru alloys; (**b**) Laurite-erlichmanite: Lr I—primary, inclusion in chrome spinels or isolated grains, laurite II—primary altered sulfides, and laurite III—secondary sulfides. The fields shown in the diagrams are drawn from [41].

4.3.1. Primary Platinum-Group Minerals

High-temperature Os-Ir-Ru alloys-I are in the form of idiomorphic inclusions in chrome spinels, and xenomorphic grains are intergrown with laurite. Single grains contain inclusions of $(Ru,Os)S_2$ of less than 10 μm in size. The alloys are enriched with Os, and its content varies from 71 to 79 wt. %; the content of Ir ranges from 20 to 28 wt. %; and the content of Ru is very low and ranges from 2 to 5 wt. %. There are some impurities of Fe and Ni. In the classification diagram, the compositions of the primary alloys correspond to osmium with low contents of Ir and Ru (Figure 8a). Dissolution microstructures are observed in some grains (Figure 9).

Laurite-Erlichmanite $(Ru,Os)S_2$

Sulfides can be divided into two groups, according to their composition and microstructural features (Table 4). The first group, laurite I, contains laurite-erlichmanite with a homogeneous composition.

Laurite-erlichmanite occurs as an inclusion in chrome spinels (10–15 μm), and in some cases, it can be intergrown with amphibole (magnesio-hastingsite hornblende) in chrome spinel or be found as individual grains (Figure 9c,d). The ratio, Ru′ = Ru/(Ru + Os), is 0.61–0.78. The content of Os is 20–33 wt. %, and Ru is 27–40 wt. %. In the classification diagram, they are in the laurite field (Figure 8b). The second group, laurite II, contains laurites of a heterogeneous composition, containing micro inclusions of (Os-Ir-Ru) II alloys and laurites, which are replaced by PGE sulfarsenides (Figure 9e–g). Laurite II (Lr II) is characterized by wide variations in the contents of Os of 1.2–49.9 wt. %, Ru of 20.8-62.2 wt. %, and the Ru/(Ru + Os) ratio of 0.44–0.99. In the classification diagram, they are in the fields of erlichmanite and laurite. In the chemical composition of laurite, a sulfur deficiency is often registered (Table 4). Insignificant amounts of irarsite IrAsS are found in the chromitites of the USO, where it replaces laurite. Irarsite forms corrosive, looped structures to replace laurite (Figure 9g,h).

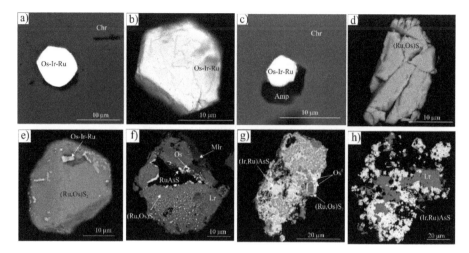

Figure 9. Back Scattered Electron (BSE) images of primary PGM, showing the textural and morphological relations of single and polyphase PGM from the Ulan-Sar'dag chromitites: (**a**) inclusion euhedral Os-Ir-Ru alloy—I in chromite, an. 1; (**b**) individual single-grain Os-Ir-Ru alloys—I, with a dissolution microstructure, an. 2; (**c**) inclusion in Cr-spinel of intergrowth laurite I and hornblende, an. 6; (**d**) individual grain of laurite I, an. 7; (**e**) grain of laurite II (an. 8), with inclusions of micro particle Os-Ir-Ru alloys—II; (**f**) grain laurite II (an. 9), associated with an unnamed phase (Ru,Ni,Os,Ir,Rh)AsS (an. 36) (laurite II is surrounded by Os-poor laurite, which grows with millerite); (**g**) substitution of laurite II by irarsite (an. 32), with remnants of laurite II (an. 10,11) and Os-Ir-Ru alloys—II; (**h**) laurite II (an. 12) surrounded by irarsite (an. 33). Abbreviations: Crsp—chrome spinel; Lr—laurite; Mlr—millerite. Notes: an. No—No analysis, as shown in Table 4.

4.3.2. Secondary Platinum-Group Minerals

The secondary PGMs are presented by (Os,Ru)S$_2$—laurite III, native Os and Ru, Os-Ir-Ru alloys-II of a variable composition, irarsite IrAsS, zaccarinite RhNiAs, and (Ru,Ni,Os,Ir,Rh)(As,S) sulfarsenides of a non-stoichiometric composition. The secondary phases are localized in the chloritized silicate intergranular space of chrome spinels. Some grains (micro particles of osmium and other phases) are very small (less than 2–3 μm), and their chemical composition can be determined only semi-quantitatively. In some cases, micro particles of a variable composition (Ru,Ir,Os,Cu,Te,Ni,Ba,S,O) can be found with secondary PGMs. Further, due to their very small size and partial coincidence with the elemental composition of the host mineral, qualitative analysis can also be hampered.

Os-Ir-Ru Alloys-II

Micro particles of Os-Ir-Ru alloys-II are the common phases and are found in a secondary mineral assemblage. They are localized mainly in laurite II, or they are a part of polyphase aggregates with Ni, Cu sulfides, sulfarsenides of Os, Ir, and Ru, zaccarinite RhNiAs, and laurite III (Figure 10a–d). Their composition varies (Table 4, Figure 8a) from native osmium to native ruthenium. Native osmium is composed of (wt. %): Os (87–92), Ir (0–12), and Ru (3–8). Native Ru is composed of (wt. %): Ru (93), Os (2.5), and Ir (2.1). The composition of the (Os-Ir-Ru) II alloys varies, with a wide range (wt. %): Os (30–74), Ir (6–32), and Ru (8–58).

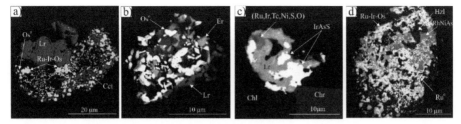

Figure 10. BSE images of secondary PGMs, showing the textural, morphological relations and assemblages of the PGM from the Ulan-Sar'dag chromitites: (**a**) intergrowth of laurite III (an. 26) and PGE bearing chalcocite Cu_2S, with micro inclusions of Os-Ir-Ru alloys- II (an. 18); (**b**) polyphase aggregate of agglomeration PGM particles, consisting of Os-rich laurite III (an.28), native Os^0 (an. 20), and Os-poor laurite III (an. 27); (**c**) composite grain of irarsite (an. 34), laurite III (an. 29), and unnamed phases (Ru, Ir, Te, Ni, S, O) within the interstitial chlorite of chromitites; (**d**) polyphase aggregate, consisting of Os-Ir-Ru alloys-II (an. 21), Ru^0 (an. 22), zaccarinite RhNiAs (an. 35), and heazlewoodite. Abbreviations: Chl—chlorite; Cct—chalcocite; Hzl—heazlewoodite; Irs—irarsite; Lr III—laurite III. Notes: an. No—No analysis, as shown in Table 4.

Laurite III

The occurrence forms of secondary laurite III are very diverse. It is mainly localized in Cr-containing chlorite, where it forms polyphase aggregates with PGE-containing chalcocite (Cu_2S) (Figure 10a), recrystallized PGM aggregates with native Os (Figure 10b), micro particles in association with Ni_3S_2, NiS and phases of a non-stoichiometric composition (Ru,Ir,Os,Cu,Te,Ni,Ba,S,O) (Figure 10c). The following features are characteristic of laurite III: a) A porous structure, with Cr-bearing minerals of the chlorite group in the voids, and b) a very small grain size (micro particles). The chemical composition corresponds to the end member of the laurite-erlichmanite solid solution: RuS_2-OsS_2 (Figure 8b). Laurite III is characterized by very low Os and Ir contents and the absence of Rh. Sometimes it contains Ni impurities. The value of Ru' varies insignificantly and amounts to 0.93–1 (Table 4); in turn, OsS_2 has low contents of Ru and Ir and is located in a recrystallized aggregate with laurite III and native Os (Figure 10b).

The (Ir,Os,Ru)AsS in the secondary association is in the form of polyphase aggregates. It replaces laurite. It contains Os and Ru in insignificant amounts. Secondary irarsite is presented as very small microparticles (less than 5 μm) in chlorite in association with Ni_3S_2 and laurite III.

(Ru,Ni,Os,Ir,Rh)(As,S) is found in the form of micro particles (7 μm) in a polyphase aggregate, consisting of laurite II and NiS (Figure 9f). There is a deficit of S and As in this phase.

Zaccarinite RhNiAs is found in polyphase aggregates in association with Ni_3S_2, secondary (Os-Ir-Ru) II alloys and native Ru (Figure 10d).

Unknown phase No. 1 (Ir,Ni,Cu,Ru,Os,Cl) is found in the intergrowth with laurite, erlichmanite, and chalcocite. This association is localized in chromite in the cracks filled with Cr-containing chlorite. It is worth noting the presence of Cl in this phase.

Unknown phase No. 2 (Os,Ir,Ru,As,S,O) is found in the millerite cracks in close association with chlorite and chrome-magnetite.

Unknown phase No. 3 (Ru,Ir,Te,Ni,Ba,S,O) is found in chlorite in the intergrowth with laurite and irarsite. The Ba and Te impurities are unusual for the platinum phases of chromitites.

Ni,Fe,Cu sulphides and arsenides. The sulphides of base metals form a dispersed impregnation mainly in serpentine-chlorite aggregate. Heazlewoodite Ni_3S_2 and millerite NiS are the predominant sulfide phases (Figure 5, Table 5). Heazlewoodite is often found in polyphase aggregates with secondary PGE minerals. It has a homogeneous composition and is identical in individual grains and in the intergrowth with platinum metal phases.

Table 5. Chemical composition of Ni, Fe, and Cu sulfides and arsenides (wt. %).

No. Mineral	2a-12 $(Fe,Ni)_9S_8$	6-12 Cu_2S	6-13 Ni_3S_2	3-13 NiS	6-13 Ni_3S_2	4-12 Ni_3S_2	3-13 Ni_3As_2
Ni	38.9		67.5	64.16	72.82	72.19	64.24
Fe	25.58						
Cu		64.89					
Co	0.87						
Os		4.34					
Ir		1.18					
Ru		8.01	5.29				
Rh			0.42				
S	33.08	19.8	25.81	33.82	27.25	27.2	
As			0.55				36.01
O		1.06					
Total	98.43	99.28	99.57	97.98	100.07	99.39	100.25
	Incl. in olivine	PGE bearing, intergrowth with PGM			PGE-free, individual grains		

5. Discussion

5.1. Chromitite Formation: Composition of Parental Melts

Chromite bodies in ophiolites are formed due to the partial melting of rocks of the upper mantle. The interaction of the melt with mantle peridotites plays an important role in the formation of podiform chromitites. In the channel filled with molten mantle, the ascending olivine-chromite-cotectic melt mixes with the silica-enriched melt, formed through the harzburgite-melt reaction. The formation of surrounding dunites along the host harzburgites is the result of a combination of olivine precipitation from "older" magma and the destruction of orthopyroxene in harzburgite, interacting with the magma channel. Thin chromite schlieren and streaks can be formed by separating the chromite from the cotectic olivine-chromite melt [9,42–46].

In the Pt/Pt*-Pd/Ir diagram (Figure 11), dunites and chromitites are within the partial melting trend. The late metamorphic and hydrothermal processes, during which Pd enrichment occurred, probably cause the high Pd/Ir ratio in some Ulan-Sar'dag chromitites. In the OSMA diagram, most of the ore chrome spinels are in the olivine-spinel equilibrium field, and some of the chrome spinels are beyond this field. This can be explained by the distortion of the magmatic system closedness, changes in the compositions during the tectonic processes and metamorphism of chrome spinels. Chromitites of the USO were formed at the 28–35% degree of partial melting (Figure 12). The data on the dunites, harzburgites, and chromitites of the northern and southern branches of the Ospa-Kitoy ophiolite are presented for comparison (Figure 12). The Mg'(Ol)–Cr'(Sp) ratio in dunites and harzburgites corresponds to the 30–40% degree of partial melting; in ore chrome spinels from the northern branch, it corresponds to 30–40%; in ore chrome spinels of the southern branch, it corresponds to 35%.

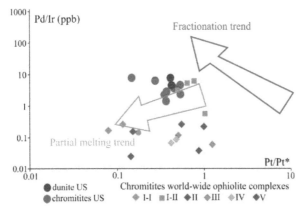

Figure 11. Plot of Pd/Ir versus Pt/Pt* for mantle peridotite and chromitites of the USO. The Pt anomaly is calculated as follows: Pt/[Pt]* = (Pt/8.3) × $\sqrt{(Rh/1.6) \times (Pd/4.4)}$ [8]. US—Ulan-Sar'dag ophiolite; ophiolite complexes from around the world: I—Wadi Al Hwanet ophiolite, Saudi Arabia [47]; II—Oman ophiolite, Semail [48]; III—Veria ophiolite, Greece [49]; IV—Shetland Ophiolite Complex, Scotland [50]; V—Ray-Iz ophiolite, Russia [51].

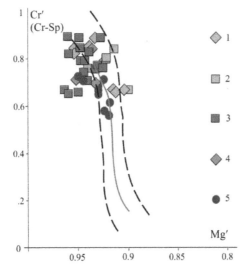

Figure 12. OSMA (olivine-spinel mantle array) is a spinel peridotite restite trend [52]. The chrome spinels are from: 1–4—the Ospa-Kitoy ophiolite: 1—dunites; 2—harzburgites; the chromitites are from: 3—the north-branch Ospa-Kitoy, 4—south-branch Ospa-Kitoy, 5—Ulan-Sar'dag ophiolite. Mg'-Mg/(Mg + Fe) in olivine; Cr'-Cr/(Cr + Al + Fe^{3+}) in chrome spinel.

The composition of the chrome spinels of the USO varies significantly, which may reflect the interaction of mantle peridotites with the melts of various compositions. The joint presence of high-Cr' and high-Al' chrome spinels is common in many ophiolite belts, but as a rule, chrome spinels with different compositions are found in different ophiolite nappes [7,9,52–59]. Their joint occurrence in one ophiolite nappe is less common; in this case, the ophiolite nappe, as a rule, contains peridotites depleted to different degrees [45,60–63]. The composition of chrome spinels of podiform chromitites that includes, as the main components, FeO, MgO, Al$_2$O$_3$, and TiO$_2$ is a function of the composition of the parental melts [9,15,16,18,21,64]. We have calculated the contents of the Al$_2$O$_3$, TiO$_2$ and FeO–MgO ratios in the parental melts, which were in equilibrium with the podiform chromitites (Table 6), according to Equations (1)–(3) [15]. Table 6 shows, for comparison, the values of these parameters for the chrome spinels of ophiolites from around the world.

$$Al_2O_{3Sp}(wt. \%) = 0.035 \times (Al_2O_3)_{melt}^{2.42}, \tag{1}$$

$$TiO_{2(melt)}(wt. \%) = TiO_{2Sp}^{0.82524} \times e^{0.20203} \tag{2}$$

$$Ln(FeO/MgO)_{Sp} = 0.47 - 1.07 \times Al'_{Sp} + 0.64 \times Fe^{3+\prime} + Ln(FeO/MgO)_{melt}, \text{ where } Al'_{Sp} = Al/(Al + Cr + Fe^{3+}) \text{ and } Fe^{3+\prime}_{Sp} = Fe^{3+}/(Al + Cr + Fe^{3+}) \tag{3}$$

The chemical composition of the chrome spinels I group (Crsp I) and the composition of the parental melt are similar in these parameters for medium-aluminous chrome spinels of the Ospa-Kitoy ophiolite (Table 6). The $(Al_2O_3)_{melt}$ value of Crsp I is 13–18 wt. %, and the Al_2O_{3Sp} / Al_2O_{3melt} ratio corresponds to the spreading trend and abyssal peridotites (Figure 13a). Despite the low TiO_2 content, Crsp I has the TiO_{2Sp}/TiO_{2melt} trend, which is calculated for spinels from the MORB-type peridotites (Figure 13b).

The chemical composition of the chrome spinels II group (Crsp II) and the composition of the parental melt are similar in these parameters for the low-Al' chrome spinels of the Ospa-Kitoy ophiolite (Table 6). The $(Al_2O_3)_{melt}$ values for the chrome spinels of groups II and III are 10–12 and 10–13 (wt. %), and the (FeO/MgO)melt ratio is 0.4–0.85 and 0.7–1, respectively. The values of the $Al_2O_{3Sp}/Al_2O_{3melt}$ ratios in chrome spinels II and III correspond to the trend of the chrome spinels from the island-arc boninites and chromitites of the Ural-Alaska complexes (Figure 13c). The $(TiO_2)_{melt}$ values for the chrome spinels I, II, and III groups overlap because of the low TiO_2 content (0–0.2 wt. %). In general, the values of $(Al_2O_3)_{melt}/(FeO/MgO)$ are similar to the high-Cr' chrome spinels from the Troodos and Zetford Mine ophiolites (Figure 13d), which were formed in a suprasubduction setting. In the discrimination relationship diagrams, the $[Al_2O_3/Fe^{2+}/Fe^{3+}]$, $[Mg'/Cr']$, and $[Al_2O_3/TiO_2]$ (Figure 14a–c) of the chrome spinels I group are in the field of the MORB–type peridotites. Chromitites were formed during the interaction of harzburgites with primitive MORB-like melts at deep levels of the upper mantle. Through the interaction of MORB-type melts, which are in equilibrium with abyssal dunites [65,66], precipitation of chrome spinels with Cr' 0.4–0.6, as a product of the melt–peridotite reaction, is possible [43,67]. Among the volcanic rocks of the USO, metabasalts, with enriched mid-ocean ridge basalt (E-MORB) geochemical characteristics, are available [68].

The chrome spinels II group is localized in the boninite field. The TiO_2 and Al_2O_3 content corresponds to the chrome spinels of the suprasubduction peridotites and overlaps with the chrome spinels from the New Caledonia island arc (Figure 14c). The chrome spinels II are formed during the reaction of the peridotites with the island-arc boninite melt in the subduction zone. The chrome spinels III group lies on the boundary of the boninite fields and magmatic complexes of the Ural-Alaskan type. Three mechanisms can be suggested for the formation of the third type of chrome spinels. The first is the interaction with high-iron low-titanium melts [81]. The second mechanism is through plastic deformations under mantle or crust-mantle conditions (low fO_2, Table 1), because of the reactions of the Mg-Fe exchange with olivine. High-iron chrome spinels are found in the structural types of chromite, including chromitites with deformation structures. When compared with chrome spinels of the Alaskan type chromitites, the chrome spinels III group from the chromitites of the USO have low TiO_2 contents, high Cr_2O_3 contents and low (mantle) fO_2 values (Table 1). Variations in TiO_2 contents, from 0 to 0.22 wt. %, may indicate a reaction with TiO_2-containing melts [82]. The third mechanism is through the partial melting of the fluid-metasomatized mantle during the interaction of andesitic melts with rocks of an overlying mantle wedge [71,83,84].

Table 6. Calculated parental melt composition of chrome spinels.

	Cr-Spinels	Al′	Cr#′	Mg′	Parental Melt (wt. %)	References
1	I group	24–60	36–74	45–74	Al_2O_3, 13–18; TiO_2, 0–0.14 FeO/MgO, 0.2–0.7	In this article
2	II group	14–23	74–81	32–48	Al_2O_3, 10–12; TiO_2, 0–0.35 FeO/MgO, 0.4–0.85	In this article
3	III group	13–27	68–81	28–35	Al_2O_3, 10–13; TiO_2, 0.08–0.13 FeO/MgO, 0.7–1	In this article
4	Ospa-Kitoy medium Al′	24—41	59–75	43–70	Al_2O_3, 12–14; TiO_2, 0.01–0.44 FeO/MgO, 0.5–1.1	[38]
5	Ospa-Kitoy low Al′	9–21	77–90	23–59	Al_2O_3, 8–11; TiO_2, 0.01–0.48 FeO/MgO, 0.5–2.4	[38]
6	MORB	35–64	29–57	57–59	Al_2O_3 13–18; TiO_2, 0.3–1.7 FeO/MgO, 0.5–0.7	[18,69];
7	BAB	61	34	75	Al_2O_3, 17.6; TiO_2, 0.4 FeO/MgO, 0.4	[18]
8	OIB	28	61	57	Al_2O_3, 12; TiO_2, 1.6 FeO/MgO, 0.6	[18]
9	IAB	15–29	61–68	58–69	Al_2O_3, 9–12; TiO_2, 0.4–0.7 FeO/MgO, 0.3–0.5	[18]
10	IABon, IAT	5–19	74–89	58–75	Al_2O_3, 6–10; TiO_2, 0.08–0.4 FeO/MgO, 0.2–0.4	[18]
11	LIP	13–35	52–72	35–61	Al_2O_3, 8–13; TiO_2, 0.2–0.5 FeO/MgO, 0.4–1.2	[18]
12	Abissal peridotite	45–77	20–50	64–77	Al_2O_3, 15–19; TiO_2, 0.08–0.1 FeO/MgO, 0.4–0.5	[70]
13	Chromite in ophiolite mantle	40–51	44–52	63–74	Al_2O_3, 14–16; TiO_2, 0.2–0.5 FeO/MgO, 0.3–0.6	[3]
14	Chromite in ophiolite mantle	21–28	67–74	56–68	Al_2O_3, 10–12; TiO_2, 0.1–0.3 FeO/MgO, 0.54–0.6	[3,48]
15	Chromite in Alaskan type	11–14	62–70	45–56	Al_2O_3, 8–9; TiO_2, 0.6–0.9 FeO/MgO, 0.4–0.5	[71]

5.2. Spinel and Olivine Geothermometers and Olivine-Spinel Oxybarometers

Several processes condition the composition of chrome spinels: partial melting, cooling of mantle peridotites and plastic deformations in the upper mantle. Many researchers have shown that, during ultramafite cooling, regardless of their formation (mantle or cumulative), (Mg $\leftrightarrow$ Fe^{2+}) exchange reactions take place between coexisting olivines and chrome spinels, as a result of which the coefficient of these elements' distribution increases in favor of chrome spinels, and consequently, the calculated temperatures of the olivine-spinel equilibrium decrease [85–87]. During plastic deformations, exchange processes and rebalancing between olivine and chrome spinels also occur, and schlieren and lenticular segregations of chromitites transform into massive chromite bodies (Figure 4c–f). Based on this, it is assumed that the obtained values for the temperature, pressure and oxygen fugacity correspond not to the formation of ultramafites and chromitites, but to the stages of the formation and transformation of these bodies: the tectonic flow of the upper mantle rocks, the effect of metasomatizing fluids and other processes.

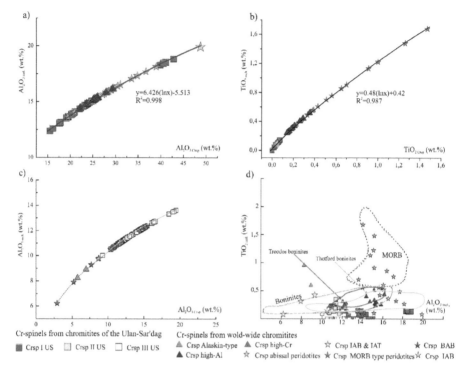

Figure 13. Plots of relations: **(a)** $Al_2O_{3\_Crsp}$–$Al_2O_{3\_melt}$; **(b)** TiO_{2_Crsp}–TiO_{2_melt}; **(c)** $Al_2O_{3\_Crsp}$–$Al_2O_{3\_melt}$; **(d)** calculated abundance of Al_2O_3–TiO_2 in melt, as in the equilibrium with the Cr spinels from the chromitites of the USO and the chrome spinels of chromitites from all over the world. The fields for the chrome spinels of boninites [72–74], Troodos boninites [75,76], Thetford boninites [77] and MORB [78–80] are shown for comparison.

We have calculated and estimated the P-T parameters with the help of an olivine-spinel (Ol-Sp) geothermometer, provided in [88]. The oxygen fugacity was determined using an Ol-Sp oxybarometer, provided in [89], in accordance with oxybarometers [90,91]. For the chrome spinels I group (high Al'), the calculated temperatures of the Ol-Sp equilibrium are T_{Ol-Sp} = 1020–920 °C, and fO_2 = (−0.7)–(−1.5) (Table 1); for the chrome spinels II group (medium-, low-Al'), the calculated temperatures are T_{Ol-Sp} = 891–1003 °C, and fO_2 = (−1.1) − (−4.4); for the chrome spinels III group, these values are T_{Ol-Sp} = 846-738 °C, and fO_2 =(−1.49) − (−3.01). For the chrome spinels containing inclusions of laurite-erlichmanite, T_{Ol-Sp} = 916–938 °C, and fO_2 = −2.4. For the chrome spinels with the olivine inclusion, T_{Ol-Sp} = 991.6 °C, and fO_2 = (−1.3). The temperature values overlap, but for the chrome spinels II group, higher temperatures are noted. The values of fO_2 are discrete. It is assumed that a more reducing environment and higher solid-phase reaction temperatures are required for the formation of the chrome spinels II group, in comparison with the chrome spinels groups III and I. The content of impurities (Al_2O_3) in olivine was used as an alternative geothermometer. For our objects, this method was limited by the small amount of preserved olivine. We were interested in the chromitites containing PGE- and PGE-free mineralization. Using the geothermometer of De Hoog and Gall for the Al content in olivine [25], the temperature was estimated by Equation (4). For the chromitites containing PGM, the temperature, according to the olivine thermometer, was 893–1332 °C, and for the PGM-free chromitites, it was 1073–1225 °C (Table 1).

$$T_{ol}(C) = 1087/(7.46 - ln Al_{ppm}) - 273 \tag{4}$$

An assessment of the temperature using an olivine thermometer and inclusions of primary high-temperature (Os-Ir-Ru)-I alloys gives the values of 1200–1300 °C, which is more consistent with the expected temperatures of the chrome spinel formation from the melt.

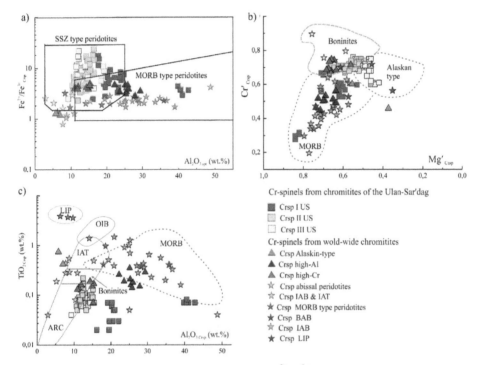

Figure 14. Tectonic discrimination diagrams: (**a**) Al_2O_3–Fe^{2+}/Fe^{3+}; (**b**) Mg'_{sp}–Cr'_{sp}; (**c**) Al_2O_3–TiO_2 from the chromitites of the USO and chrome spinels of chromitites from all over the world. The chrome spinels' composition fields for peridotites from different geodynamic settings are drawn from [18].

5.3. Distribution of PGE in Mantle Peridotites and Chromitites

The form of PGE distribution in the mantle peridotites and chromitites of ophiolites reflects the mantle conditions of chromitite formation and PGE mineralization. In the process of partially melting the mantle source, the PGEs fractionate. The melt is enriched in Rh, Pt, and Pd (PPGE), since PPGE is incompatible with Os, Ir, and Ru (IPGE). The mantle restite will be enriched in IPGE [17,92,93]. This process is confirmed by the form of PGE distribution in the Ulan-Sar'dag peridotites (Figure 15a,b), for which $\sum IPGE > \sum PPGE$.

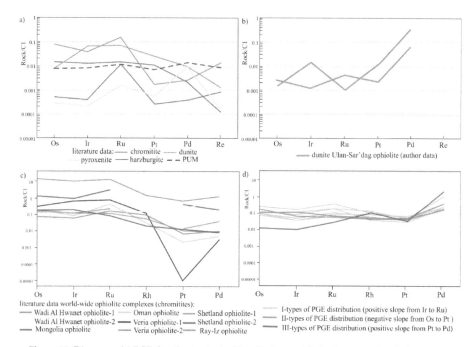

Figure 15. Diagrams: (**a**) PGE distribution in the Ulan-Sardag peridotite literature data [94]; (**b**) PGE distribution in the Ulan-Sar'dag peridotite author data; (**c**) PGE pattern of chromitites of ophiolite complexes from all over the world: The Wadi Al Hwanet ophiolite, Saudi Arabia [47]; Oman ophiolite, Semail [48]; Veria ophiolite, Greece [49]; Shetland Ophiolite Complex, Scotland [50]; Ray-Iz ophiolite, Russia [51]; Mongolian ophiolite [95]; 1—chromitites are enriched in Os-Ir-Ru; 2—chromitites are enriched in Pt-Pd; (**d**) PGE patterns in the Ulan-Sar'dag chromitites.

In the case of a low degree of partial melting, there are no obvious differences between the PPGE and IPGE contents, and the PGE distribution is flat. At a high degree of partial melting, PPGE is depleted in restite mantle rocks, relative to IPGE, and the total PGE contents become lower. In this case, the distributions have a negative slope. The PGE distribution in the mantle peridotite of Ulan-Sar'dag demonstrates: (1) a positive Ru and Pd picks; (2) a flat type of distribution, with a negative slope towards Pd; (3) a positive slope of Ir, Rh, and Pt and a negative slope of Ru (Figure 15b). Podiform chromitites are characterized by a fractionated form of PGE distribution (Figure 15c), which indicates a high degree of partial melting (about 20%–24%) of the mantle source. Three types of PGE distribution are observed in Ulan-Sar'dag chromitites (Figure 15d): (1) a negative slope from Os to Ir and Ru to Pt and a positive slope from Ir to Ru and Pt to Pd; (2) a negative slope of the distribution curve from Os to Pt and a positive slope from Pt to Pd; (3) low contents of Os, Ir and Ru and an enrichment in Rh (Pd is uncharacteristic of chromitites). Chondrite-normalized PGE relations in chromitites of the USO are similar to those of chromitites of ophiolites (Figure 15c,d) formed in a suprasubduction environment from all over the world [7,11,57,63,96–98]. The pronounced positive Ru anomaly is due to the predominant laurite phase in the chromitites of the USO. It is known that Ru has a maximum affinity with S, and the predominance of IPGE sulfides, in turn, indicates a high fugacity of sulfur in the melt, in contrast to the parental melts for chromitites of the northern and southern branches of the Ospa-Kitoy ophiolite (Figure 16a–c).

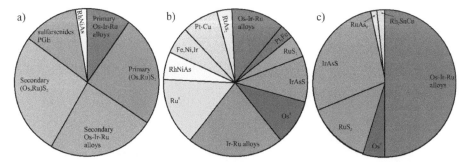

Figure 16. PGE assemblage (**a**) Ulan-Sar'dag chromitites; (**b**) the northern branch of the SEPES ophiolite; (**c**) the southern branch of the SEPES ophiolite [36].

The experimental data show that Os, Ir, and Ru are concentrated by trapping submicroscopic clusters of these elements in the metallic state during chromite crystallization [99–101]. Tsoupas [49] believes that an IPGE enrichment in chromitites can be associated with post-magmatic processes during a long period of deformations, beginning from plastic deformations in the asthenospheric mantle and ending with brittle deformations in the crust. This is confirmed by wide variations in the contents of IPGE, PPGE and the IPGE/PPGE ratio from 0.03 to 2.15 (Table 3) in the studied chromitites. Negative Pt anomalies in chromitites are closely related to the unique properties of Pt itself. The distribution coefficients of the alloy/sulfide liquid at 1000 °C are 1–2 for Pd and more than 1000 for Pt. Thus, the distribution coefficient of the alloy/sulfide melt for Pt is 1000 times greater than it is for Pd [102]. This is confirmed by the distribution diagrams, where a Pt negative anomaly is clearly visible (Figure 15c,d). An extreme Pd enrichment in one of the chromitites is most likely associated with late magmatic processes and exposure to the reduced fluid. Chrome spinels has high FeO contents and low fO_2 values. There are no Pd phases in this chromite. We believe that PGM is concentrated in recrystallized dunite, which requires further study. Detailed studies on the metaperidotites, metagabbros, and metavolcanogenic sedimentary rocks of the USO showed that some rocks have signs of significant exposure to a high-temperature fluid phase, which leads to rock metasomatism [68,103].

5.4. Sequence of the Formation and Transformation of the Platinum-Group Mineral Assemblage

Based on the chemical and textural features of PGM and associations with the magmatic and hydrothermal minerals of chromitites, several stages of PGE mineralization were distinguished (Figure 17).

- I—Magmatic Stage

At the magmatic stage, under the upper mantle conditions, euhedral–subhedral high-temperature Os-Ir-Ru I alloys and laurite I are formed (Figures 8 and 9a–d), which are captured by chromite grains [5,11,22,104–106]. The high osmium content in the primary (Os-Ir-Ru) I alloys is caused by an early crystallization of laurite-erlichmanite (Figure 9c). According to the experimental data, laurite without an Os impurity crystallizes from the melt at a high temperature (T = 1200–1300 °C), P = 5–10 kbar and low log sulfur fugacity (fS_2), from (−0.39) to 0.07 [24,106–109]. A decrease in temperature and increase in fS_2 leads to a replacement of Ru by Os, and as a result, laurite rich in Os is formed. The predominance of laurite-erlichmanite over solid (Os-Ir-Ru) solutions in the chromitites of the USO is a distinctive feature, in comparison with the PGE mineralization of the northern and southern branches of the Ospa-Kitoy ophiolite (Figure 16a–c). In combination with the high and medium Al' chrome spinels, this indicates the formation of chromitites and PGE mineralization of the USO because of the interaction between the initially S-saturated tholeiite magma and depleted harzburgites. Sulfarsenides and arsenides of Ru and Ir are formed from the residual fluid phase at the late magmatic stage (Figure 10g,h). With magmatic system cooling, volatile components, such as S and As, accumulate with the formation of the residual fluid phase. There is a partial replacement of laurite by irarsite with the formation of laurite II.

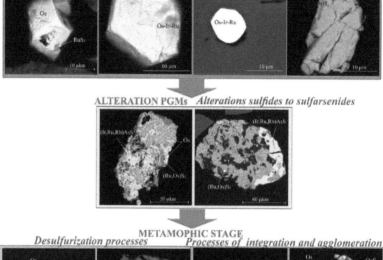

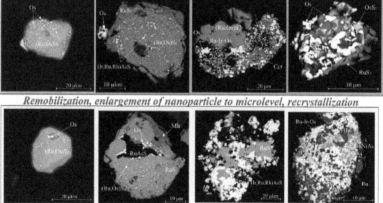

Figure 17. Scheme of the alteration and transformation of platinum-group minerals.

- II—Stage of Serpentinization and Exposure of Fluid

The PGMs in chromitites demonstrate signs of PGE remobilization (Figure 10a–d). The most intense changes of PGMs occur at the stage of the serpentinization of ultramafites. A fluid–rock interaction occurs with the participation of reduced gases (H_2, CH_4) and the H_2O of mantle origin. Dehydrating rocks of the subducting slab, as well as mantle-reduced fluids penetrating along the fault zones in tectonically weakened sectors, serve as the fluid source. At this stage, the following platinum–metal phases are formed: native Os and Ru, (Ir-Ru) alloys, phases of a variable (Os-Ir-Ru) composition, and newly formed laurite III, IrAsS, RhNiAs and RuAs (Figures 9f and 10a–d). Remobilized secondary PGMs form polyphase aggregates in the serpentine-chlorite matrix (Figure 10c) in association with

Ni sulfides, sulfarsenides, and arsenides. The joint occurrence of PGMs with Ni sulfides and the presence of such elements as Ni, Fe, Te, Cu, Co, As, and Sb in the platinum phases indicate PGE mobility in a fluid-saturated medium. The processes of the redistribution and concentration of PGE, including refractory Os, Ir, and Ru, occur at relatively low temperatures, reducing conditions corresponding to the formation of nickel sulfides, sulfarsenides, and arsenides, and low-temperature PGE-bearing intermetallides [110]. At the initial stage, the penetration of fluid through the permeable zones into laurite-erlichmanite led to the desulfurization of sulfides, a deviation from stoichiometry, the appearance of microdefects in the crystal lattice, and the formation of nanopores on the grain surface. These processes led to the formation of microporous structures and the separation of native Os and Ru (Figure 9e,f) [108,109,111,112]. According to the experimental data, congruent RuS_2 → native Ru decomposition occurs at $T = 300\ °C$, $\log fS_2 = (-20)$, and $P = 0.5\ kbar$ [113–118]. The (Os-Ir-Ru) phases of a variable composition are probably the products of changes in $(Os,Ir,Ru)AsS$, since Ir has a maximum affinity with As, and irarsite therefore survives for the longest. At the same stage, secondary (newly formed) laurite III can be formed. They grow over primary laurite-erlichmanite or are confined to chloritization zones, and as a rule, they are in association with nickel and copper sulfides and sulfarsenides (Figure 10a). Irarsites IrAsS can be formed during serpentinization. In this case, irarsites are in close association with Ni_3S_2, forming joint aggregates. The microstructural features of such aggregates indicate their simultaneous formation (Figure 10c). The physical–chemical modeling of the forms of PGE transport in the fluid systems indicates the formation of carbonyl, chloride, hydrosulfide, and bisulfide complexes, in the form of which they are transported, and the formation of secondary PGM occurs. When PGEs are transported by bisulfide complexes, and As and Sb appear in the system, PGE solubility decreases, the composition of the solution changes, and the system deviates from equilibrium. Sulfur released from bisulfide complexes reacts with Ni to form Ni_3S_2.

- III—Stage of Ophiolite Obduction. Regional Metamorphism.

As the ophiolite rises to the surface, the rocks undergo repeated processes of serpentinization under the influence of metamorphogenic fluids with an increased activity of O_2, As, and Sb [48]. In chromitites, chrome spinels change into chrome magnetites or magnetites. There are no clear criteria for distinguishing remobilized PGMs under the crustal conditions (under the influence of metamorphogenic fluids). We believe that non-stoichiometric platinum-metal phases, containing Cu, Te, As, Sb, and O, could be formed under the crustal conditions. These elements can be transported by aqueous solutions, with a subsequent re-deposition [119,120]. These events are mainly controlled by the Eh-pH conditions, and these minerals can be formed directly in the supergenic environment. Under the conditions of a changing temperature, varying Eh-pH in the Os-S-O-H system, low $f(S_2)$ and exposure to an oxidizing high-temperature fluid at a temperature of about 500 °C [121], Os becomes more mobile than other PGEs, which leads to the further redistribution and redeposition of osmium. At this stage, the following PGE-containing phases can be formed: (Ru,Ir,Te,Ni,Ba,S,O), (Os,Ir,Ru,As,S,O), and (Ir,Ni,Cu,Ru,Os,Cl). Most often, such phases are localized in the micro voids of early PGMs (laurite, irarsite, etc.). Another process of PGM changing is the enlargement and agglomeration of PGE nanoparticles to the micro level. During progressive metamorphism (epidote-amphibolite and amphibolite facies) and/or a thermal event (introduction of granite intrusion), the changed P-T conditions affect the stability of nanoparticles. As the temperature rises to 590–650 °C, the nanoparticles in the sulfide matrix become unstable [122], which leads to their coalescence (Figure 10b) and enlargement (to micron sizes). The accessory mineralization of the crust-metamorphogenic stage is represented by the products of the changes in Ni, Fe, and Cu sulfides and sulfarsenides, with the appearance of oxygen-containing phases of a non-stoichiometric composition.

6. Conclusions

(1) High- and medium-Al′ chrome spinels were formed through the interaction of mantle peridotites with tholeiite melts in a spreading setting. High-Cr chrome spinels were formed

during the interaction of mantle peridotites with boninite melts in suprasubduction environments. The predominance of PGE sulfides over high-temperature Os-Ir-Ru alloys indicates their formation from S-saturated magma, which is typical of tholeiite melts.

(2) The formation temperatures of magmatic PGM–chromite association are estimated at 1000–1200 °C. The temperatures of olivine–spinel equilibrium, reflecting the formation of chromitites and tectonic deformation processes, range from 1000 to 740 °C, and the *log oxygen* fugacity $f(O_2)$ is low, ranging from (−0.76) to (−4.4), which indicates the upper mantle conditions, as well as the effect of reduced mantle fluids.

(3) Platinum-group mineralization in the Ulan-Sar'dag chromitites reflects a long history of formation and transformation, a change in the fluid conditions from magmatic to metamorphic ones. Primary PGMs (Os-Ir-Ru alloys-I, laurite I) were formed under the condition of a high fugacity of sulfur. The physical–chemical conditions were as follows: $T < 1200$–1300 °C, and $\log f(S_2) > (-2)/(-1)$. Under the influence of reduced fluids on chromitites, the desulfurization of laurite and the formation of secondary PGMs (native Os and Ru, Os-Ir-Ru alloys-II, laurite III, IrAsS, and RhNiAs) occur in association with serpentine, chlorite, nickel sulfides and arsenides. This association can be formed at $T = 300$–700 °C, $\log f(S_2) = (-20)$, and P = 0.5 kbar.

(4) At the stage of ophiolite obduction and exposure to a metamorphogenic fluid, especially osmium, a further redistribution of PGE occurs. New phases of a non-stoichiometric composition (PGE + Cu, Te, Ba, As, Sb, O, and Cl) are formed. During progressive metamorphism, an enlargement and agglomeration of PGE nanoparticles to the microlevel occur.

Author Contributions: Conceptualization, O.N.K. and E.V.A.; methodology, D.K.B.; software, D.K.B. and E.V.A.; validation, S.M.Z. and E.V.A.; formal analysis, O.N.K.; investigation, O.N.K. and E.V.A.; resources, O.N.K. and S.M.Z.; data curation, O.N.K.; writing—original draft preparation, O.N.K. and E.V.A.; writing—review and editing, O.N.K., S.M.Z. and E.V.A.; visualization, E.V.A. and D.K.B.; supervision, S.M.Z.; project administration, O.N.K.; funding acquisition, O.N.K. All authors have read and agreed to the published version of the manuscript.

Funding: This research was funded by the Ministry of Science and Higher Education of the Russian Federation and Russian Fond Basic Research Grant No. 16-05-00737a and 19-05-00764a. Work is done on state assignment of IGM SB RAS.

Acknowledgments: The work was carried out at the Analytical Center for multi-elemental and isotope research, SB RAS. The authors are grateful of I. Ashchepkov for constructive comments and suggestions when preparing an article.

Conflicts of Interest: The authors declare no conflicts of interest.

References

1. Zhou, M.F.; Robinson, P.T. Origin and tectonic environment of podiform chromite deposits. *Econ. Geol.* **1997**, *92*, 259–262. [CrossRef]
2. Arai, S.; Matsukage, K. Petrology of a chromitite micropod from Hess Deep, equatorial Pacific: A comparison between abyssal and alpine-type podiform chromitites. *Lithos* **1998**, *43*, 1–14. [CrossRef]
3. Dönmez, C.; Keskin, S.; Günay, K.; Çolakoğlu, A.O.; Çiftçi, Y.; Uysal, İ.; Türkel, A.; Yıldırım, N. Chromite and PGE geochemistry of the Elekdağ Ophiolite (Kastamonu, Northern Turkey): Implications for deep magmatic processes in a supra-subduction zone setting. *Ore Geol. Rev.* **2014**, *57*, 216–228. [CrossRef]
4. Ahmed, A.H.; Arai, S. Platinum group minerals in podiform chromitites of the Oman ophiolite. *Can. Mineral.* **2003**, *41*, 597–616. [CrossRef]
5. Uysal, I.; Sadiklar, M.B.; Tarkian, M.; Karsli, O.; Aydin, F. Mineralogy and composition of the chromitites and their platinum-group minerals from Ortaca (Mugla-SW Turkey): Evidence for ophiolitic chromitite genesis. *Mineral. Petrol.* **2005**, *83*, 6–13. [CrossRef]
6. Thalhammer, O.A.R.; Prochaska, W.; Mühlhans, H.W. Solid inclusions in chromspinels and platinum group element concentration from the Hochgrössen and Kraubath Ultramafic Massifs (Austria). *Contrib. Mineral. Petrol.* **1990**, *105*, 66–80. [CrossRef]

7. Melcher, F.; Grum, W.; Simon, G.; Thalhammer, T.V.; Stumpfl, E.F. Petrogenesis of the ophiolitic giant chromite deposits of Kempirsai, Kazakhstan: A study of solid and fluid inclusions in chromite. *J. Petrol.* **1997**, *38*, 1419–1458. [CrossRef]

8. Garuti, G.; Fershtater, G.; Bea, F.; Montero, P.G.; Pushkarev, E.V.; Zaccarini, F. Platinum-group element distribution in mafic–ultramafic complexes of central and southern Urals: Preliminary results. *Tectonophysics* **1997**, *276*, 181–194. [CrossRef]

9. Zhou, M.F.; Sun, M.; Keays, R.R.; Kerrich, R.W. Controls on platinum-group elemental distribution of podiform chromitites: A case study of high-Cr and high-Al chromitites from chinese orogenic belts. *Geochimica et Cosmochim. Acta* **1998**, *62*, 677–688. [CrossRef]

10. Gervilla, F.; Proenza, J.A.; Frei, R.; González-Jiménez, J.M.; Garrido, C.J.; Melgarejo, J.C.; Meibom, A.; Díaz-martínez, R.; Lavaut, W. Distribution of platinum-group elements and Os isotopes in chromite ores from Mayarí-Baracoa Ophiolilte Belt (eastern Cuba). *Contrib. Mineral. Petrol.* **2005**, *150*, 589–607. [CrossRef]

11. Uysal, I.; Tarkian, M.; Sadıklar, M.B.; Sen, C. Platinum group-element geochemistry and mineralogy of ophiolitic chromitites from the Kop Mountains, Northeastern Turkey. *Can. Mineral.* **2007**, *45*, 355–377. [CrossRef]

12. Proenza, J.A.; Zaccarini, F.; Escayola, M.; Cábana, C.; Shalamuk, A.; Garuti, G. Composition and textures of chromite and platinum-group minerals in chromitites of the western ophiolitic belt from Córdoba Pampeans Ranges, Argentine. *Ore Geol. Rev.* **2008**, *33*, 32–48. [CrossRef]

13. Zaccarini, F.; Pushkarev, E.; Garuti, G. Platinum-group element mineralogy and geochemistry of chromitite of the Kluchevskoy ophiolite complex, central Urals (Russia). *Ore Geol. Rev.* **2008**, *33*, 20–30. [CrossRef]

14. Barnes, S.J.; Naldrett, A.J.; Gorton, M.P. The origin of the fractionation of platinum-group elements in terrestrial magmas. *Chem. Geol.* **1985**, *53*, 303–323. [CrossRef]

15. Maurel, C.; Maurel, P. Étude expérimentale de la distribution de l'aluminium entre bain silicaté basique et spinelle chromifère. Implications pétrogénétiques: Teneur en chrome des spinelles. *Bull. Minéralogie* **1982**, *105*, 197–202. [CrossRef]

16. Dick, H.J.B.; Bullen, T. Chromium-spinel as a petrogenetic indicator in abyssal and alpine-type peridotites and spatially associated lavas. *Contrib. Mineral. Petrol.* **1984**, *86*, 54–76. [CrossRef]

17. Gueddari, K.; Piboule, M.; Amosee, J. Differentiation of platinum-group elements (PGE) and of gold during partial melting of peridotites in the lherzolitic massifs of the Betico-Rifean range (Ronda and Beni Bousera). *Chem. Geol.* **1996**, *134*, 181–197. [CrossRef]

18. Kamenetsky, V.; Crawford, A.J.; Meffre, S. Factors controlling chemistry of magmatic spinel: An empirical study of associated olivine, Cr-spinel and melt inclusions from primitive rocks. *J. Petrol.* **2001**, *42*, 655–671. [CrossRef]

19. Ahmed, A.H.; Arai, S. Unexpectedly high-PGE chromitite from the deeper mantle section of the northern Oman ophiolite and its tectonic implications. *Contrib. Mineral. Petrol.* **2002**, *143*, 263–278. [CrossRef]

20. Ahmed, A.H. Diversity of platinum-group minerals in podiform chromitites of the late Proterozoic ophiolite, Eastern Desert, Egypt: Genetic implications. *Ore Geol. Rev.* **2007**, *33*, 31–45. [CrossRef]

21. Rollinson, H. The geochemistry of mantle chromitites from the northern part of the Oman ophiolite: Inferred parental melt composition. *Contrib. Mineral. Petrol.* **2008**, *156*, 273–288. [CrossRef]

22. Akmaz, R.M.; Uysal, I.; Saka, S. Compositional variations of chromite and solid inclusions in ophiolitic chromitites from the southeastern Turkey: Implications for chromitite genesis. *Ore Geol. Rev.* **2014**, *58*, 208–224. [CrossRef]

23. Ballhaus, C.; Berry, R.; Green, D. High pressure experimental calibration of the olivine-orthopyroxene-spinel oxygen geobarometer: Implication for the oxidation state of the upper mantle. *Contrib. Mineral. Petrol.* **1991**, *107*, 27–40. [CrossRef]

24. Andrews, D.R.A.; Brenan, J.M. Phase-equilibrium constraints of the magmatic origin of laurite and Os-Ir alloy. *Can. Mineral.* **2002**, *40*, 1705–1716. [CrossRef]

25. De Hoog, J.C.M.; Gall, L. Trace element geochemistry of mantle olivine and its application to geothermometry. *Ofioliti* **2005**, *30*, 182–183.

26. Korolyuk, V.N.; Pokhilenko, L.N. Electron probe determination of trace elements in olivine: Thermometry of depleted peridotites. *Russ. Geol. Geophys.* **2016**, *57*, 1750–1758. [CrossRef]

27. Kuznetsov, A.P.; Kukushkin, Y.N.; Makarov, D.F. The use of nickel matte as a collector of precious metals in the analysis of poor products. *J. Anal. Chem. USSR* **1974**, *29*, 2156–2160. (In Russian)

28. Beklemishev, M.K.; Kuzmin, N.M.; Zolotov, Y.A. Extraction and extraction-kinetic determination of Os using aza analogs of dibenzo-18-crown-6. *J. Anal. Chem. USSR* **1989**, *2*, 356–362. (In Russian)

29. Shlenskaya, V.I.; Khvostova, V.P.; Kadyrova, G.I. Kinetic methods for the determination of osmium and ruthenium (review). *J. Anal. Chem. USSR* **1973**, *28*, 779–784. (In Russian)

30. Rao, N.V.; Ravana, P.V. Kinetic-catalytic determination of osmium. *Mikrochim. Acta* **1981**, *76*, 269–276. [CrossRef]

31. Belichenko, V.G.; Butov, Y.P.; Boos, R.G.; Vratkovskaya, S.V.; Dobretsov, N.L.; Dolmatov, V.A.; Zhmodik, S.M.; Konnikov, E.G.; Kuzmin, M.I.; Medvedev, V.N.; et al. *Geology and Metamorphism of Eastern Sayan*; Nauka: Novosibirsk, Russia, 1988. (In Russian)

32. Dobretsov, N.L.; Konnikov, E.G.; Dobretsov, N.N. Precambrian ophiolitic belts of Southern Siberia (Russia) and their metallogeny. *Precambr. Res.* **1992**, *58*, 427–446. [CrossRef]

33. Zhmodik, S.; Kiseleva, O.; Belyanin, D.; Damdinov, B.; Airiyants, E.; Zhmodik, A. PGE mineralization in ophiolites of the southeast part of the Eastern Sayan (Russia). In Proceedings of the 12th International Platinum Symposium, Abstracts, Russia, 11–14 August 2014; Anikina, E.V., Ariskin, A.A., Barnes, S.-J., Barnes, S.J., Borisov, A.A., Evstigneeva, T.L., Kinnaird, J.A., Latypov, R.M., Li, C., Maier, W.D., et al., Eds.; Institute of Geology and Geochemistry UB RAS: Yekaterinburg, Russia, 2014; pp. 221–225.

34. Kuzmichev, A.B. *The Tectonic History of the Tuva–MongolianMassif: Early Baikalian, late Baikalian, and Early Caledonian Stages*; Probel Publishing House: Moscow, Russia, 2004; 192p. (In Russian)

35. Kuzmichev, A.B.; Larionov, A.N. Neoproterozoic island arcs of East Sayan: Duration of magmatism (from U-Pb zircon dating of volcanic clastics). *Russ. Geol. Geophys.* **2013**, *54*, 34–43. [CrossRef]

36. Kiseleva, O.N.; Zhmodik, S.M.; Damdinov, B.B.; Agafonov, L.V.; Belyanin, D.K. Composition and evolution of PGE mineralization in chromite ores from the Il'chir ophiolite complex (Ospa-Kitoi and Khara-Nur areas, East Sayan). *Russ. Geol. Geophys.* **2014**, *55*, 259–272. [CrossRef]

37. Sklyarov, E.V.; Kovach, V.P.; Kotov, A.B.; Kuzmichev, A.B.; Lavrenchuk, A.V.; Perelyaev, V.I.; Shipansky, A.A. Boninites and ophiolites: Problems of their relations and petrogenesis of boninites. *Geol. Geophys.* **2016**, *57*, 127–140. (In Russian) [CrossRef]

38. Kiseleva, O.; Zhmodik, S. PGE mineralization and melt composition of chromitites in Proterozoic ophiolite complexes of Eastern Sayan, Southern Siberia. *Geosci. Front.* **2017**, *8*, 721–731. [CrossRef]

39. Skopintsev, V.G. *Geological Structure and Mineral Resources of the Upper Rivers Gargan, Urik, Kitoy, Onot*; Results of prospecting works on the site of Kitoy (East Sayan); Report of the Samartin and Kitoy parties; Buryatia Publishing House: Ulan-Ude, Russia, 1995; Book 1; 319p. (In Russian)

40. Pavlov, N.V.; Kravchenko, G.G.; Chuprynina, I.I. *Chromites from the Kempirsai Pluton*; Nauka: Moscow, Russia, 1968. (In Russian)

41. Cabri, L.J. The platinum group minerals. In *The Geology, Geochemistry, Mineralogy and Mineral Beneficiation of Platinum Group Elements*; Published for the Geological Society of CIM; Canadian Institute of Mining, Metallurgy and Petroleum: Montreal, QC, Canada, 2002; Volume 54, pp. 13–131.

42. Kelemen, P.B. Reaction between ultramafic rock and fractionating basaltic magma I. Phase relations, the origin of the calcalkaline magma series, and the formation of discordant dunite. *J. Petrol.* **1990**, *31*, 51–98. [CrossRef]

43. Arai, S.; Yurimoto, H. Podiform chromitites from the Tari-Misaka ultramafic complex, southwestern Japan, as melt-mantle interaction products. *Econ. Geol.* **1994**, *89*, 1279–1288. [CrossRef]

44. Zhou, M.-F.; Robinson, P.T. High-chromium and high-aluminum podiform chromitites, western China: Relationship to partial melting and melt/rock interaction in the upper mantle. *Int. Geol. Rev.* **1994**, *36*, 678–686. [CrossRef]

45. Zhou, M.-F.; Robinson, P.; Malpas, J.; Li, Z. Podiform chromites in the Luobusa Ophiolite (Southern Tibet): Implications for melt-rock interaction and chromite segregation in the upper mantle. *J. Petrol.* **1996**, *37*, 3–21. [CrossRef]

46. Arai, S. Control of wall-rock composition on the formation of podiform chromitites as a result of magma/peridotite interaction. *Resour. Geol.* **1997**, *47*, 177–187.

47. Ahmed, A.H.; Harbi, H.M.; Habtoor, A.M. Compositional variations and tectonic settings of podiform chromitites and associated ultramafic rocks of the Neoproterozoic ophiolite at Wadi Al Hwanet, northwestern Saudi Arabia. *J. Asian Earth Sci.* **2012**, *56*, 118–134. [CrossRef]

48. Prichard, H.M.; Lord, R.A.; Neary, C.R. A model to explain the occurrence of platinum- and palladium- rich ophiolite complexes. *J. Geol. Soc.* **1996**, *153*, 323–328. [CrossRef]

49. Tsoupas, G.; Economou-Eliopoulos, M. High PGE contents and extremaly abundant PGE-minerals hosted in chromitites from Veria ophiolite complex, northern Greece. *Ore Geol. Rev.* **2008**, *33*, 3–19. [CrossRef]

50. O'Driscoll, B.; Day, J.M.D.; Walker, R.J.; Daly, J.S.; McDonough, W.F.; Piccoli, P.M. Chemical heterogeneity in the upper mantle recorded by peridotites and chromitites from the Shetland Ophiolite Complex. Scotland. *Earth Planet. Sci. Lett.* **2012**, *333*, 226–237. [CrossRef]

51. Gurskaya, L.I.; Smelova, L.V.; Kolbantsev, L.R.; Lyakhnitskaya, V.D.; Lyakhnitsky, Y.S.; Shakhova, S.N. *Platinoids of Chromite-Bearing Massifs of the Polar Urals*; Publishing House SPb Card Factory VSEGEI: St. Petersburg, Russia, 2005; 306p. (In Russian)

52. Arai, S. Characterization of spinel peridotites by olivine—Spinel compositional relationships: Review and interpretation. *Chem. Geol.* **1994**, *113*, 191–204. [CrossRef]

53. Leblanc, M.; Violette, J.F. Distribution of Aluminium-rich and Chromium-rich chromite pods in ophiolite peridotites. *Econ. Geol.* **1983**, *78*, 293–301. [CrossRef]

54. Zhou, M.F.; Bai, W.J. Chromite deposits in China and their origin. *Miner. Depos.* **1992**, *27*, 192–199. [CrossRef]

55. Leblanc, M. Chromite and ultramafic rock compositional zoning through a paleotransform fault, Poum, New Caledonia. *Econ. Geol.* **1995**, *90*, 2028–2039. [CrossRef]

56. Graham, I.T.; Franklin, B.J.; Marshall, B. Chemistry and mineralogy of podiform chromitite deposits, southern NSW, Australia: A guide to their origin and evolution. *Mineral. Petrol.* **1996**, *37*, 129–150. [CrossRef]

57. Economou-Ellopoulos, M. Platinum-group element distribution in chromite ores from ophiolite complexes: Implications for their exploration. *Ore Geol. Rev.* **1996**, *11*, 363–381. [CrossRef]

58. Proenza, J.; Gervilla, F.; Melgarejo, J.C.; Bodinier, J.L. Al-and Cr-rich chromitites from the Mayari–Baracoa Ophiolitic Belt (Eastern Cuba): Consequence of interaction between volatile-rich melts and peridotite in suprasubduction mantle. *Econ. Geol.* **1999**, *94*, 547–566. [CrossRef]

59. Ahmed, A.H.; Arai, S.; Attia, A.K. Petrological characteristics of podiform chromitites and associated peridotites of the Pan African Proterozoic ophiolite complexes of Egypt. *Miner. Depos.* **2001**, *36*, 72–84. [CrossRef]

60. Thayer, P.T. Principal features and origin of podiform chromite deposits, and some observations on the Guleman-Soridag district, Turkey. *Econ. Geol.* **1964**, *59*, 1497–1524. [CrossRef]

61. Thayer, T.P. *Chromite Segregations as Petrogenetic Indicators*; 1 (Special Publications); The Geological Society of South Africa: Johannesburg, South Africa, 1970; pp. 380–389.

62. Uysal, I.; Tarkian, M.; Sadiklar, M.B.; Zaccarini, F.; Meisel, T.; Garuti, G.; Heidrich, S. Petrology of Al- and Cr-rich ophiolitic chromitites from the Muğla, SW Turkey: Implications from composition of chromite, solid inclusions of platinum-group mineral, silicate, and base-metal mineral, and Os-isotope geochemistry. *Contrib. Mineral. Petrol.* **2009**, *158*, 659–674. [CrossRef]

63. Xiong, F.; Yang, J.; Liu, Z.; Guo, G.; Chen, S.; Xu, X.; Li, Y.; Liu, F. High-Cr and high-Al chromitite found in western Yarlung-Zangbo suture zone in Tibet. *Acta Petrol. Sin.* **2013**, *29*, 1878–1908.

64. González-Jiménez, J.M.; Proenza, J.A.; Gervilla, F.; Melgarejo, J.C.; Blanco-Moreno, J.A.; Ruiz-Sánchez, R.; Griffin, W.L. High-Cr and high-Al chromitites from the Sagua de Tánamo district, Mayarí-Cristal ophiolitic massif (eastern Cuba): Constraints on their origin from mineralogy and geochemistry of chromian spinel and platinum-group elements. *Lithos* **2011**, *125*, 101–121. [CrossRef]

65. Kelemen, P.B.; Shimizu, N.; Salters, V.J.M. Extraction of mid-ocean-ridge basalt from the upwelling mantle by focused flow of melt in dunite channels. *Nature* **1995**, *375*, 747–753. [CrossRef]

66. Arai, S. Role of dunite in genesis of primitive MORB. *Proc. Jpn. Acad.* **2005**, *B 81*, 14–19. [CrossRef]

67. Arai, S.; Miura, M. Podiform chromitites do form beneath mid-ocean ridges. *Lithos* **2015**, *232*, 143–149. [CrossRef]

68. Kiseleva, O.N.; Airiyants, E.V.; Belyanin, D.K.; Zhmodik, S.M. *Geochemical Features of Peridotites and Volcanogenic-Sedimentary Rocks of the Ultrabasic-Basitic Massif of Ulan-Sar'dag (East Sayan, Russia)*; The Bulletin of Irkutsk State University: Irkutsk, Russia, 2019. (In Russian)

69. Wilson, M. *Igneous Petrogenesis*; Unwin Hyman: London, UK, 1989.

70. Jonson Kevin, T.M.; Dick Henry, J.B. Open System Melting and Temporal and Spatial Variation of Peridotite and Basalt at the Atlantis II Fracture Zone. *J. Geophys. Res.* **1992**, *97*, 9219–9241. [CrossRef]

71. Garuti, G.; Pushkarev, E.V.; Thalhammer, O.A.R.; Zaccarini, F. Chromitites of the Urals (Part 1): Overview of chromite mineral chemistry and geotectonic setting. *Ofioliti* **2012**, *37*, 27–53.
72. Jenner, G.A. Geochemistry of high-Mg andesites from Cape Vogel, Papua New Guinea. *Chem. Geol.* **1981**, *33*, 307–332. [CrossRef]
73. Kamenetsky, V.S.; Sobolev, A.V.; Eggins, S.M.; Crawford, A.J.; Arculus, R.J. Olivine enriched melt inclusions in chromites from low-Ca boninites, Cape Vogel, Papua New Guinea: Evidence for ultramafic primary magma, refractory mantle source and enriched components. *Chem. Geol.* **2002**, *83*, 287–303. [CrossRef]
74. Walker, D.A.; Cameron, W.E. Boninite primary magmas: Evidence from the Cape Vogel Peninsula, PNG. *Contrib. Mineral. Petrol.* **1983**, *83*, 150–158. [CrossRef]
75. Cameron, W.E. Petrology and origin of primitive lavas from the Troodos ophiolite, Cyprus. *Contrib. Mineral. Petrol.* **1985**, *89*, 239–255. [CrossRef]
76. Flower, M.F.J.; Levine, H.M. Petrogenesis of a tholeiite–boninite sequence from Ayios Mamas, Troodos ophiolite: Evidence for splitting of a volcanic arc? *Contrib. Mineral. Petrol.* **1987**, *97*, 509–524. [CrossRef]
77. Page, P.; Barnes, S.J. Using trace elements in chromites to constrain the origin of podiform chromitites in the Thetford Mines Ophiolite, Québec, Canada. *Econ. Geol.* **2009**, *104*, 997–1018. [CrossRef]
78. Shibata, T.; Thompson, G.; Frey, F.A. Tholeiitic and alkali basalts from the mid Atlantic ridge at 43° N. *Contrib. Mineral. Petrol.* **1979**, *70*, 127–141. [CrossRef]
79. Le Roex, A.P.; Dick, H.J.B.; Gulen, L.; Reid, A.M.; Erlank, A.J. Local and regional heterogeneity in MORB from the mid-Atlantic ridge between 54.5° S and 51° S: Evidence for geochemical enrichment. *Geochim. Cosmochim. Acta* **1987**, *51*, 541–555. [CrossRef]
80. Presnall, D.C.; Hoover, J.D. High pressure phase equilibrium constraints on the origin of mid-ocean ridge basalts. *Geochem. Soc. Spec. Pap.* **1987**, *1*, 75–89.
81. Rollinson, H. Chromite in the mantle section of the Oman ophiolite: A new genetic model. *Isl. Arc* **2005**, *14*, 542–550. [CrossRef]
82. Barnes, S.J.; Kunilov, V.Y. Spinels and Mg ilmenites from the Noril'sk and Talnakh intrusions and other mafic rocks of the Siberian flood basalt province. *Econ. Geol.* **2000**, *95*, 1701–1717. [CrossRef]
83. Burns, L.E. The Borger Range ultramafic and mafic complex, south-central Alaska: Cumulative fractionates of island-arc volcanics. *Can. J. Earth Sci.* **1985**, *22*, 1020–1038. [CrossRef]
84. Volchenko, Y.A.; Ivanov, K.S.; Koroteev, V.A.; Auge, T. Structural-substantial evolution of the Urals platiniferous belt's complexes in the time of Uralian type chromite-platinum deposits formation. Part I. *Lithosphere* **2007**, *3*, 3–27.
85. Irvine, T.N. Chromian spinel as a petrogenetic indicator: Part II. Petrologic applications. *Can. J. Earth Sci.* **1967**, *4*, 71–103. [CrossRef]
86. Henry, D.; Medaris, L. Application of pyroxene and olivine-spinel geothermometers to spinel peridotites in South-western Oregon. *Am. J. Sci.* **1980**, *280*, 211–231.
87. Bedard, J.H. A new projection scheme and differentiation index for Cr-spinels. *Lithos* **1997**, *42*, 37–45. [CrossRef]
88. Ashchepkov, I.V. Program of the mantle thermometers and barometers: Usage for reconstructions and calibration of PT methods. *Vestn. Otd. Nauk Zemle* **2011**, *3*, NZ6008. [CrossRef]
89. Ashchepkov, I.V.; Pokhilenko, N.P.; Vladykin, N.V.; Rotman, A.Y.; Afanasiev, V.P.; Logvinova, A.M.; Kostrovitsky, S.I.; Pokhilenko, L.N.; Karpenko, M.A.; Kuligin, S.S.; et al. Reconstruction of mantle sections beneath Yakutian kimberlite pipes using monomineral thermobarometry. *Geol. Soc. Spec. Publ.* **2008**, *293*, 335–352. [CrossRef]
90. O'Neill, H.S.C.; Wal, V.J. The olivine orthopyroxene-spinel oxygen geobarometer, the nickel precipitation curve, and the oxygen fugacity of the Earth's upper mantle. *J. Petrol.* **1987**, *28*, 1169–1191. [CrossRef]
91. Taylor, W.R.; Kammerman, M.; Hamilton, R. New thermometer and oxygen fugacity sensor calibrations for ilmenite and chromium spinel-bearing peridotitic assemblages. In Proceedings of the 7th International Kimberlite Conference, Cape Town, South Africa, 11–17 April 1998; Extended Abstracts. pp. 891–901.
92. Lorand, J.P.; Keays, R.R.; Bodiner, J.R. Copper- and noble metal enrichment across the asthenosphere-lithosphere mantle diapiris: The Lanzo lherzolite massif. *J. Petrol.* **1993**, *34*, 1111–1140. [CrossRef]

93. Brugmann, G.E.; Armdt, N.T.; Hoffmann, A.W.; Tobschall, H.J. Nobel metal abundances in Komatiite suites from Alexo, Ontario and Gorgona Island, Colombia. *Geochim. Cosmochim. Acta* **1987**, *51*, 2159–2169. [CrossRef]

94. Wang, K.-L.; Chu, Z.; Gornova, M.A.; Dril, S.; Belyaev, V.A.; Lin, K.-Y.; O'Reilly, S.Y. Depleted SSZ type mantle peridotites in Proterozoic Eastern Sayan ophiolotes in Siberia. *Geodyn. Tectonophys.* **2017**, *8*, 583–587. [CrossRef]

95. Agafonov, L.V.; Lkhamsuren, J.; Kuzhuget, K.S.; Oidup, C.K.B. *Platinum-Group Element Mineralization of Ultramafic-Mafic Rocks in Mongolia and Tuva*; Tomurtogoo, O., Ed.; Ulaanbaatar Publishing House: Ulaanbaatar, Mongolia, 2005. (In Russian)

96. Page, N.J.; Engin, T.; Singer, D.A.; Haffty, J. Distribution of platinum-group elements in the Bati Kef chromite deposit, Güleman-Elaziğ area, Eastern Turkey. *Econ. Geol.* **1984**, *79*, 177–184. [CrossRef]

97. Yaman, S.; Ohnenstetter, M. Distribution of platinum-group elements of chromite deposits within ultramafic zone of Mersin ophiolite (south Turkey). *Bull. Geol. Congr. Turk.* **1991**, *6*, 253–261.

98. Garuti, G.; Pushkarev, E.V.; Zaccarini, F. Diversity of chromite-PGE mineralization in ultramafic complexes of the Urals. In Proceedings of the Platinum-Group Elements—From Genesis to Beneficiation and Environmental Impact: 10th International Platinum Symposium, Oulu, Finland, 8–11 August 2005; Geological Survey of Finland: Esbo, Finland, 2005.

99. Ballhaus, C.; Sylvester, P. PGE enrichment processes in the Merensky reef. *J. Petrol.* **2000**, *41*, 454–561. [CrossRef]

100. Matveev, S.; Ballhaus, C. Role of water in the origin of podiform chromititedeposits. *Earth Planet. Sci. Lett.* **2002**, *203*, 235–243. [CrossRef]

101. Sattari, P.; Brenan, J.M.; Horn, I.; McDonough, W.F. Experimental constraints in the sulfide-and chromite-silicate melt partitioning behaviour of rhenium and platinum-group elements. *Ecol. Geol.* **2002**, *97*, 385–398. [CrossRef]

102. Fleet, M.E.; Crocket, J.H.; Lin, M.H.; Stone, W.E. Laboratory partitioning of platinum-group elements and gold with application to magmatic sulfide-PGE deposits. *Lithos* **1999**, *47*, 127–142. [CrossRef]

103. Kiseleva, O.N.; Airiyants, E.V.; Belyanin, D.K.; Zhmodik, S.M. Geochemical and mineralogical indicators (Cr-spinelides, Platinum Group Minerals) of the geodynamic settings of formation of maficultramafic Ulan Saridag massif (Eastern Sayan). In Proceedings of the EGU General Assembly Conference Abstracts, Vienna, Austria, 8–13 April 2018.

104. Prichard, H.M.; Tarkian, M. Platinum and palladium minerals from two PGElocalities in the Shetland ophiolite complex. *Can. Mineral.* **1988**, *26*, 979–990.

105. Garuti, G.; Zaccarini, F.; Economou-Eliopoulos, M. Paragenesis and composition of laurite from chromitites of Othrys (Greece): Implications for Os-Ru fractionation in ophiolitic upper mantle of the Balkan peninsula. *Miner. Depos.* **1999**, *34*, 312–319. [CrossRef]

106. Brenan, J.M.; Andrews, D. High-temperature stability of laurite and Ru-Os-Ir alloy and their role in PGE fractionation in mafic magmas. *Can. Mineral.* **2001**, *39*, 341–360. [CrossRef]

107. Ballhaus, C.; Bockrath, C.; Wohlgemuth-Ueberwasser, C.; Laurenz, V.; Berndt, J. Fractionation of the noble metals by physical processes. *Contrib. Mineral. Petrol.* **2006**, *152*, 667–684. [CrossRef]

108. Bockrath, C.; Ballhaus, C.; Holzheid, A. Stabilities of laurite RuS_2 and monosulphide liquid solution atmagmatic temperature. *Chem. Geol.* **2004**, *208*, 265–271. [CrossRef]

109. Finnigan, C.S.; Brenan, J.M.; Mungall, J.E.; McDonough, W.F. Experiments and models bearing on the role of chromite as a collector of platinum group minerals by local reduction. *J. Petrol.* **2008**, *49*, 1647–1665. [CrossRef]

110. Dick, H.J.B. Terrestrial nickel–iron from the josephinite peridotite, its geologic occurrence, associations and origin. *Earth Planet. Sci. Lett.* **1974**, *24*, 291–298. [CrossRef]

111. Zaccarini, F.; Proenza, J.A.; Ortega-Gutiérrez, F.; Garuti, G. Platinum group minerals in ophioliticchromitites from Tehuitzingo (Acatlán complex, southern Mexico): Implications for postmagmatic modification. *Miner. Petrol.* **2005**, *84*, 147–168. [CrossRef]

112. Garuti, G.; Proenza, J.A.; Zaccarini, F. Distribution and mineralogy of platinum-group elements in altered chromitites of the Campo Formoso layered intrusiyn (Bahia State, Brazil): Control by magmatic and hydrothermal processes. *Miner. Petrol.* **2007**, *89*, 159–188. [CrossRef]

113. Stockman, H.W.; Hlava, P.F. Platinum-group minerals in Alpine chromitites from southwestern Oregon. *Econ Geol.* **1984**, *79*, 492–508. [CrossRef]

114. Nilsson, L.P. Platinum-group mineral inclusions in chromitite from Osthammeren ultramafic tectonite body, south central Norway. *Mineral. Petrol.* **1990**, *42*, 249–263. [CrossRef]

115. Bowles, J.F.W.; Gize, A.P.; Vaughan, D.J.; Norris, S.J. Development of platinum-group minerals in laterites—Initial comparison of organic and inorganic controls. *Trans. Inst. Min. Metall. (Sect. B Appl. Earth Sci.)* **1994**, *103*, 53–56.

116. Garuti, G.; Zaccarini, F. In situ alteration of platinum-group minerals at low temperature: Evidence from serpentinized and weathered chromitite of the Vourinos complex, Greece. *Can. Mineral.* **1997**, *35*, 611–626.

117. Bai, W.; Robinson, P.T.; Fang, Q.; Yang, J.; Yan, B.; Zhang, Z.; Hu, X.-F.; Zhou, M.-F.; Malpas, J. The PGE and base-metal alloys in the podiform chromitites of the Luobusa ophiolite, southern Tibet. *Can. Mineral.* **2000**, *38*, 585–598. [CrossRef]

118. Evans, B.V.; Hattori, K.; Barronet, A. Serpentinite: What, Why, Where? *Elements* **2013**, *9*, 99–106. [CrossRef]

119. Bowles, J.F.W. The development of platinum-group minerals in laterites. *Econ. Geol.* **1986**, *81*, 1278–1285. [CrossRef]

120. Bowles, J.F.W.; Lyon, J.C.; Saxton, J.M.; Vaughan, D.J. The origin of Platinum Group Minerals from the Freetown intrusions, Sierra Leone, inferred from osmium isotope systematics. *Econ. Geol.* **2000**, *95*, 539–548. [CrossRef]

121. Xiong, Y.; Wood, A. Experimental quantifycation of hydrothermal solubility of platinum- group elements with special reference to porphyry copper environments. *Mineral. Petrol.* **2000**, *68*, 1–28. [CrossRef]

122. González-Jiménez, J.M.; Reich, M.; Camprubí, T.; Gervilla, F.; Griffin, W.L.; Colás, V.; O'Reilly, S.Y.; Proenza, J.A.; Pearson, N.J.; Centeno-García, E. Thermal metamorphism of mantle chromites and the stability of noble-metal nanoparticles. *Contrib. Mineral. Petrol.* **2015**, *170*, 15. [CrossRef]

 *minerals*

Article

Platinum Group Elements in Arsenopyrites and Pyrites of the Natalkinskoe Gold Deposit (Northeastern Russia)

Raisa G. Kravtsova, Vladimir L. Tauson, Artem S. Makshakov *, Nikolay V. Bryansky and Nikolay V. Smagunov

A.P. Vinogradov Institute of Geochemistry, Siberian Branch of Russian Academy of Sciences, Favorskiy str 1A, 664033 Irkutsk, Russia; krg@igc.irk.ru (R.G.K.); vltauson@igc.irk.ru (V.L.T.); tridigron@yandex.ru (N.V.B.); nicksm@igc.irk.ru (N.V.S.)
* Correspondence: artem_m@mail.ru

Received: 3 March 2020; Accepted: 30 March 2020; Published: 31 March 2020

Abstract: The peculiarities of the distribution and binding forms of platinum group elements (Pt, Pd, Ru, Rh, Os and Ir) in the arsenopyrites and pyrites of the Natalkinskoe gold ore deposit (Northeastern Russia) were examined using atomic absorption spectrometry with analytical data selections for single crystals (AAS-ADSSC), a "phase" chemical analysis (PCA) based on AAS of different size-fractions of minerals, scanning electron microscopy with energy dispersive X-ray spectrometry (SEM-EDX) and laser ablation inductively coupled plasma mass spectrometry (LA-ICP-MS). The arsenopyrites and pyrites of the Natalkinskoe gold deposit were found to concentrate not only Au but also platinum group elements (PGEs) such as Pt, Pd, Ru and Rh. The PCA showed that the highest contents (in ppm) were found in the monofractions of arsenopyrite—Pt up to 128, Pd up to 20, Ru up to 86 and Rh up to 21—and comparably lower in monofractions of pyrite—Pt to 29, Pd to 15, Ru to 58 and Rh to 5.9. The AAS-ADSSC method revealed two forms of uniformly distributed Pt, Pd and Ru corresponding to the chemically bound element in the structure of the mineral and in the superficial non-autonomous phase (NAP). The superficially bound form dominates over the structural form and presumably exists in a very thin surface layer of the crystal (~100–500 nm). The maximum contents of these PGE, chemically bound in the structure of arsenopyrite, reached values of (in ppm) 48, 5.9 and 48; and in pyrite structure, 68, 5.2 and 34 for Pt, Pd and Ru respectively. The contents of Pt, Pd and Ru related to NAP on the surface of the crystal were significantly higher and amounted (in ppm) for arsenopyrite to 714, 114 and 1083; and for pyrite 890, 62 and 690 for Pt, Pd and Ru, respectively. Preliminary results for the Rh form in arsenopyrite crystals suggest that the surface-related form (154–678 ppm) is more abundant than the structural form (17–45 ppm). Data from studying the surfaces of sulphide minerals by SEM-EDX and LA-ICP-MS confirmed the presence of Pt, Pd, Ru and Rh on the surface of arsenopyrite and pyrite crystals. These methods generated primary data on the content of Os and Ir in arsenopyrite and pyrite in the surface layer. The maximum content of Os and Ir found in arsenopyrites was up to 0.7 wt%. PGE-enriched fluids (up to ~3 ppm Pt) may exist in the gold ore deposit. It is assumed that there is a common mechanism of impurities uptake associated with the active role of the crystal surface and surface defects for gold-bearing arsenopyrites and pyrites. The surface enrichment is due to peculiarities in the crystal growth mechanism through the medium of NAP and the dualism of the element distribution coefficient in the system of mineral–hydrothermal solution, which is higher for NAP, compared to the volume of the crystal. Although mineral forms of Pt, Pd, Ru, Rh, Os and Ir have not been found at the Natalkinskoe gold deposit, their existence in the form of nano-scale particles is not excluded. This follows from the evolutionary model of surficial NAPs, assuming their partial transformation and aggregation with the formation of nano- and micro-sized autonomous phases of trace elements. The presence of PGE in the ores and the possibility of their extraction significantly increase the quality and value of the extracted raw gold materials at the Natalkinskoe deposit, and adds to the list of known platiniferous ore formations.

Minerals **2020**, *10*, 318

Keywords: Natalkinskoe gold deposit; arsenopyrite; pyrite; platinum group elements; distribution; binding forms

1. Introduction

Studying the platinum content of gold ore deposits and platinum group elements (PGEs) in general in the rocks of carbon-containing and black shale formations is currently a promising area of research, both in Russia and abroad. Many researchers see these deposits as a promising new mineral resource, although views on the presence and potential of extracting PGE from the ores of these deposits are somewhat contradictory [1–13]. The Natalkinskoe deposit (Northeastern Russia, Magadan region) is one of the largest gold deposits in Russia, and so is of particular interest. In addition to gold, there are quite high contents of Pt and Pd in the ores of the deposit. There are few data on other PGEs (Ru, Rh, Ir and Os). PGE contents and distribution studies are necessary to complete the most important tasks in the geochemistry of endogenous ore formation, such as the study of the material composition and the identification of genesis features of deposits, including giant and unique deposits. These studies are of great practical importance. The presence of Pt in ores and the possibility of its extraction would significantly complement the list of already known platiniferous ore formations and significantly increase the quality and value of the gold ore raw materials mined.

Work to detect platinum content at the Natalkinskoe deposit began in 1990 by a research group headed by V.I. Goncharov from the North–East Interdisciplinary Scientific Research Institute of the Far-East Branch of the Russian Academy of Sciences (Magadan, Russia) [1–3]. Much geochemical and mineralogical work has been performed over the years, but in terms of contents of PGE " ... the obtained analytical results remain ambiguous, and often contradictory" [3] (p. 132). The reliability of previously obtained high Pt content in ores (tens of ppm) and Pd (several ppm) [3,14] is still questionable. It was concluded that "Minerals-concentrators of platinum group elements are not found in the deposit" [3] (p. 136) nor in native mineral forms of PGE.

Subsequently, and until now, questions relating to this issue have been raised repeatedly, but there has been no clear response. Plyusnina and co-authors [14] (p. 840) concluded that " ... the industrial content of precious metals (Au and Pt) at the Natalkinskoe deposit is caused by metasomatic alteration and sulphidization of carbonaceous rocks". Unlike Au, the involvement of the carbonaceous substance in the accumulation of Pt has not been established by the authors of this work: the connection between this metal and the organic component of rocks and ores has not been found. It was assumed that sulphide minerals concentrate not only Au [3,15], but also Pt [14]. Later arsenopyrite was found to act as an Au, Pt and Pd concentrator at the Natalkinskoe deposit [11]. According to preliminary data obtained at that time, Pt contents in monofractions of this sulphide ranged from 23 to 62 ppm, and Pd from 2.3 to 9.5 ppm. Two main non-mineral forms of Pt and Pd, structural and surface-related, were identified. The nature of high contents of non-mineral Pt and Pd forms in arsenopyrite was found to be mainly superficial. Distinct mineral forms of Pt and Pd were not observed.

Considering all the above, the study of the binding forms and concentration levels of PGE in the most common sulphide minerals at the Natalkinskoe deposit, arsenopyrite and pyrite, seems to be an interesting and important task. This paper presents new data on the content of Pt and Pd in arsenopyrite and pyrite. The concentration levels of Ru, Rh, Ir and Os in these minerals are estimated for the first time. It is shown that the ambiguity of the results of previous studies may have been caused by the presence of hidden (so-called "invisible") forms of PGEs.

2. Research Objects

The Natalkinskoe gold ore deposit, where this research was conducted, is located in the territory of Northeastern Russia (see Figure 1, inset). It is part of the Omchak ore-placer cluster and one of the largest gold reserves in Russia. Structurally, the deposit is confined to the boundary part of the estimated granite pluton in the Tenka deep fault zone and is associated with the collision development stage of the Yana–Kolyma fold system. The primary objects of the study are arsenopyrites and pyrites selected from the most industrially significant vein, streaky-vein and veinlet-disseminated ores of the Natalkinskoe deposit, which belong to the productive gold-bearing quartz–pyrite–arsenopyrite (with native gold) mineral association.

The deposit falls under the category of orogenic gold–quartz ore formations, and it is characterized by a complex, long-term nature of development and, according to most researchers, by metamorphogenic-hydrothermal genesis. The ore hosted rocks are identified as late Palaeozoic (upper Permian), and ore mineralization is presumably Mesozoic (from late Jurassic to late Cretaceous) [3]. The ores of the Natalkinskoe gold deposit, unique in scale, despite the diversity, form a deposit with uniform internal structure consisting of zones of quartz, carbonate–quartz, sulphide–quartz veins and veinlets, surrounded by a wide halo of sulphidized rocks. They are generally characterized by the same elemental and mineral composition, differing only in the degree of their manifestation: the quantitative ratios. The host volcanogenic sedimentary strata include diamictites, argillites and siltstone shales with layers of sandstones and gravelites with high carbon content. The ores of the deposit, as already mentioned, demonstrate relatively high contents of Pt and Pd. Commercial ores emerged in the interaction of host rocks with low- and moderate-salinity water–bicarbonate fluids in the salinity range of 3–12 wt% eq. NaCl, at temperatures of 360–280 °C and pressures of approximately 2.4–1.1 kbar [16]. More details on the geology, mineralogy, geochemistry and conditions of formation of the Natalkinskoe giant gold deposit are available [1–4,11,14–28].

To study the distribution features and forms of occurrence of PGEs in arsenopyrites and pyrites, 35 large-volume ore samples were selected, each weighing up to 10 kg. The samples were taken from quarries and boreholes of different horizons of the deposit from those most rich in the number of sulphide mineral veins, veinlets and metasomatites classed as gold vein, streaky-vein and veinlet-disseminated ores (the areas named "Geological", "North–West", "Central" and "South–East"). The schematic geological map of the Natalkinskoe gold deposit with areas from which samples were taken is shown in Figure 1, ore types in Figure 2.

Non-metallic minerals such as quartz, carbonates, feldspar, sericite, chlorite and carbonaceous matter (micrographite) were found in the specimens characterizing these samples. Arsenopyrite and pyrite amounted to 95–99% of the ore minerals, with the former predominating. Ore minerals, along with arsenopyrite and pyrite, the amount of which in some areas reached ~7%, included galena, sphalerite, chalcopyrite, rutile and native gold (up to 1%). Native gold with a fineness from 750‰, to 900‰, less frequently electrum, mainly of large grain size, was located mainly in the vein quartz and intergrowths with sulphide minerals. Small quantities of pyrrhotite, sulphoarsenides of nickel and cobalt and ilmenite were identified. More than 70 minerals were found at the deposit in total [3,19,20].

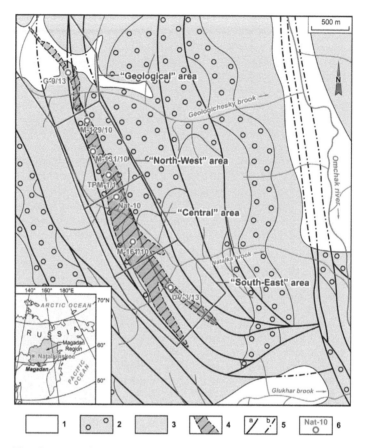

Figure 1. The schematic geological map of the Natalkinskoe gold deposit with areas from which ore samples were taken. Constructed by the authors of the present paper using the data [3,19,21,24,26]. Legend: 1—Quaternary sediments; 2 and 3—upper Permian ore hosted volcanogenic sedimentary rocks: 2—diamictites, 3—argillites and siltstone shales with layers of sandstones and gravelites with high carbon content; 4—ore deposit; 5—faults (a—reliable, b—estimated); 6—sampling points and their numbers. The inset map shows the location of the Natalkinskoe deposits in the territory of Northeastern Russia.

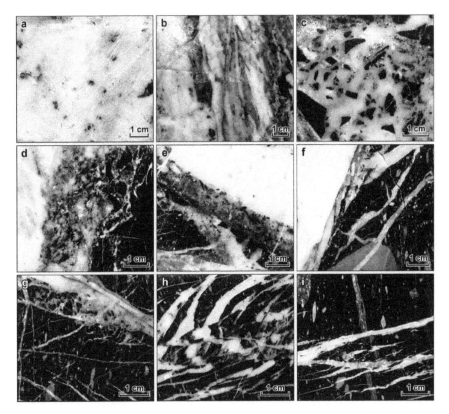

Figure 2. Types of ores in the Natalkinskoe deposit (a–c—vein, d–f—streaky-vein, g–i—veinlet-disseminated): (**a**) a massive quartz vein with a small number of fragments containing siltstones, nests and inclusions of arsenopyrite ("North–West" area, borehole DH-20/11n, interval 566–569 m); (**b**) quartz vein of laminated texture (bands are composed of completely silicificated siltstones) with the impregnations of small grains of arsenopyrite ("Central" area, quarry, elevation 810 m); (**c**) massive quartz–sulphide vein of breccia structure with fragments of host diamictites and enclaves of arsenopyrite and pyrite crystals ("North–West" area, quarry, elevation 900 m); (**d**) siltstones dissected by a net of quartz and carbonate–quartz veinlets with nests and individual crystals of arsenopyrite and pyrite, on the border with a massive quartz vein ("Central" area, mine, elevation 600 m); (**e**) silicificated diamictites dissected by quartz and carbonate–quartz veins with enclaves of arsenopyrite and pyrite crystals on the border with a massive quartz vein ("South–East" area, borehole DH50/12n, interval 452.0–454.9 m); (**f**) massive quartz vein on the border with diamictites, dissected by the carbonate–quartz veinlets, with fragments of sandy argillites of gravel size with enclaves of crystals and clusters of arsenopyrite and pyrite ("South–East" area, borehole DH50/12n, interval 460.8–463.8 m); (**g**) diamictites with fragments of argillites of gravel size dissected by a dense net of quartz and carbonate–quartz veins, with enclaves of arsenopyrite and pyrite crystals ("Central" area, borehole DH30/29, interval 273.5–276.5 m); (**h**) diamictites with enclaves of crystals, clusters and crosscuts of arsenopyrite and pyrite, from the zone of intense quartz and quartz–carbonate streaking ("South–East" area, borehole DH50/12n, interval 428.1–431.0 m); (**i**) diamictites dissected by transverse quartz and quartz–carbonate veins with fragments of argillites of gravel size and enclaves of arsenopyrite and pyrite crystals and clusters ("South–East" area, borehole DH50/12n, interval 528.4–531.4 m).

3. Research Methods

All ore samples were analyzed for Au and PGEs using atomic absorption spectrometry (AAS) and inductively coupled plasma mass spectrometry (ICP-MS) [29–31]. Twenty-two monofractions of arsenopyrite and 14 monofractions of pyrite were selected from the 7 ore samples richest in amount of sulphides, Au and PGE. They were all subjected to AAS analysis for Au, as were the ore samples [29,30]. All sulphide minerals were found to be gold-bearing. The level of Au content in arsenopyrite was 1.4–1383 ppm, in pyrite 0.8–158 ppm.

Further study of these gold-bearing arsenopyrites and pyrites for PGE was carried out using "phase" chemical analysis (PCA) based on AAS, coupled with analytical data selections for single crystals (ADSSC) [32,33], scanning electron microscopy with energy dispersive X-ray spectrometry (SEM-EDX) and laser ablation inductively coupled plasma mass spectrometry (LA-ICP-MS). These methods best characterize the peculiarities of minerals such as internal parts and surface layers, microinclusions, the so-called "invisible" impurity elements, their forms, distribution and concentration levels.

This set of methods was applied because PGEs in many deposits are represented by at least two forms—a solid solution in a carrier mineral and the submicroscopic inclusions of a PGE-rich material in an invisible form, undetectable by optical and scanning electron microscopy [34]. The results of the experiment and theoretical analysis suggest that not only Au, but also other noble metals (PGE and Ag) can form chemically bound forms of impurities in pyrite and other sulphides and chalcogenides [35]. The presence of these forms (at least one of them—a solid solution of PGE in sulphide or sulphoarsenide matrix) is identified even in low-temperature (~100 °C) platinum-bearing associations [36].

Current methods of "phase" chemical analysis for the determination of PGE in monofractions of arsenopyrite and pyrite, as well as the analysis of individual crystals of these minerals by AAS-ADSSC, allow a reliable determination of the contents of only three PGEs: Pt, Pd and Ru. The Rh analysis is under development, and the data are preliminary. All PGE—Pt, Pd, Ru, Rh, Ir and Os—can be analyzed using SEM-EDX and LA-ICP-MC methods. Restrictions are related only to their detection limits. Requirements for the selection of the test material are discussed below.

Monomineral samples of different sizes (0.07–0.14, 0.14–0.2, 0.2–0.25, 0.25–0.5, 0.5–1.0 and 1.0–2.0 mm) are taken for "phase" chemical analysis: individual and cluster crystals of arsenopyrite represented by pseudo-orthorhombic and monoclinic prisms; individual and cluster crystals of pyrite of cubic, pentagon-dodecahedral and cuboctahedral habits (Figure 3).

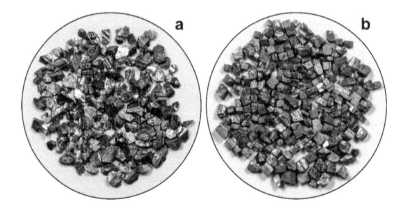

Figure 3. Monomineral fractions: (**a**) arsenopyrite, (**b**) pyrite.

In order to study PGEs in arsenopyrites and pyrites using the AAS-ADSSC method, euhedral individuals with well-defined morphology were selected from monomineral fractions of different sizes, from 0.25 to 2 mm. The crystals of arsenopyrite were predominantly shaped as pseudo-orthorhombic prisms, close to a slightly distorted parallelepiped, and less frequently as monoclinic prisms. Selected pyrite crystals generally had the shape of a cube or parallelepiped, although the samples were often complicated by pentagonal dodecahedron and octahedron facets (Figure 4). Shape is the characteristic required for calculation of the specific surface area of average crystal in size sample using the form coefficient for the true polyhedron. In our case, the coefficient was six for cubes and parallelepipeds. A total of 415 arsenopyrite crystals and 234 pyrite crystals were selected and studied.

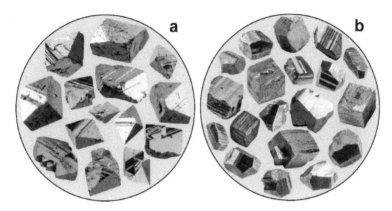

Figure 4. Crystals: (**a**) arsenopyrite, (**b**) pyrite.

Twenty arsenopyrite crystals and 14 pyrite crystals were selected and examined by SEM-EDX and LA-ICP-MS in order to study the PGE distribution in the surface layer. Euhedral pseudo-orthorhombic crystals of arsenopyrite and cubic crystals of pyrite 0.5–1 mm in size were used to study the surface of these minerals by SEM-EDX. Larger (1–2 mm) crystals of arsenopyrite and pyrite with at least one clearly manifested face belonging to certain simple form were selected for the LA-ICP-MS study (see Figure 4).

3.1. "Phase" Chemical Analysis

PCA was carried out with 10–20 mg samples with the view to determine the PGE in monofractions of arsenopyrite and pyrite. Monofractions were ground to powder and decomposed with Aqua Regia during heating. After decomposition, the material was treated with concentrated HCl and evaporation to dry state to remove nitric acid residues and convert the salts to chloride form. After cooling, the samples were diluted with 2M HCl to the condition required for the AAS method with electrothermal atomization on the M-503 device (Perkin-Elmer, Waltham, MA, USA) using graphite atomizer HGA-72 (Perkin-Elmer, Waltham, MA, USA) [29,30]. Au was determined in all the samples, along with Pt, Pd and Ru [29]. The accuracy of determination was ±10–12 rel. %.

3.2. AAS-ADSSC Method

The study of the so-called "invisible" uniformly distributed impurity component of PGE in arsenopyrites and pyrites was carried out using the AAS-ADSSC method. The method was developed by V.L. Tauson and co-authors for the study of structural and surface-bound forms of gold in ore minerals [32,33,37,38]. The method was further refined and successfully used to study structural and surface-bound forms of Pt, Pd, Ru and Rh [11,38,39]. The technique is designed to determine the content of the structural impurity of the element with an uncertainty of ±30 rel. % [40,41]. The determination of Pt, Pd, Ru and Rh in solutions obtained through the decomposition of individual crystals of arsenopyrite and pyrite was carried out following their preliminary extraction, concentration and separation from the matrix. Tristyrylphosphine, $(C_6H_5CH–CH)_3P$, was used as an extracting agent. Extraction was carried out using hydrochloric acid solutions (0.5 M HCl). The extracting agent concentration was

0.05 M (in toluene), and the phase contact time was 30 min. The ratio of aqueous to organic phases was 2:1. The extraction was carried out in static mode at room temperature and without labilising additives. The organic phase was used to measure element concentrations. Measurements were performed by AAS with electrochemical atomization on an M-503 (Perkin-Elmer, Waltham, MA, USA) device with graphite atomizer HGA-72 (Perkin-Elmer, Waltham, MA, USA) [29,30].

The data obtained were statistically processed according to the regulations of the distribution of different binding forms of the element [33,38]. We divided the dataset (>30–40 crystals with evenly distributed element) into the intervals of crystal masses (sizes), chosen to be as narrow as possible, although statistically representative, and determined the average crystal mass in every size fraction ($\overline{m}$), average size ($\overline{r}$), specific surface area ($\overline{S}_{sp}$) and element content ($\overline{C}$). The more size fractions and number of crystals in the final samples, the more reliable the results obtained. When constructing the dependences $\overline{C} = f(\overline{S}_{sp})$, we usually obtained a number of points best approximated with an exponent. The extrapolation of these curves to a zero-specific surface, that is, to a virtual infinite crystal, gave the bulk element content. In our model, this was equal to the structurally bound element content (C_{str}) because all other possible bulk forms were eliminated at the stage of initial dataset processing. The superficially bound element content ($\overline{C}_{sur}$) characterizes an average crystal among all size samples, that is, the surface-related excess content of the element, and can be calculated with the equation given in [38]. It is important to note that in such an approach, the amount of the material is normalized to the whole crystal, and this allows a comparison of the contribution of surface and bulk related forms of the element to its total content.

3.3. SEM-EDX Method

The study of the "invisible" impurity content of the element on the surface of arsenopyrite and pyrite crystals by SEM-EDX was carried out on a Quanta 200 scanning electron microscope (FEI Company, Hillsboro, OR, USA) with energy-dispersive accessory EDAX (FEI Company, Hillsboro, OR, USA) with nitrogen-free cooling for X-ray spectral microanalysis. The maximum spatial resolution of the device with a tungsten cathode and a standard detector of secondary and backscattered electrons was 3.5 nm. In elemental analysis, the resolution maximally amounted to 2 μm. The EDAX accessory with EdaxGenesis software enables not only semiquantitative but also quantitative analysis of elements in a wide range (from Be to U) with a resolution of 127 eV and a minimum detection limit of 0.5 wt% for the elements considered. Microphotography and crystal surface analyses were performed under high-vacuum conditions, without sputtering and at an accelerating voltage of 30 kV. Elemental analysis was performed on a flat area with a side of at least 2 μm. The contents were calculated using the three-correction ZAF method (atomic number, absorption and fluorescence). The ZAF method is based on measuring the X-ray intensity of the i-th element of the sample (J_i) and the standard of the known composition containing this element ($J_{(i)}$) under the same conditions, minus the intensity of the X-ray background radiation, followed by normalization by 100%. To study the chemical composition of the surface of arsenopyrite and pyrite crystals using the SEM-EDX method, the most "clean" areas, with virtually no inclusions or significant defects visible at maximum magnification were selected. It seems meaningless to use polished samples while examining natural surfaces. However, it is possible to minimize fluctuations of the probe current by selection of smooth-faced sections and their positioning in the SEM device. Sections with as little roughness as possible (2–3 nm, well comparable with ordinary polished surfaces) were selected using atomic force microscopy on a SMM-2000 scanning probe microscope (Proton-MIET, Moscow, Russia).

3.4. LA-ICP-MS Method

The study of the surface layer of arsenopyrite and pyrite crystals was continued using the LA-ICP-MS method. A quadrupole inductively coupled plasma mass spectrometer NexION 300D (Perkin-Elmer, Waltham, MA, USA) was used in combination with a laser ablation system NWR-213 (New Wave Research, Fremont, CA, USA) with a laser wavelength of 213 nm. The working and carrier gases were Ar and He, at 99.999% purity. The laser burned a continuous groove (track) no longer than 1.4 mm on the surface of the crystal. The diameter of the laser beam was 100 μm, the frequency was 10 Hz and the energy on the surface of the sample was 0.4 J/cm^2. The laser movement speed was 70 μm/s, and the background measuring time was 20 s. The calibration graphs were constructed in compliance with the NISTSRM610, NISTSRM612, NISTSRM614, BHVO-2G, TB-1G and NKT-1G international standards. It was possible to verify the results for several elements (Au, Pd and Pt) using the in-house sulphide reference sample, which included highly homogeneous ferrous greenockite (α-CdS) crystals synthesized hydrothermally at 500 °C and 1 kbar pressure [42]. The instrument error determined for PGE did not exceed 10%. The contents of the elements in arsenopyrite and pyrite were calculated according to the depth of the track. The ablation depth was controlled by a Quanta 200 scanning electron microscope (FEI Company, Hillsboro, OR, USA). An analysis depth from 0.5 to 1 μm per laser pass was obtained for all samples. For one of the arsenopyrite crystals (sample M-129/10) selected for a detailed study, the laser parameters were adjusted in such a way as to achieve the burned layer of ~100 nm in depth in one pass. Depth was controlled using atomic force microscopy with a SMM-2000 scanning probe microscope (Proton-MIET, Moscow, Russia). In this case, the diameter of the laser beam was 5 μm, the frequency 20 Hz, and the energy on the surface of the sample was 10 J/cm^2 with similar parameters. The instrument error for Pt and Pd did not exceed 30%.

4. Results and Discussion

This section presents data for the study of distribution features and forms of occurrence of PGE (Pt, Pd, Ru, Rh, Ir and Os) in arsenopyrite and pyrite from the Natalkinskoe gold deposit. The data were obtained through local (SEM-EDX and LA-ICP-MS) and, so to speak, "semi-local" analyses (PCA and ADSSC). That is, the methods are not exactly local but were by no means bulk analysis, with a resolution at the level of different size fractions (PCA) or the size of individual crystals (ADSSC). The combination of local analysis and statistics for the compositions of individual monocrystals of different sizes is a unique feature of our approach.

4.1. Pt, Pd, Ru and Rh Content in Monomineral Fractions of Arsenopyrite and Pyrite According to PCA

All studied arsenopyrites and pyrites from the deposit were found to be concentrators not only of Au, but also of such elements as Pt, Pd and Ru. Data on their content are given in Table 1. The highest contents (in ppm) were observed in arsenopyrite: Pt up to 128, Pd up to 20 and Ru up to 86, with lower values identified in pyrite: Pt up to 29, Pd up to 15 and Ru up to 58. The first, although preliminary, results on the content in the monofractions of the studied sulphides of another platinum group element, Rh, have been obtained (Table 2). Maximum contents of this element, up to 21 ppm, as well as for Pt, Pd and Ru are found in arsenopyrite, and lower levels up to 5.9 ppm were found in pyrite. Not only gold-bearing arsenopyrites, but also pyrites were shown to act as concentrators of these elements. It can be argued that the most platinum-rich ores have a sulphide–quartz composition. The data on PGE content in bulk ore samples are in favor of this argument. Relatively high contents (in ppm) were observed for Pt (up to 0.260) and Pd (up to 0.029), and elevated contents for Os (up to 0.003), Ru (up to 0.004), Rh (up to 0.0012) and Ir (up to 0.0006).

Table 1. Platinum, palladium and ruthenium content in different-sized monofractions of arsenopyrite and pyrite selected from vein, streaky-vein and veinlet-disseminated ores of the Natalkinskoe deposit.

Fraction No.	Sample No.*	Characteristics of Samples for Selection of Monofractions	Fraction (mm)	Pt (ppm)	Pd (ppm)	Ru (ppm)
		Arsenopyrite				
1	G-9/13	Veinlet-disseminated ore type. Diamictites dissected by a dense net of quartz and carbonate–quartz veinlets with inclusions of arsenopyrite and pyrite crystals. Arsenopyrite crystals shaped as pseudo-orthorhombic and monoclinic prisms predominate.	0.5–1	60	12	44
2			0.25–0.5	70	20	55
3			0.2–0.25	128	5.4	25
4			0.14–0.2	64	14	53
5			0.07–0.14	63	17	86
6	TPM-1/1	Streaky-vein ore type. Veins and veinlets of quartz and carbonate–quartz in diamictites, with inclusions of crystals and nodules of arsenopyrite, less frequently—pyrite. The crystals of arsenopyrite are shaped as pseudo-orthorhombic and monoclinic prisms.	0.25–0.5	40	4.0	31
7			0.14–0.25	73	11	54
8			0.07–0.14	120	6.7	67
9	M-129/10	Vein ore type. Sulphidized quartz vein on contact with diamictites dissected by a dense net of quartz and carbonate–quartz veins. Sulphides are mainly represented by arsenopyrite dominated by crystals in the form of pseudo-orthorhombic and monoclinic prisms.	0.5–1	23	3.5	18
10			0.25–0.5	25	3.9	21
11			0.2–0.25	27	4.1	24
12	M-131/10	Streaky-vein ore type. Offset subparallel quartz and carbonate–quartz veins and veinlets in diamictite. Sulphides are mainly represented by arsenopyrite. The crystals are shaped as pseudo-orthorhombic and monoclinic prisms.	0.5–1	24	2.8	16
13			0.25–0.5	26	2.3	24
14			0.14–0.25	29	3.6	17
15	M-161/10	Vein ore type. Massive sulphide–quartz vein with a small number of fragments of diamictites, with nodes and inclusions of arsenopyrite, rarely pyrite. Arsenopyrite crystals shaped as monoclinic prisms dominate.	1–2	27	3.7	19
16			0.5–1	47	6.3	33
17			0.25–0.5	62	9.5	31
18			0.14–0.25	44	6.2	28
19	Nat-10	Veinlet-disseminated ore type. Diamictites interspersed with pyrite, rarely arsenopyrite, dissected by a dense net of quartz and carbonate veins from hair-thick to 5 mm. The crystals of arsenopyrite are shaped as pseudo-orthorhombic and monoclinic prisms.	0.2–0.25 **	20	3.4	20
				20	1.9	17
20	UV-3/13	Veinlet-disseminated ore type. Diamictites interspersed with arsenopyrite and pyrite, dissected by a dense net of quartz and carbonate veins. Arsenopyrite crystals shaped as pseudo-orthorhombic and monoclinic prisms predominate.	0.5–1	15	3.5	3.7
21			0.25–0.5 **	5.9	0.7	5.4
22			0.2–0.25	8.0	0.4	4.0
				10	0.7	4.2

Table 1. *Cont.*

Fraction No.	Sample No.*	Characteristics of Samples for Selection of Monofractions	Fraction (mm)	Pt (ppm)	Pd (ppm)	Ru (ppm)
		Pyrite				
23	TPM-1/1	Streaky-vein ore type. Veins and veinlets of quartz and carbonate–quartz in diamictites, with inclusions of crystals and nodules of arsenopyrite, less frequently—pyrite. Pyrite forms cubic and pentagon-dodecahedral crystals.	0.25–0.5	1.0	0.5	4.4
24			0.14–0.25	3.3	0.3	2.9
25			0.07–0.14	1.1	0.4	4.2
26	M-161/10	Vein ore type. Massive sulphide–quartz vein with a small number of fragments of diamictites, with nodules and inclusions of arsenopyrite, rarely pyrite. Cubic pyrite crystals predominate.	0.5–1	14	12	24
27			0.25–0.5	25	15	33
28			0.2–0.25	27	14	41
29	Nat-10	Veinlet-disseminated ore type. Diamictites interspersed with pyrite, rarely arsenopyrite, dissected by a dense net of quartz and carbonate veins from hair-thick to 5 mm. Pyrite forms cubic, pentagon-dodecahedral, rarely cuboctahedral crystals and their combined forms.	1–2 **	25	4.7	21
				10	1.0	28
30			0.5–1 **	15	1.0	57
				15	1.0	39
31			0.25–0.5 **	12	2.1	58
				10	5.8	48
32			0.2–0.25 **	10	2.8	22
				13	3.6	11
33			0.14–0.2	18	4.9	17
34	UV-3/13	Veinlet-disseminated ore type. Diamictites interspersed with arsenopyrite and pyrite, dissected by a dense net of quartz and carbonate veins. Pyrite forms cubic, pentagon-dodecahedral, rarely cuboctahedral crystals.	0.5–1 **	15	1.0	16
				8.0	0.5	2.8
35			0.25–0.5 **	11	1.0	28
				17	1.0	16
36			0.2–0.25	29	1.7	14

Note: The data for the "phase" chemical analysis are presented here and in Table 2. Here: elev.—elevation relative to sea level. * Ore sampling location: G-9/13—"Geological" area, borehole DH329n, interval 151.6–154.6 m (elev. 750 m); TPM-1/1—"North-West" area, quarry (elev. 860 m); M-129/10—"North-West" area, ore zone 33, surface (elev. 920 m); M-131/10—"North-West" area, ore zone 33, surface (elev. 900 m); M-161/10—"Central" area, mine (elev. 600 m, crosscut 11); Nat-10—"Central" area, quarry (elev. 790 m); UV-3/13—"South-East" area, borehole DH70/5n, interval 160.1–163.1 m (elev. 590 m). ** Measurements of Pt, Pd and Ru were carried out on two parallel samples.

Table 2. The content of rhodium in the differently sized monofractions of arsenopyrite and pyrite selected from vein and streaky-vein ores of the Natalkinskoe deposit.

Fraction No.	Sample No.	Monofraction (mm)	Rh (ppm)
		Arsenopyrite	
1		0.5–1	13
2	M-131/10	0.25–0.5	10
3		0.14–0.25	11
4		1–2	9.0
5	M-161/10	0.5–1	21
6		0.25–0.5	21
7		0.14–0.25	12
		Pyrite	
8		0.25–0.5	5.9
9	TPM-1/1	0.14–0.25	3.4
10		0.07–0.14	4.8

Note: Ore sampling location and brief characteristics are presented in Table 1.

4.2. Results of the Study of Pt, Pd, Ru and Rh in Arsenopyrites and Pyrites by AAS-ADSSC

Data on the so-called "invisible" uniformly distributed impurity of Pt, Pd and Ru and the ratio of their structural and surface-bound forms in arsenopyrites and pyrites of the Natalkinskoe deposit are presented in Figures 5–7 and Tables 3–5. In all cases, highly deterministic ($R^2 = 0.82$–1.0) dependences of the average element content in the size sample on the specific surface area of the average crystal in it were obtained. The highest R^2 values were set for arsenopyrite: Pt, 0.92–0.99; Pd, 0.93–1.0 and Ru, 0.97–1.0 and were slightly lower for pyrite: Pt, 0.82–0.99; Pd, 0.94–0.99 and Ru, 0.91–0.99.

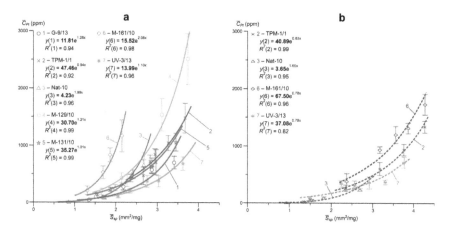

Figure 5. Dependences of the average content of evenly distributed platinum on the specific surface area of an average crystal in size fractions of arsenopyrite (**a**) and pyrite (**b**) from vein, streaky-vein and veinlet-disseminated ores of the Natalkinskoe deposit. The expressions for approximate curves are shown here and in Figures 6–8, in which the pre-exponential factor (in bold) is an estimate of the content of the structural form of the element. The contents of structurally and surface-bound forms of Pt, Pd, Ru and Rh in arsenopyrite and pyrite are given in Tables 3–6. Points 1–7 here and in Figures 6–8 correspond to the ore samples described in Table 1.

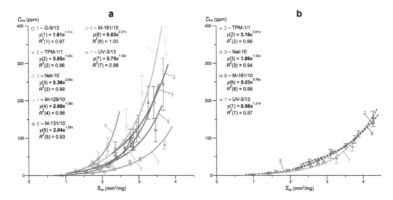

Figure 6. Dependences of the average content of evenly distributed palladium on the specific surface area of an average crystal in size fractions of arsenopyrite (**a**) and pyrite (**b**) from vein, streaky-vein and veinlet-disseminated ores of the Natalkinskoe deposit.

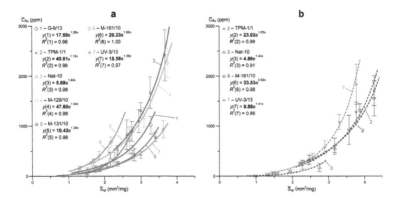

Figure 7. Dependences of the average content of evenly distributed ruthenium on the specific surface area of an average crystal in size fractions of arsenopyrite (**a**) and pyrite (**b**) from vein, streaky-vein and veinlet-disseminated ores of the Natalkinskoe deposit.

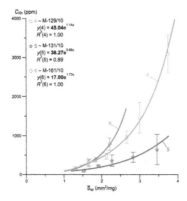

Figure 8. Dependences of the average content of evenly distributed rhodium on the specific surface area of an average crystal in size fractions of arsenopyrite from vein and streaky-vein ores of the Natalkinskoe deposit.

Table 3. Results of platinum analysis of single crystal size selections of arsenopyrite and pyrite from vein, streaky-vein and veinlet-disseminated ores of the Natalkinskoe deposit.

Sample No.	Number of Crystals (Initial–Final Sample)	Characteristics of the Final Sample						Contents (ppm)			
		Number of Crystals	Interval of Masses (mg)	$\bar{m}$ (mg)	$\bar{r}$ (mm)	S_{sp} (mm²/mg)	$C \pm \Delta$ (ppm)	$\bar{C}_{tot}$	$\bar{C}_{ev}$	$\bar{C}_{str}$	$\bar{C}_{sur}$
Arsenopyrite											
G-9/13	66–48	8	0.12–0.18	0.15	0.292	3.411	710 ± 109				
		12	0.19–0.28	0.24	0.339	2.873	579 ± 56				
		11	0.29–0.69	0.55	0.448	2.189	281 ± 54	184	145	12	140
		7	0.79–1.32	1.14	0.572	1.722	100 ± 30				
		10	1.34–3.64	2.34	0.727	1.355	55 ± 9				
TPM-1/1	76–58	12	0.10–0.14	0.12	0.271	3.672	1488 ± 259				
		10	0.15–0.20	0.17	0.304	3.262	1052 ± 130				
		6	0.21–0.22	0.22	0.33	2.97	713 ± 221				
		5	0.23–0.23	0.23	0.336	2.945	852 ± 98	844	728	48	714
		10	0.24–0.28	0.26	0.35	2.827	727 ± 115				
		6	0.29–0.33	0.31	0.371	2.664	477 ± 156				
		9	0.34–0.46	0.41	0.406	2.412	502 ± 77				
M-129/10	42–35	9	0.10–0.12	0.11	0.262	3.745	2519 ± 443				
		7	0.13–0.24	0.19	0.314	3.114	1543 ± 282				
		6	0.27–0.37	0.31	0.371	2.664	802 ± 87	669	536	31	581
		8	0.48–0.73	0.59	0.459	2.143	438 ± 86				
		5	1.01–1.45	1.3	0.597	1.645	202 ± 58				
M-131/10	53–39	9	0.10–0.17	0.14	0.284	3.456	1067 ± 157				
		6	0.18–0.32	0.24	0.339	2.873	649 ± 136				
		8	0.34–0.52	0.43	0.412	2.369	436 ± 67	320	288	35	244
		7	0.54–1.05	0.88	0.524	1.872	236 ± 35				
		9	1.07–2.15	1.58	0.637	1.541	152 ± 28				
M-161/10	40–31	7	0.40–0.62	0.58	0.456	2.151	1168 ± 278				
		8	0.66–1.01	0.9	0.529	1.866	838 ± 118				
		9	1.03–1.39	1.2	0.582	1.694	532 ± 114	565	499	16	494
		7	1.42–4.31	2.58	0.751	1.312	216 ± 75				

Table 3. *Cont.*

Sample No.	Number of Crystals (Initial–Final Sample)	Number of Crystals	Characteristics of the Final Sample					Contents (ppm)			
			Interval of Masses (mg)	m (mg)	r (mm)	S_{sp} (mm²/mg)	$C \pm \Delta$ (ppm)	C_{tot}	C_{ev}	C_{str}	C_{sur}
Arsenopyrite											
Nat-10	60–39	8	0.21–0.41	0.3	0.366	2.679	669 ± 85	135	78	4.2	89
		8	0.42–0.56	0.49	0.431	2.275	397 ± 102				
		9	0.58–0.98	0.74	0.495	1.987	272 ± 51				
		8	1.17–2.31	1.65	0.646	1.518	125 ± 38				
		6	3.73–14.33	9.21	1.147	0.857	17 ± 14				
UV-3/13	78–50	11	0.11–0.24	0.18	0.33	3.63	688 ± 119	252	188	14	190
		8	0.25–0.35	0.31	0.396	3.035	375 ± 97				
		8	0.36–0.51	0.41	0.434	2.756	375 ± 104				
		9	0.52–0.75	0.67	0.512	2.348	210 ± 41				
		7	0.76–1.10	0.99	0.583	2.06	115 ± 31				
		7	1.19–2.16	1.7	0.698	1.72	90 ± 22				
Pyrite											
TPM-1/1	43–33	8	0.10–0.12	0.11	0.28	4.276	1350 ± 100	574	534	41	462
		9	0.13–0.18	0.16	0.317	3.768	940 ± 100				
		6	0.21–0.33	0.27	0.378	3.175	610 ± 110				
		10	0.41–1.65	0.67	0.512	2.348	272 ± 54				
M-161/10	58–45	12	0.10–0.12	0.11	0.28	4.276	1730 ± 180	990	940	68	890
		10	0.13–0.16	0.15	0.311	3.869	1320 ± 90				
		8	0.17–0.21	0.19	0.336	3.565	1200 ± 150				
		10	0.22–0.31	0.27	0.378	3.175	940 ± 57				
		5	0.34–1.13	0.65	0.507	2.373	380 ± 140				
Nat-10	54–34	8	0.30–0.55	0.45	0.448	2.676	239 ± 21	56	33	3.7	35
		8	0.60–1.96	1.46	0.663	1.806	97 ± 17				
		7	2.24–2.71	2.51	0.795	1.511	56 ± 11				
		6	2.84–4.14	3.65	0.9	1.332	31 ± 12				
		5	4.48–16.2	11.5	1.32	0.909	13 ± 5				

Table 3. *Cont.*

Sample No.	Number of Crystals (Initial–Final Sample)	Interval of Masses (mg)	$\bar{m}$ (mg)	$\bar{r}$ (mm)	$\bar{S}_{sp}$ (mm²/mg)	C±Δ (ppm)	Contents (ppm)			
							$\bar{C}_{tot}$	$\bar{C}_{ev}$	$\bar{C}_{str}$	$\bar{C}_{sur}$
				Pyrite						
UV-3/13	79–52	0.10–0.18	0.16	0.317	3.768	716 ± 97				
		0.20–0.29	0.24	0.363	3.294	372 ± 42				
		0.31–0.39	0.35	0.412	2.91	360 ± 107	322	208	37	189
		0.41–0.58	0.52	0.47	2.549	316 ± 75				
		0.60–1.00	0.78	0.538	2.227	342 ± 54				
		1.05–3.60	2.69	0.813	1.474	84 ± 31				

Note: Here and in Tables 4–6; $\bar{m}$, the average mass of the crystal in size fraction; $\bar{r}$, the average size of the crystal in size fraction; $\bar{S}_{sp}$, the specific surface area of average crystal in size fraction; $C ± Δ$, the average content of evenly distributed element ± one standard deviation; $\bar{C}_{tot}$, average total content, $\frac{\sum C_i m_i}{\sum m_i}$; $\bar{C}_{ev}$, average evenly distributed element content for all size fractions; C_{str}, structurally bound form content (C extrapolation to zero $\bar{S}_{sp}$); $\bar{C}_{sur}$, average content of surface-associated form.

Table 4. Results of palladium analysis of single crystal size selections of arsenopyrite and pyrite from vein, streaky-vein and veinlet-disseminated ores of the Natalkinskoe deposit.

Sample No.	Number of Crystals (Initial–Final Sample)	Interval of Masses (mg)	$\bar{m}$ (mg)	$\bar{r}$ (mm)	$\bar{S}_{sp}$ (mm²/mg)	C±Δ (ppm)	Contents (ppm)			
							$\bar{C}_{tot}$	$\bar{C}_{ev}$	$\bar{C}_{str}$	$\bar{C}_{sur}$
				Arsenopyrite						
G-9/13	66–42	0.12–0.18	0.15	0.292	3.411	195 ± 43				
		0.19–0.28	0.25	0.345	2.857	97 ± 27				
		0.29–0.69	0.57	0.453	2.16	27 ± 6	40	27	1.6	26
		0.79–1.32	1.05	0.556	1.766	26 ± 6				
		1.34–3.64	2.42	0.735	1.339	9.5 ± 3.5				

Table 4. *Cont.*

Sample No.	Number of Crystals (Initial–Final Sample)	Number of Crystals	Interval of Masses (mg)	$\bar{m}$ (mg)	$\bar{r}$ (mm)	S_{sp} (mm²/mg)	$\bar{C} \pm \Delta$ (ppm)	Contents (ppm)			
								C_{tot}	C_{ev}	C_{str}	C_{sur}
Arsenopyrite											
TPM-1/1	88–63	10	0.10–0.14	0.12	0.271	3.672	245 ± 69	142	114	5.9	114
		11	0.15–0.19	0.17	0.304	3.262	196 ± 45				
		8	0.20–0.22	0.21	0.324	3	132 ± 30				
		10	0.23–0.24	0.23	0.336	2.945	117 ± 27				
		9	0.25–0.29	0.27	0.353	2.769	101 ± 23				
		7	0.30–0.34	0.32	0.373	2.609	75 ± 20				
		8	0.35–0.65	0.44	0.416	2.36	73 ± 18				
M-129/10	41–31	9	0.10–0.12	0.11	0.262	3.745	361 ± 39	97	83	2.6	82
		6	0.13–0.24	0.2	0.321	3.091	221 ± 47				
		5	0.27–0.37	0.31	0.371	2.664	133 ± 46				
		6	0.48–0.73	0.6	0.461	2.125	54 ± 19				
		5	1.01–1.45	1.25	0.59	1.671	20 ± 4				
M-131/10	52–38	8	0.10–0.17	0.14	0.284	3.456	122 ± 31	32	30	2	27
		6	0.18–0.32	0.26	0.35	2.827	77 ± 10				
		8	0.34–0.52	0.43	0.412	2.369	52 ± 10				
		8	0.54–1.05	0.87	0.523	1.886	25 ± 6				
		8	1.07–2.15	1.58	0.637	1.541	10 ± 2				
M-161/10	40–24	8	0.40–0.62	0.58	0.456	2.151	83 ± 18	39	31	0.6	31
		5	0.66–0.98	0.85	0.518	1.894	46 ± 18				
		6	1.00–1.32	1.12	0.569	1.734	34 ± 6				
		5	1.38–4.31	2.87	0.777	1.262	11 ± 7				
Nat-10	60–39	9	0.21–0.41	0.27	0.353	2.769	96 ± 18	12	7.7	0.4	8.1
		8	0.42–0.56	0.51	0.438	2.257	33 ± 9				
		8	0.58–0.98	0.77	0.501	1.956	18 ± 5				
		7	1.17–2.31	1.64	0.646	1.527	9.8 ± 4.2				
		7	3.73–14.33	8.42	1.113	0.883	1.9 ± 0.8				

Characteristics of the Final Sample

Table 4. *Cont.*

Sample No.	Number of Crystals (Initial–Final Sample)	Number of Crystals	Interval of Masses (mg)	$\bar{m}$ (mg)	$\bar{r}$ (mm)	$\bar{S}_{sp}$ (mm²/mg)	$\bar{C}\pm\Delta$ (ppm)	Contents (ppm)			
								$\bar{C}_{tot}$	$\bar{C}_{ev}$	$\bar{C}_{str}$	$\bar{C}_{sur}$
Arsenopyrite											
UV-3/13	76–58	12	0.11–0.24	0.17	0.324	3.705	82 ± 11	21	18	0.8	25
		10	0.25–0.35	0.31	0.396	3.035	44 ± 5				
		9	0.35–0.51	0.41	0.434	2.756	35 ± 6				
		10	0.52–0.75	0.68	0.514	2.331	17 ± 3				
		8	0.76–1.10	0.99	0.583	2.06	10 ± 2				
		9	1.19–2.16	1.61	0.685	1.749	6.7 ± 1.1				
Pyrite											
TPM-1/1	43–33	10	0.10–0.12	0.11	0.28	4.276	156 ± 17	58	53	3.2	50
		8	0.13–0.18	0.16	0.317	3.768	97 ± 11				
		6	0.21–0.33	0.29	0.378	3.099	58 ± 12				
		9	0.41–1.65	0.7	0.519	2.309	25 ± 5				
M-161/10	54–44	10	0.10–0.12	0.11	0.28	4.276	142 ± 11	73	67	5.2	62
		9	0.13–0.17	0.15	0.311	3.869	111 ± 8				
		7	0.18–0.22	0.2	0.342	3.509	80 ± 6				
		9	0.24–0.31	0.28	0.383	3.143	62 ± 3				
		9	0.34–1.13	0.54	0.476	2.518	37 ± 7				
Nat-10	60–41	8	0.29–0.51	0.42	0.438	2.741	35 ± 6	8.7	6.3	1.1	4.9
		7	0.55–1.63	1.24	0.628	1.908	14 ± 3				
		8	1.72–2.57	2.17	0.757	1.584	11 ± 3				
		9	2.66–4.14	3.23	0.864	1.387	7.9 ± 0.8				
		9	4.48–16.2	10.66	1.287	0.932	2.7 ± 0.5				
UV-3/13	79–56	10	0.10–0.18	0.17	0.324	3.705	77 ± 11	22	15	1	18
		9	0.20–0.29	0.25	0.368	3.25	45 ± 12				
		11	0.31–0.39	0.35	0.412	2.91	37 ± 4				
		12	0.41–0.58	0.51	0.467	2.566	25 ± 2				
		9	0.60–1.00	0.83	0.55	2.187	18 ± 3				
		5	1.05–3.60	3.2	0.862	1.393	4.2 ± 1.2				

Characteristics of the Final Sample

Table 5. Results of ruthenium analysis of single crystal size selections of arsenopyrite and pyrite from vein, streaky-vein and veinlet-disseminated ores of the Natalkinskoe deposit.

Sample No.	Number of Crystals (Initial–Final Sample)	Characteristics of the Final Sample						Contents (ppm)			
		Number of Crystals	Interval of Masses (mg)	m (mg)	r (mm)	S_{sp} (mm²/mg)	$C \pm \Delta$ (ppm)	C_{tot}	C_{ev}	C_{str}	C_{sur}
					Arsenopyrite						
G-9/13	64–42	12	0.12–0.20	0.16	0.296	3.286	1034 ± 177				
		9	0.22–0.33	0.27	0.353	2.769	547 ± 76				
		6	0.40–0.69	0.57	0.453	2.16	339 ± 149	283	193	18	190
		7	0.79–1.34	1.1	0.565	1.741	160 ± 53				
		8	1.39–3.64	2.49	0.742	1.327	82 ± 26				
TPM-1/1	85–61	9	0.10–0.14	0.12	0.271	3.672	2424 ± 512				
		12	0.15–0.19	0.17	0.304	3.262	1678 ± 316				
		10	0.20–0.22	0.21	0.324	3	1321 ± 159	1318	1129	41	1083
		9	0.23–0.24	0.23	0.336	2.945	1108 ± 225				
		8	0.25–0.29	0.28	0.358	2.746	904 ± 197				
		6	0.30–0.34	0.32	0.373	2.609	832 ± 298				
		7	0.35–0.65	0.45	0.42	2.352	530 ± 86				
M-129/10	43–33	9	0.10–0.12	0.11	0.262	3.745	2269 ± 285				
		6	0.13–0.24	0.2	0.321	3.091	1179 ± 196				
		5	0.27–0.37	0.3	0.366	2.679	812 ± 63	656	537	48	520
		7	0.48–0.73	0.61	0.464	2.118	516 ± 63				
		6	1.01–1.45	1.27	0.592	1.656	231 ± 59				
M-131/10	40–29	6	0.10–0.17	0.15	0.292	3.411	789 ± 92				
		5	0.18–0.32	0.26	0.35	2.827	432 ± 121				
		7	0.34–0.52	0.44	0.416	2.36	263 ± 33	179	165	10	162
		6	0.54–1.05	0.9	0.529	1.866	129 ± 30				
		5	1.07–2.15	1.7	0.65	1.505	63 ± 15				
M-161/10	40–33	9	0.40–0.62	0.56	0.451	2.179	667 ± 107				
		8	0.66–1.01	0.91	0.53	1.852	422 ± 75				
		9	1.03–1.39	1.2	0.582	1.694	309 ± 42	331	297	20	290
		7	1.42–4.31	2.62	0.755	1.305	166 ± 33				

Table 5. *Cont.*

Sample No.	Number of Crystals (Initial–Final Sample)	Number of Crystals	Interval of Masses (mg)	$\bar{m}$ (mg)	$\bar{r}$ (mm)	$\bar{S}_{sp}$ (mm²/mg)	C±Δ (ppm)	Contents (ppm)			
								C_{tot}	$\bar{C}_{ev}$	C_{str}	$\bar{C}_{sur}$
Arsenopyrite											
Nat-10	59–35	5	0.21–0.41	0.30	0.336	2.679	729 ± 262	128	79	5.7	79
		8	0.42–0.56	0.50	0.434	2.260	375 ± 103				
		8	0.58–0.98	0.80	0.508	1.936	187 ± 51				
		8	1.17–2.31	1.65	0.646	1.518	124 ± 49				
		6	3.73–14.33	9.07	1.141	0.861	24 ± 10				
UV-3/13	78–58	10	0.11–0.24	0.18	0.33	3.63	941 ± 172	329	238	19	218
		9	0.25–0.35	0.32	0.4	3	428 ± 69				
		10	0.36–0.51	0.4	0.431	2.787	372 ± 65				
		10	0.52–0.75	0.66	0.509	2.355	245 ± 49				
		9	0.76–1.10	0.98	0.581	2.067	211 ± 56				
		10	1.19–2.16	1.63	0.688	1.742	103 ± 20				
Pyrite											
TPM-1/1	43–29	8	0.10–0.12	0.11	0.28	4.276	1720 ± 300	686	574	23	577
		8	0.13–0.18	0.16	0.317	3.768	1280 ± 210				
		6	0.21–0.33	0.27	0.378	3.175	605 ± 160				
		7	0.41–1.65	0.73	0.527	2.283	235 ± 90				
M-161/10	58–43	10	0.10–0.12	0.11	0.28	4.276	1560 ± 350	850	700	34	690
		10	0.13–0.16	0.15	0.311	3.869	1320 ± 130				
		6	0.17–0.21	0.19	0.336	3.565	930 ± 290				
		10	0.22–0.31	0.27	0.378	3.175	540 ± 70				
		7	0.34–1.13	0.58	0.488	2.464	330 ± 90				
Nat-10	58–40	6	0.29–0.55	0.42	0.438	2.741	189 ± 86	48	35	4.9	30
		10	0.60–1.72	1.23	0.627	1.918	109 ± 22				
		7	1.81–2.66	2.32	0.774	1.549	58 ± 15				
		9	2.71–4.48	3.58	0.895	1.343	33 ± 5				
		8	5.78–16.2	9.91	1.256	0.955	14 ± 4				

Table 5. *Cont.*

Sample No.	Number of Crystals (Initial–Final Sample)	Characteristics of the Final Sample						Contents (ppm)			
		Number of Crystals	Interval of Masses (mg)	$\bar{m}$ (mg)	$\bar{r}$ (mm)	$\bar{S}_{sp}$ (mm²/mg)	$\bar{C}\pm\Delta$ (ppm)	$\bar{C}_{tot}$	$\bar{C}_{ev}$	$\bar{C}_{str}$	$\bar{C}_{sur}$
					Pyrite						
UV-3/13	79–55	10	0.10–0.18	0.16	0.317	3.768	1840 ± 400	454	272	9.9	330
		11	0.20–0.29	0.24	0.363	3.294	1085 ± 149				
		10	0.31–0.39	0.36	0.416	2.884	577 ± 129				
		9	0.41–0.58	0.52	0.47	2.549	384 ± 74				
		9	0.60–1.00	0.83	0.558	2.187	248 ± 39				
		6	1.05–3.60	2.9	0.834	1.439	67 ± 23				

Table 6. Results of rhodium analysis of size selections of arsenopyrite single crystals from vein and streaky-vein ores of the Natalkinskoe deposit.

Sample No.	Number of Crystals (Initial–Final Sample)	Characteristics of the Final Sample						Contents (ppm)			
		Number of Crystals	Interval of Masses (mg)	$\bar{m}$ (mg)	$\bar{r}$ (mm)	$\bar{S}_{sp}$ (mm²/mg)	$\bar{C}\pm\Delta$ (ppm)	$\bar{C}_{tot}$	$\bar{C}_{ev}$	$\bar{C}_{str}$	$\bar{C}_{sur}$
M-129/10	43–31	9	0.10–0.12	0.11	0.262	3.745	3108 ± 491	801	715	45	678
		5	0.13–0.24	0.21	0.324	3.000	1473 ± 394				
		6	0.27–0.37	0.31	0.371	2.664	973 ± 159				
		6	0.48–0.73	0.60	0.461	2.125	483 ± 78				
		5	1.01–1.45	1.29	0.595	1.647	296 ± 85				
M-131/10	38–28	5	0.10–0.17	0.14	0.284	3.456	653 ± 391	211	190	36	154
		4	0.18–0.31	0.26	0.350	2.827	453 ± 130				
		4	0.34–0.52	0.49	0.431	2.275	368 ± 146				
		7	0.54–1.04	0.83	0.514	1.910	224 ± 34				
		8	1.07–2.15	1.64	0.650	1.527	103 ± 19				
M-161/10	40–28	8	0.40–0.62	0.54	0.446	2.210	782 ± 155	360	309	17	299
		7	0.66–1.01	0.93	0.534	1.840	387 ± 55				
		7	1.03–1.39	1.26	0.592	1.669	320 ± 89				
		6	1.42–4.31	2.65	0.757	1.297	158 ± 46				

Extrapolation of these dependences to zero specific surface area ($\overline{S}_{sp}$, mm^2/mg, Tables 3–5), i.e., on a conditionally infinite crystal in the first case of arsenopyrite, and in the second pyrite, yields the following estimates of the contents of the structural form of the elements (C_{str}, ppm): for arsenopyrite Pt, 4.2–48; Pd, 0.4–5.9 and Ru, 5.7–48; for pyrite Pt, 3.7–68; Pd, 1.0–5.2 and Ru 4.9–34 (Figures 5–7). The average contents of surface-bound form of elements ($\overline{C}_{sur}$, ppm) are given in Tables 3–5 and for arsenopyrite amount to the following values: Pt, 89–714; Pd, 8.1–114 and Ru, 79–1083; for pyrite Pt, 35–890; Pd, 4.9–62 and Ru, 30–690.

By now, the range of PGE studied by the AAS-ADSSC method has expanded. The first results for the forms of occurrence of Rh, one more element of the platinum group in crystals of arsenopyrite, were obtained. These data are still limited and preliminary. The results of the study of the content of evenly distributed Rh and the ratio of its structural and surface-bound forms in arsenopyrites from vein and streaky-vein ores of the Natalkinskoe deposit are presented in Figure 8 and in Table 6. In all cases, as well as for Pt, Pd and Ru, highly deterministic ($R^2 = 0.89$–1.0) dependences of the average Rh content in the size fraction on the specific surface area of the average arsenopyrite crystal in it were obtained. Extrapolation of these dependences to a zero-specific surface, i.e., to a conditionally infinite crystal, yielded estimates of the contents of the structural form of this element of 17–45 ppm. The average grades of the surface-associated forms were 154–678 ppm for arsenopyrite (see Table 6).

The data presented (PCA and AAS-ADSSC) suggest that arsenopyrite also acts as a concentrator mineral for Rh. Structural and surface-bound forms in arsenopyrite (as well as in pyrite), as the main uniformly distributed forms, are characteristic not only of Pt, Pd and Ru, but also Rh. The data obtained confirms that arsenopyrites and pyrites are major mineral concentrators of PGE at the Natalkinskoe gold ore deposit. It can also be assumed that the size dependence of the content of evenly distributed Pt, Pd, Ru and Rh, and the confinement of a significant part to the surface of the crystals, is not a unique phenomenon for sulphide minerals of ore deposits. Structurally and superficially bound forms, as the main uniformly distributed forms, may be typical of both the remaining PGE (Os and Ir) and of other noble metals—Au and Ag [11,37–39,41–45].

New data for Pt and Pd forms in arsenopyrites, both structural and surface-bound, confirmed earlier conclusions [11]. The uniformly distributed Pt, Pd and Ru and, as we believe, Rh in arsenopyrites and pyrites are forms of elements which are chemically bound in the structure of these minerals and in nanoscale nonautonomous phases (NAPs) on the surface of arsenopyrite and pyrite crystals. The latter usually dominates over the former according to the content of these elements. Recent experimental and natural data [37,38,46] suggest that Pt, Pd, Ru and Rh, as well as Au, can be incorporated into NAP formed on the surface of arsenopyrite and pyrite crystals in a very thin surface layer (~100–500 nm). These elements are absorbed by NAP from the hydrothermal solution even more efficiently than gold, which is incompatible in pyrite.

There is currently high demand for studies of the surface layer of sulphide minerals as possible hosts for "invisible" forms of trace elements, the nature of their distribution and concentration levels. Nanoscale NAPs enriched with Pt, Pd, Ru and Rh form a part of the so-called impurity component on the surface of arsenopyrite and pyrite crystals. The importance of its investigation is obvious. It is assumed that there is a common mechanism of absorption of all noble metals for the studied sulphides due to the active role of the surface during the growth of the crystal from a supersaturated hydrothermal solution [38,39]. This mechanism is closely related to the conditions of ore formation. This is an important point that must be taken into account in practical terms, for example, in the concentration of ores that include sulphides, in our case arsenopyrite and pyrite.

The structurally bound form of PGE, primarily Pt, despite a lower content compared to the surface form, is also extremely important. As shown earlier, data on the structural form of Au, Ag and Pt make it possible to make comparative estimates of the contents of these elements in the ore-forming fluids that form gold deposits. It is the only form of the element occurrence that can be used as an indicator of element activity in hydrothermal solution [37,46]. For example, the experimental data on the distribution and segregation of trace elements in crystals of ore minerals grown in hydrothermal systems were used to assess the content of Au and Pt in the ore-forming fluid (C_{aq}) of the Natalkinskoe and Degdekan gold deposits. The content of the element in the structural form in natural pyrite, and the distribution coefficient (D_{str}) for the same form between pyrite and hydrothermal solution according to experimental data, were used for estimating element content in the ore-forming fluid [38]. In Table 7 these results are supplemented by the data from the present work on samples M-161/10, TPM-1/1 and Nat-10.

Table 7. Evaluation of gold and platinum grade in ore-forming fluids by pyrite composition using experimental distribution coefficients D_{str}.

Sample No.	Deposit	Number of Crystals (Initial–Final Sample)	Element	C_{str} (ppm)	$\overline{C}_{sur}$ (ppm)	C_{aq} (ppm)	Au/Pt$_{aq}$
M-163/10	Degdekan	95–59	Au	0.2	0.5	2.1	21
		92–71	Pt	2.0	7.1	0.1	
DG-10/14		67–44	Au	0.1	1.1	1.3	6.5
		34–22	Pt	5.1	102	0.2	
UV-3/13	Natalkinskoe	80–52	Au	0.3	1.4	2.9	1.6
		79–52	Pt	37	189	1.8	
M-161/10		58–40	Au	0.2	3.6	1.8	0.6
		58–45	Pt	68	890	3.2	
TPM-1/1		43–30	Au	0.3	3.4	3.2	1.6
		43–33	Pt	41	462	2.0	
Nat-10		60–39	Au	1.8	2.7	18	90
		54–34	Pt	3.6	35	0.2	

Note: C_{str}, structurally bound element content; $\overline{C}_{sur}$, superficially bound element content; C_{aq}, element content in the ore-forming fluid.

The average value of D_{str} is 0.1 for Au and 21 for Pt [41,46]. The evaluation of Au and Pt content in the ore-forming fluid showed that Au content in the fluids of the Degdekan and Natalkinskoe gold deposits was mainly at the level of 1.3–3.2 ppm, and it reached a maximum value of 18 ppm for the Nat-10 sample. Pt grade ranged from 0.1 to 3.2 ppm and, on average, was clearly elevated at the Natalkinskoe deposit compared to Degdekan. The fluid of the Natalkinskoe deposit was rich in both elements compared to the fluid of the Degdekan deposit. Au and Pt contents in solution were also comparable for most samples from the Natalkinskoe deposit (with the exception of the anomalous Au in sample Nat-10), whereas Au dominated over Pt in the Degdekan deposit. This is another argument in favor of studying platinum-bearing gold deposits, of which the Natalkinskoe deposit is no exception. Information about the Pt content in the fluid is of particular value in view of the complex behavior of Pt and the heterogeneity of its distribution under the capture of gas–liquid inclusions, the composition of which is determined by LA-ICP-MS as a proxy of the ore fluid composition [47].

4.3. Study of the Chemical Composition of the Surface of Crystals of Arsenopyrite and Pyrite Using the SEM-EDX Method

Nearly isometric pseudo-orthorhombic crystals of arsenopyrite and cubic crystals of pyrite without visible inclusions of ore minerals and with a minimum number of non-metallic microinclusions (mainly quartz and carbonate) were selected for the study (Figure 9). The so-called "clean" areas virtually lacking visible inclusions and significant defects were then selected to study the composition of the "invisible" impurity component on the surface of these crystals (Figure 10). The size (in µm) of such areas was 10×10, 20×20, and it was at least 30×30. Sections with low roughness were selected using the scanning probe microscopy technique (see above).

Minerals **2020**, 10, 318

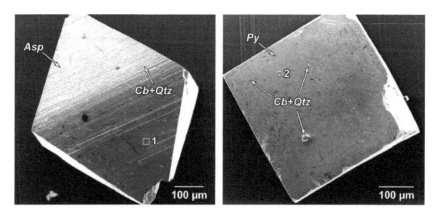

Figure 9. Crystals of arsenopyrite (Asp) and pyrite (Py) with microinclusions of carbonate (Cb) and quartz (Qtz). Surface areas of arsenopyrite (1) and pyrite (2) crystals, where measurements were carried out. Their enlarged fragments are shown in Figure 10.

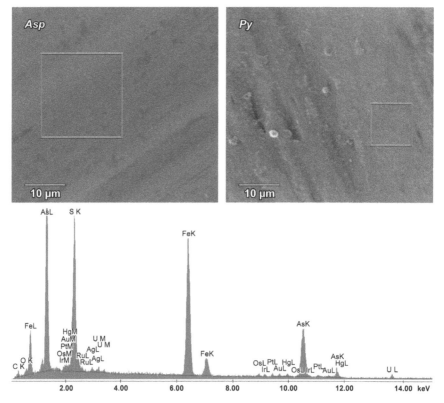

Figure 10. Enlarged fragments of arsenopyrite (Asp) and pyrite (Py) crystals with "clean" areas, where measurements were carried out (see Table 8, sample M-129/10, *N* 3, *n* 3 and Table 9, sample M-129/10, *N* 1, *n* 2). The bottom of the Figure 9 shows the X-ray energy spectra of arsenopyrite (see Table 8, *N* 3, *n* 3).

Faces with striation of the same crystallographic type were selected during scanning in order to eliminate the effect of capturing various impurities by faces of different crystallographic types during crystal growth: for arsenopyrite—prism {141}, for pyrite—cube {100}.

The surfaces of arsenopyrites and pyrites selected from the Natalkinskoe ore deposit (samples G-9/13, TPM-1/1, M-129/10, M-131/10, M-161/10 and Nat-10) clearly demonstrated the peculiarities of its elemental composition. On the "clean" surface of arsenopyrite, PGEs were thus represented by the following elements (in wt%): Pt (up to 2.1), Ru (1.4), Ir (up to 0.7) and Os (up to 0.7). Other ore elements were also determined (in wt%): U (up to 2.0), Hg (up to 1.1), Au (up to 1.0), rarely Ag (up to 0.7) and Cu (up to 0.6) (Table 8). The only PGE present on the "clean" surface of pyrite was Pt (up to 2.9 wt%), and other ore elements were represented by As (up to 2.2 wt%) and Cu (up to 0.7 wt%) (Table 9).

The SEM-EDX method confirmed the presence of Au and Pt on the surface of arsenopyrite crystals from the Natalkinskoe deposit, and high contents of them were detected earlier [11]. It is the first time such PGEs as Ru, Os, Ir and a number of ore elements—Ag, Hg, Cu and U—were detected on the surface of arsenopyrite. Pt and As are shown for the first time to be constantly present on the surface of pyrite crystals. Comparing the data from the AAS-ADSSC and SEM-EDX methods regarding the surface contents of Au, Pt and Ru, it is necessary to recall that the data on $\overline{C}_{sur}$ of AAS-ADSSC method are related to the entire average crystal, whereas the SEM-EDX method virtually defines local content of the element at the local area ("point") on the surface. It would be more correct to compare these results with the content of the element within NAP, which can be calculated under certain assumptions [46]. According to such calculations, Au in NAP on pyrite amounts to 0.3–0.5 wt% [38], which is comparable by an order of magnitude to the data in Tables 8 and 9 on Au and Pt. Increased platinum on pyrite (see Table 9) is qualitatively consistent with the fact that, according to AAS-ADSSC, $\overline{C}_{sur}$ Pt > $\overline{C}_{sur}$ Au (see Table 7).

Table 8. Chemical composition (in wt%) of arsenopyrite crystal surfaces with micron-size rare inclusions of non-metallic minerals. Samples from vein ores of the Natalkinskoe deposit.

Sample No.	N	n	Fe	As	S	C	O	Al	Si	Hg	Au	Ag	Cu	Ti	Pt	Os	Ir	Ru	U
	1	1	**31.6**	**36.0**	**23.4**	2.8	2.0	1.2	0.2	0.7	0.7	0.2	-	-	*0.4*	-	-	-	0.8
		2	**32.4**	**35.6**	**20.2**	3.4	2.5	1.3	0.2	1.1	0.9	0.3	-	-	*0.6*	-	-	-	1.5
		3	**31.9**	**36.8**	**20.8**	3.7	1.5	0.6	0.1	1.0	0.9	0.3	-	-	*0.8*	-	-	-	1.6
M-129/10	2	1	**30.0**	**36.3**	**20.4**	4.3	2.5	1.2	*0.1*	1.1	0.9	*0.1*	-	-	*1.1*	-	-	-	2.0
		2	**28.9**	**39.0**	**20.6**	4.6	2.5	0.5	*0.2*	0.9	0.8	*0.2*	-	-	*0.4*	-	-	-	1.4
	3	1	**32.6**	**41.8**	**16.1**	3.9	1.5	-	-	1.0	0.7	-	-	-	0.5	0.6	*0.4*	-	0.9
		2	**32.1**	**43.4**	**14.2**	2.3	1.4	-	-	0.9	1.0	0.2	-	-	0.9	0.6	*0.7*	1.0	1.3
		3	**32.4**	**43.1**	**15.4**	1.1	1.0	-	-	0.9	0.9	0.7	-	-	0.9	0.7	*0.7*	1.0	1.2
	4	1	**34.0**	**40.2**	**15.1**	4.6	2.7	-	-	*0.4*	0.3	-	-	-	*0.4*	0.2	0.3	1.4	*0.4*
		2	**34.7**	**41.5**	**13.8**	4.8	1.8	-	-	*0.4*	0.4	-	-	-	*0.4*	0.2	0.2	1.2	*0.6*
M-161/10	5	1	**30.8**	**41.2**	**17.1**	3.9	3.2	1.0	0.1	-	-	-	0.6	-	2.1	-	-	-	-
		2	**31.9**	**41.7**	**19.0**	1.9	2.1	1.4	0.1	-	-	*0.1*	0.5	0.2	1.1	-	-	-	-
	6	1	**33.3**	**42.4**	**19.0**	1.0	1.1	0.6	0.2	-	-	0.3	0.2	0.2	0.6	0.2	0.3	0.3	0.3
		2	**33.2**	**42.9**	**19.1**	0.6	0.8	0.5	0.2	-	-	*0.1*	0.2	*0.1*	1.2	0.2	0.3	0.3	0.3
		3	**33.1**	**43.2**	**19.1**	0.4	0.5	0.5	*0.1*	-	-	0.6	0.6	*0.1*	0.8	0.2	0.3	0.2	0.3

Note: Pd, Rh, F, Cl, Pb, Zn, Mo, W, Bi and REE were not detected. Analytical lines: Fe$K\alpha$, As$K\alpha$, S$K\alpha$, C$K\alpha$, O$K\alpha$, Al$K\alpha$, Si$K\alpha$, Hg$L\alpha$, Au$L\alpha$, Ag$L\alpha$, Cu$K\alpha$, Ti$K\alpha$, Pt$L\alpha$, Os$L\alpha$, Ir$L\alpha$, Ru$L\alpha$ and U$L\alpha$. Here and in Table 9: *N*—the serial number of the crystal, *n*—the serial number of the measurement area on the crystal surface; hyphen—element is not detected; italics—the elements content below the detection limit (<0.5 wt%), so their presence is problematic but without taking them into account the sum total does not reach 100%; the main composition is in bold.

Table 9. Chemical composition (in wt%) of pyrite crystal surfaces with micron-size rare inclusions of non-metallic minerals. Samples from vein ores of the Natalkinskoe deposit.

Sample No.	N	n	Fe	S	C	O	Al	Si	Ca	Na	As	Cu	Ti	Pt
M-129/10	1	1	38.5	43.1	5.3	7.7	1.2	0.2	-	-	1.3	0.7	0.3	1.7
		2	36.9	39.2	7.0	10.7	1.3	0.3	-	-	1.2	0.7	0.4	2.3
		3	41.0	47.9	2.9	2.6	1.4	0.2	-	-	1.7	0.5	0.2	1.6
	2	1	39.8	44.1	5.1	6.8	1.1	0.2	-	-	1.0	-	0.2	1.7
		2	35.9	42.7	5.6	10.2	0.8	1.1	-	-	1.0	0.6	0.1	2.0
		3	38.8	47.0	3.3	6.2	0.9	0.9	-	-	1.1	0.3	0.1	1.4
	3	1	37.0	38.3	4.3	13.2	0.9	0.7	0.2	0.3	1.8	0.6	-	2.7
		2	39.5	48.4	2.3	3.0	0.8	0.1	0.1	0.2	2.2	0.5	-	2.9
M-161/10	4	1	37.4	37.5	3.1	16.4	1.3	1.0	-	-	1.1	0.4	0.1	1.7
		2	40.5	46.5	3.6	5.6	1.4	0.1	-	0.2	1.4	0.2	-	0.5
	5	1	41.7	47.2	2.6	3.3	0.9	0.1	-	-	2.0	0.3	-	1.9
		2	40.7	47.5	3.0	4.0	0.9	0.2	-	-	2.0	0.2	-	1.5

Note: Au, Ag, Hg, Pd, Os, Ir, Ru, Rh, U, F, Cl, Pb, Zn, Mo, W, Bi and REE were not detected. Analytical lines: FeK_α, SK_α, CK_α, OK_α, AlK_α, SiK_α, CaK_α, NaK_α, AsK_α, CuK_α, TiK_α and PtL_α.

4.4. Study of the Chemical Composition of the Surface of Arsenopyrite and Pyrite Crystals by LA-ICP-MS

When studying the surface layers of arsenopyrite and pyrite crystals using the LA-ICP-MS method, attention was largely focused on Pt and Pd. Ten arsenopyrite crystals and six pyrite crystals were analyzed (Tables 10 and 11). For all crystals of arsenopyrite (samples M-129/10, M-131/10, M-161/10 and TPM-1/1) and pyrite (samples M-161/10 and TPM-1/1) a track depth of 0.5–0.6 μm per laser pass was obtained. Unfortunately, we do not know the NAP thickness of the studied crystals. If we focus on the previously obtained data [44,48], the thickness of NAP on pyrite is ≤0.5 μm, and for arsenopyrite it is, according to preliminary data, much less [11]. The first pass of the laser thus best represents the composition of NAP, although significant "contamination" by the material underlying NAP is likely. The parameters of the laser were adjusted for one of the arsenopyrite crystals selected for detailed study (see Table 10, sample M-129/10, crystal 2), in such a way that a layer of ~100 nm thick would be burned in one pass (see above). In this case, the maximum depth of the laser-dig groove was 600 nm. Figure 11 shows graphs of the detailed distribution of Pt and Pd contents in the surface layer of Crystal 2 from the M-129/10 sample.

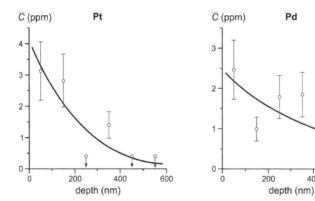

Figure 11. Distribution of the average content ($\overline{C}$) of platinum and palladium in the surface layer of arsenopyrite crystal (sample M-129/10, Crystal 2) according to the results of LA-ICP-MS analysis (the Natalkinskoe deposit).

Table 10. Results of LA-ICP-MS analysis of the surface of arsenopyrite crystals for Pt and Pd (the Natalkinskoe deposit).

Sample No.	Crystal No. (Size, mm)	Track No.	Contents of Elements (ppm)	
			Pt	Pd
M-129/10	1 (1.0)	1	2.9	1.8
		2	1.4	1.8
	2 (1.1)	1	**3.1**	**2.5**
			2.8	**1.0**
			-	**1.8**
			1.4	**1.8**
			-	-
			-	**1.0**
M-131/10	1 (1.3)	1	0.4	0.7
		2	0.4	0.5
	2 (1.3)	1	-	0.1
		2	-	0.2
	3 (1.0)	1	<0.4	0.1
		2	-	0.1
M-161/10	1 (1.0)	1	0.4	0.4
		2	23	3.8
		3	0.4	0.2
		4	<0.4	0.1
		5	6.2	1.4
	2 (1.2)	1	<0.4	<0.1
		2	-	0.6
TPM-1/1	1 (1.3)	1	0.4	0.8
	2 (1.5)	1	0.4	1.2
	3 (1.5)	1	<0.4	0.9
		2	0.4	4.8

Note: Here and in Tables 11–13:hyphen—element not detected; sign "<"—below detection limit. Bold—results of a detailed study of Pt and Pd contents to the depth of 600 nm with a step of 100 nm (see Figure 11).

Table 11. Results of LA-ICP-MS analysis of the surface of pyrite crystals for Pt and Pd (the Natalkinskoe deposit).

Sample No.	Crystal No. (Size, mm)	Track No.	Contents of Elements (ppm)	
			Pt	Pd
M-161/10	1 (1.0)	1	1.6	5.9
		2	0.4	0.1
		3	-	0.2
		4	-	0.1
	2 (1.2)	1	<0.4	0.9
		2	<0.4	0.2
		3	<0.4	0.4
	3 (1.5)	1	-	0.2
		2	-	0.4
TPM-1/1	1 (1.8)	1	0.4	3.8
		2	-	0.1
	2 (1.7)	1	-	0.7
		2	-	0.1
	3 (1.9)	1	0.4	0.1
		2	-	0.7

A generalization of the results of previous studies [11] and the new data suggest that the increased content of Pt and Pd is confined to the surface layer of not only arsenopyrite crystals, but also pyrite. It can also be assumed that the tendency for a Pt and Pd concentration in the surficial layer at <500 nm thick is common for both minerals (see Figure 11 and Table 11). Several exceptions can be noted in Table 10, however, which may be due to inhomogeneity in the impurity distribution because of the evolution of NAP during crystal growth [39].

Determination of the contents of a wider range of elements belonging to the group of PGEs—Pt, Pd, Rh and Ir—was carried out in the example of studying the surfaces of arsenopyrite and pyrite crystals taken from the UV-3/13 sample (Figures 12 and 13). The surface layer of these crystals was studied to a depth of 6 μm (depth per pass is 1 μm). Ru and Os contents were not determined due to their absence in the standards we used. According to LA-ICP-MS analysis, the surface layers of arsenopyrite and pyrite, along with Pt and Pd, concentrated PGEs such as Rh and Ir. The contents of elements (ppm) in the surface layers of arsenopyrite were Pt up to 5.2, Pd up to 8.4, Rh up to 8.9 and Ir up to 0.2, and in the surface layers of pyrite were Pt up to 13, Pd up to 896, Rh up to 6.7 and Ir up to 4.1. The maximum contents of Pt, Pd and Ir were established in pyrite, and Rh was established in arsenopyrite. Like Pt and Pd, all the highest contents of Rh and Ir were confined to the first removed layer (1 μm thick). There was a distinct tendency for the content of all these elements to decrease with depth (Tables 12 and 13).

High Pd content and Pd/Pt ratios on the pyrite surface (see Table 13) contradict the AAS-ADSSC and SEM-EDX data presented above, according to which the Pt content was more than an order of magnitude higher than Pd. The reason for this contradiction requires special research and may be related to the existence of an unidentified surface form of Pd (for example, Pd nanoparticles), which is inhomogeneously distributed over crystals. The SEM-EDX method was also local (locality is even higher than for LA-ICP-MS), and therefore cannot serve to confirm this hypothesis. In the AAS-ADSSC method, crystals with such an excess of Pd over the average content simply will not be included in the final sample.

Figure 12. Arsenopyrite Crystals 1 and 2 (sample UV-3/13) from ores of the Natalkinskoe deposit. Tracks 1–9, numbers of laser passes over the crystal surface. The results of LA-ICP-MS analysis are presented in Table 12.

Figure 13. Pyrite Crystals 1 and 2 (sample UV-3/13) from ores of the Natalkinskoe deposit. Tracks 1–8, numbers of laser passes over the crystal surface. The results of LA-ICP-MS analysis are presented in Table 13.

Table 12. Results of LA-ICP-MS analysis (in ppm) of the surface of arsenopyrite crystals (1 and 2) for Pt, Pd, Rh and Ir (sample UV-3/13) (the Natalkinskoe deposit).

n	Pt	Pd	Rh	Ir	Pt	Pd	Rh	Ir	Pt	Pd	Rh	Ir
						Crystal 1						
		Track 1				Track 2				Track 3		
1	0.7	1.6	7.6	0.2	<0.7	5.2	1.6	-	<0.7	7.9	2.3	0.1
2	-	<0.2	2.5	<0.1	0.9	2.2	1.0	-	-	3.4	0.5	-
3	-	0.3	0.2	-	-	7.6	0.3	-	-	3.5	1.6	<0.1
4	-	0.4	<0.2	-	-	3.2	0.5	<0.1	-	2.9	0.4	-
5	-	0.2	0.3	-	-	2.0	0.4	<0.1	-	1.4	0.3	-
6	-	<0.2	0.2	-	-	2.0	0.6	-	-	1.5	-	-
						Crystal 2						
		Track 4				Track 5				Track 6		
1	<0.7	0.8	1.2	0.1	5.2	2.9	0.6	-	0.7	5.8	7.5	-
2	-	0.2	0.5	0.1	-	1.2	0.2	-	-	4.3	0.7	-
3	-	0.2	0.2	-	-	<0.2	0.2	-	-	1.7	0.5	-
4	-	0.3	0.5	-	-	1.4	0.7	-	-	1.5	0.2	-
5	-	<0.2	<0.2	-	-	0.4	0.4	-	-	1.3	2.1	-
6	-	-	<0.2	-	-	0.9	<0.2	-	-	0.7	0.2	-
		Track 7				Track 8				Track 9		
1	1.3	8.2	1.9	-	<0.7	2.8	6.4	<0.1	1.5	8.4	8.9	-
2	-	0.6	1.7	-	-	1.5	0.5	-	-	3.7	3.6	-
3	<0.7	<0.2	0.8	-	<0.7	1.8	0.8	-	-	2.5	3.7	-
4	-	0.4	0.9	-	-	<0.2	0.3	-	-	3.7	1.1	-
5	-	<0.2	0.4	-	-	0.2	0.9	-	-	3.6	1.5	-
6	-	0.4	<0.2	-	-	<0.2	-	-	-	2.7	0.7	<0.1

Note: Here and in Table 13: *n*—numbers of laser passes along the track (depth per pass is 1.0 μm).

Table 13. Results of LA-ICP-MS analysis (in ppm) of the surface of pyrite crystals (1 and 2) for Pt, Pd, Rh and Ir (sample UV-3/13) (the Natalkinskoe deposit).

n	Pt	Pd	Rh	Ir	Pt	Pd	Rh	Ir	Pt	Pd	Rh	Ir	Pt	Pd	Rh	Ir
								Crystal 1								
	Track 1				**Track 2**				**Track 3**				**Track 4**			
1	2.5	118	0.9	0.5	3.6	135	1.6	0.4	1.8	134	0.3	1.1	2.2	70	5.1	0.4
2	1.0	54	-	0.3	1.1	108	<0.2	0.4	1.1	89	0.3	0.5	2.2	109	1.7	0.2
3	0.5	26	0.7	0.1	<0.7	54	0.3	0.6	<0.7	41	-	0.2	1.1	47	0.4	-
4	0.3	10	-	-	0.9	56	<0.2	<0.2	-	18	<0.2	<0.2	<0.7	63	0.7	-
5	-	7.7	1.8	-	0.7	28	<0.2	-	-	14	-	-	<0.7	31	0.5	0.2
6	-	2.6	-	-	0.9	39	<0.2	-	-	17	<0.2	-	-	34	0.2	-
								Crystal 2								
	Track 5				**Track 6**				**Track 7**				**Track 8**			
1	11	453	2.4	1.8	13	855	2.9	3.1	13	896	6.7	4.1	7.7	693	1.2	3.0
2	5.3	820	0.5	3.6	9.1	845	1.1	4.1	8.4	624	2.4	1.9	5.3	715	0.4	3.0
3	2.7	503	1.2	0.7	5.9	521	0.9	1.8	2.8	378	0.5	1.5	8.3	461	0.4	2.8
4	1.1	101	0.4	0.5	4.7	365	0.2	1.3	3.8	210	1.6	0.8	0.9	301	<0.2	0.7
5	0.8	50	0.2	0.4	3.1	206	1.0	0.6	2.5	130	0.2	0.5	2.6	135	0.6	0.5
6	0.5	35	<0.2	0.5	1.3	89	0.3	0.7	0.9	112	0.2	0.5	2.2	143	<0.2	1.1

The maximum PGE contents at the Natalkinskoe deposit are currently known to be associated with sulphide minerals. High levels of PGE in the sulphide gravity concentrate of this deposit ore were established by sample decomposition using fluoro oxidants [49] and amounted to (in ppm) Pt, 12.9–92.8; Pd, 0.62–2.97; Ru, 0.11–3.27; Rh, 0.95–1.54 and Ir, 0.1–0.3. The speculation about the possible role of sulphide minerals as Pt concentrators at the Natalkinskoe gold deposit was previously made by L.P. Plyusnina and co-authors [14]; however, microprobe analysis did not reveal high content of noble metals in conventional preparations of arsenopyrite and pyrite [3]. Arsenopyrite concentrates of not only Au, but also Pt and Pd, were later demonstrated at the Natalkinskoe deposit [11]. The new data confirmed and supplemented the earlier conclusion. The highest contents of PGE (in ppm) observed in arsenopyrite were Pt up to 128, Pd up to 20, Ru up to 86 and Rh up to 21, and they were less high in pyrite: Pt up to 29, Pd up to 15, Ru up to 58 and Rh up to 5.9. AAS-ADSSC found that the high content of PGE largely was due to the existence of non-autonomous phases, causing their surface concentration. The fact that the nature of high contents of PGE in arsenopyrite and pyrite is mainly superficial was confirmed by the data obtained by SEM-EDX and (to some extent) LA-ICP-MS. If the data from the first two methods are comparable (taking into account the above considerations), the latter gives different results, which can be explained by the following reasons. First, there is some dispersion of the material on the surface after the first and each subsequent laser pass, which increases the error of determination. Second, the analysis of surface areas of several crystals does not have the necessary statistical reliability for determining the average contents, as is the case in the analytical technology of AAS-ADSSC. It is quite possible to meet with an "empty" crystal or, conversely, a crystal with the surface strongly enriched with impurity. This is all the more likely because the heterogeneity of the distribution of PGE is well known. For example, in pyrite from the intercumulus sulphide phases of the main sulphide zone of Great Dyke (Zimbabwe), the ion-microprobe analysis revealed a strong heterogeneity in the Pt distribution—from 0.4 to 244 ppm in different grains of the same sample [50]. Thirdly, significant errors can occur due to the lack of external calibration on such surface structures, for example, deposited on the crystal films of sulphide material with a certain content of admixture of the studied element. In the current state of the study of the surface composition of sulphides by LA-ICP-MS, although the history of such studies seems to date back 12 years [51], it is possible to speak only about the general tendency consisting of the enrichment of PGE of a surface of pyrite and arsenopyrite crystals; however, this trend is fully confirmed by other methods.

The reason for this phenomenon involves the features of the crystal growth mechanism and the dualism of the element distribution coefficient in the mineral-hydrothermal solution system, which is an order of magnitude higher for NAP compared to the crystal volume [38,39]. Non-autonomous phases are extremely important, but they are still rarely considered components of synthetic materials and minerals, the geochemical role of which lies in their ability to accumulate impurity elements in ultra-high contents [52]. Such phases occur on the surface of a growing crystal due to the chemical modification and structural reconstruction of its surface layer, which is in local equilibrium with a supersaturated solution. NAP contain defects, chemical bonds and valence states of elements unusual for crystal volume, for example AsV, AsIII, AsII and FeIII on pyrite [41], which provides additional opportunities for the incorporation of minor and trace elements, especially incompatible, into their structures. Since the stability field of arsenopyrite is adjacent to the pyrite field, according to the principle of continuity of phase formation on mineral surfaces and the correspondence of chemical forms of elements on them [53], the surface NAP on arsenopyrite can have a defective pyrite structure. Since platarsite Pt(As,S)$_2$ has the structure of pyrite $Pa3$, we should therefore expect an increase in the solubility of Pt in the surficial phase compared to the volume of the crystal.

Gold and PGE nanoparticles are increasingly, especially recently, found in ores of different genesis deposits [7,8,54–56]. Although no PGE mineral forms were identified at the Natalkinskoe deposit, the need to continue research in this direction is obvious. This may be demonstrated by the Degdekan deposit, another gold project in the northeast of Russia, which is in the territory of Pre Kolyma. Its ores, classed as black shale formation, yielded the first findings of a number of platinum group minerals of submicron dimensions: native osmium, routheniridosmine, osmirid, ruthenosmirid, laurite, iridarsenite, ruthenium arsenide and arsenide of osmium and iridium [8]. The studies by [5,6] established the existence of different mineral forms of platinum metals, including native platinum, phases of Pt–Cu–Fe system, sperrylite, cooperite, and minerals of the Pd–Bi system in gold ores of the Sukhoi Log deposit (Lena ore district, Irkutsk region, Russia). Their size mainly ranges within 1–10 µm. Later, native Pt nanoparticles, with a dominant size of 0.5–20 nm, were found in the concentrates of insoluble carbonaceous matter of shales containing ore mineralization [7]. It follows that there is a high probability of nanoscale mineral forms of PGE existing at the Natalkinskoe deposit, according to the model of surface NAP evolution in sulphide minerals, suggesting their partial transformation and aggregation with the formation of nano- and microinclusions of separate (autonomous) phases of trace elements [39]. The possibility of finding nano-sized particles of PGE in sulphides is confirmed by experimental data. For example, nanometer-size PtS$_2$ inclusions in synthetic Pt-containing pyrite were detected with high-resolution transmission electron microscopy [57].

5. Conclusions

Arsenopyrites and pyrites of the Natalkinskoe gold deposit selected from vein, streaky-vein and veinlet-disseminated ores belonging to the most productive gold-rich hydrothermal mineral formations concentrated not only Au, but also platinum group elements. The highest contents (in ppm) were detected in monofractions of arsenopyrite: Pt up to 128, Pd up to 20, Ru up to 86 and Rh up to 21, and less high in monofractions of pyrite: Pt up to 29, Pd up to 15, Ru up to 58 and Rh up to 5.9.

The original AAS-ADSSC analytical technology in arsenopyrites and pyrites from the deposit established that there were two forms of uniformly distributed Pt, Pd, Ru and Rh—structural and surface-bound chemically bound in the structure of these minerals and in non-autonomous phases located on the surface of arsenopyrite and pyrite crystals. The surficially bound form dominates and probably exists in a very thin surface layer of the crystal (~100–500 nm). This phenomenon arises due to the features of the crystal growth mechanism and the dualism of the element distribution coefficient in the mineral-hydrothermal solution system, which is an order of magnitude higher for NAP compared to the crystal volume. It is assumed that for gold-bearing arsenopyrites and pyrites there is a common mechanism of impurity absorption associated with the active role of the crystal surface and surface defects. Structural forms of Pt, Pd, Ru and Rh, despite a lower content of the associated element compared to the surface, are also extremely important. They are a reliable indicator of element activity in ore-forming fluids forming gold deposits and a criterion of their potential ore content.

Restoration of the ore-forming fluid composition using the content of the structural admixture of Au and Pt in pyrite and experimentally established coefficients of their distribution in the pyrite–hydrothermal solution system showed that ore-forming fluid that deposited pyrite locally contained comparable concentrations of gold and platinum. Unlike gold, which mostly occurred in visible native form, the majority of PGEs are in so-called hidden or invisible forms. Previously, they were not recognized due to the inability to establish the nature of the carrier without the use of methods to determine the surficial and structural forms of the element.

The data resulting from studying the surface of sulphide minerals from the Natalkinskoe deposit using SEM-EDX and LA-ICP-MS methods confirmed the surface presence on arsenopyrite and pyrite crystals of Pt, Pd, Ru and Rh and such PGE as Os and Ir, as well as other ore trace elements including Ag, Hg, Pb, Cu and U. The maximum contents of invisible forms of PGE, as in the case of gold, are confined to the surface of crystals. An early assumption that surface-bound forms, as the main uniformly distributed form of elements in ore minerals, can be characteristic not only for Au, Pt and Pd was confirmed [11]. They are typical of other PGE (Ru, Rh, Os and Ir) and, in general, for ore trace elements.

The contradiction of LA-ICP-MS data with other methods with respect to the distribution of Pt and Pd in pyrite was revealed. This is assumed to be due to the presence of an unknown surface form of Pd, inhomogeneously distributed over crystals (Pd nanoparticles or any of its compounds). Although no proper mineral forms of PGE (Pt, Pd, Ru, Rh, Ir and Os) were established at the Natalkinskoe gold deposit, the probability of nano- or micro-sized mineral forms of PGE at this deposit is very high. This conclusion is based on the model of the evolution of surface NAP, assuming their partial transformation and aggregation with the formation of nano- and microinclusions of separate (autonomous) phases of trace elements.

The results obtained through the study of PGEs in sulphide minerals are of interest both theoretically and practically. Data on the nature of distribution, concentration level and forms of Pt, Pd, Ru, Rh, Ir and Os in arsenopyrites and pyrites significantly supplement existing ideas concerning peculiarities in the composition of gold mineralization, and they serve as a criterion for assessing the potential ore content of fluids forming gold deposits. The association of maximum contents of invisible forms of PGEs with the surface of sulphide mineral crystals can be used to develop methods of PGE recovery, primarily Pt, without destruction of the mineral structure, through transformation and dissolution of the surface layer only. This is an important point that must be taken into account in the concentration of ores, which includes sulphides, and in our case arsenopyrite and pyrite. The presence of PGE in ores and the possibility of their extraction significantly enhances the value of the gold ore mined, thereby increasing the prospects of the deposit under study.

Author Contributions: Conceptualization, R.G.K.; methodology, R.G.K. and V.L.T.; validation, R.G.K.; formal analysis, V.L.T., A.S.M., N.V.B. and N.V.S.; investigation, R.G.K. and V.L.T.; resources, R.G.K. and A.S.M.; writing—original draft preparation, R.G.K. and V.L.T.; writing—review and editing, R.G.K., V.L.T., A.S.M. and N.V.B.; visualization, R.G.K., A.S.M. and N.V.S. All authors have read and agreed to the published version of the manuscript.

Funding: The work was carried out within the framework of the execution of the state task for the projects IX.125.3.4. (0350-2019-0003) and IX.130.3.1. (0350-2019-0010) with the financial support of the RFBR (projects Nos. 20-05-00142 and 18-05-00077) and using scientific equipment of the shared research centre "Isotopic-geochemical investigations" IGC SB RAS (Irkutsk, Russia) and the shared research centre "Ultramicroanalysis" LIN SB RAS (Irkutsk, Russia).

Acknowledgments: The authors thank T.M. Pastushkova, I.Yu. Voronova, V.N. Vlasova, G.A. Shcherbakova, S.V. Lipko and K.Yu. Arsent'ev for their assistance in analytical studies. We would like to express our special gratitude to all geological services of the JSC "RiM" (Magadan, Russia), its chief geologist R.N. Ovsov and leading geologist E.M. Nikitenko, for comprehensive assistance in conducting fieldwork in the period from 2008 to 2016.

Conflicts of Interest: The authors declare no conflict of interest.

References

1. Goncharov, V.I.; Voroshin, S.V.; Sidorov, V.A. Platinum mineralization of gold ore deposits in black shale sediments of the North–East of Russia: Challenges and prospects. In *Platinum of Russia. Possibilities of Development of the Raw Material Base of Platinum Metals (Third meeting of the Scientific-methodical Council of the program "Platinum of Russia")*; Orlov, V.P., Ed.; Geoinformmark: Moscow, Russia, 1995; Volume 2, Part 2; pp. 156–161. (In Russian)
2. Goncharov, V.I.; Sidorov, V.A.; Pristavko, V.A. Platinum metal mineralization of the Natakinskoye gold ore deposit: Research findings. *Kolyma* **2000**, *2*, 49–53. (In Russian)
3. Goncharov, V.I.; Voroshin, S.V.; Sidorov, V.A. *Natalkinskoe Gold Lode Deposit*; NEISRI FEB RAS: Magadan, Russia, 2002; p. 250. (In Russian)
4. Voroshin, S.V.; Sidorov, V.A.; Tyukova, E.E. Geology, geochemistry and prospects of platinum mineralization at the Natakinskoye gold ore deposit (North–East of Russia). In *Platinum of Russia. Possibilities of Development of the Raw Material Base of Platinum Metals (Third meeting of the Scientific-Methodical Council of the Program "Platinum of Russia")*; Orlov, V.P., Ed.; Geoinformmark: Moscow, Russia, 1995; Volume 2, Part 2; pp. 161–176. (In Russian)
5. Distler, V.V.; Mitrofanov, G.L.; Nemerov, V.K.; Kovalenker, V.A.; Mokhov, A.V.; Semeikina, L.K.; Yudovskaya, M.A. Forms of occurrence of platinum group metals and their genesis in the Sukhoi Log gold ore deposit (Russia). *Geologiya Rudnykh Mestorozhdenij* **1996**, *38*, 467–484. (In Russian)
6. Distler, V.V.; Yudovskaya, M.A.; Razvozzhaeva, E.A.; Mokhov, A.V.; Trubkin, N.V.; Mitrofanov, G.L.; Nemerov, V.K. New data on PGE mineralization in gold ores of the Sukhoi Log deposit, Lensk gold-bearing district, Russia. *Dokl. Earth Sci.* **2003**, *393*, 1265–1267.
7. Nemerov, V.K.; Razvozzhaeva, E.A.; Spiridonov, A.M.; Sukhov, B.G.; Trofimov, B.A. Nanodispersed state of metals and their migration in carbonaceous natural media. *Dokl. Earth Sci.* **2009**, *425*, 334–337. [CrossRef]
8. Goryachev, N.A.; Sotskaya, O.T.; Goryacheva, E.M.; Mikhalitsyna, T.I.; Man'shin, A.P. The first discovery of platinum group minerals in black shale gold ores of the Degdekan deposit, Northeast Russia. *Dokl. Earth Sci.* **2011**, *439*, 902–905. [CrossRef]
9. Khanchuk, A.I.; Plyusnina, L.P.; Nikitenko, E.M.; Kuzmina, T.V.; Barinov, N.N. The noble metal distribution in the black shales of the Degdekan gold deposit in northeast Russia. *Russ. J. Pac. Geol.* **2011**, *5*, 89–96. [CrossRef]
10. Liu, J.; Chen, Y.; Liao, Z.; Zhan, Y.; Guan, Y. Progress in platinum group element (PGE) in black shale series. *Appl. Mech. Mater.* **2013**, *353–356*, 1183–1186. [CrossRef]
11. Kravtsova, R.G.; Tauson, V.L.; Nikitenko, E.M. Modes of Au, Pt, and Pd occurrence in arsenopyrite from the Natalkinskoe deposit, NE Russia. *Geochem. Int.* **2015**, *53*, 964–972. [CrossRef]
12. Pasava, J.; Ackerman, L.; Halodova, P.; Pour, O.; Durisova, J.; Zaccarini, F.; Aiglsperger, T.; Vymazalova, A. Concentrations of platinum-group elements (PGE), Re and Au in arsenian pyrite and millerite from Mo–Ni–PGE–Au black shales (Zunyi region, Guizhou Province, China): Results from LA-ICPMS study. *Eur. J. Mineral.* **2017**, *29*, 623–633. [CrossRef]
13. Vasil'eva, I.E.; Shabanova, E.V.; Goryacheva, E.M.; Sotskaya, O.T.; Labusov, V.A.; Neklyudov, O.A.; Dzyuba, A.A. Determination of precious metals in geological samples from four gold ore deposits of the north–east of Russia. *J. Anal. Chem.* **2018**, *73*, 539–550. [CrossRef]
14. Plyusnina, L.P.; Khanchuk, A.I.; Goncharov, V.I.; Sidorov, V.A.; Goryachev, N.A.; Kuz'mina, T.V.; Likhoidov, G.G. Gold, platinum, and palladium in ores of the Natalkinskoe deposit, upper Kolyma region. *Dokl. Earth Sci.* **2003**, *391A*, 836–840.
15. Volkov, A.V.; Genkin, A.D.; Goncharov, V.I. The forms of the presence of gold in the ores of the Natalka and May deposits (Northeast Russia). *Tikhookeanskaya Geol.* **2006**, *25*, 18–29. (In Russian)
16. Goryachev, N.A.; Vikent'eva, O.V.; Bortnikov, N.S.; Prokof'ev, V.Y.; Alpatov, V.A.; Golub, V.V. The world-class Natalka gold deposit, northeast Russia: REE patterns, fluid inclusions, stable oxygen isotopes, and formation conditions of ore. *Geol. Ore Depos.* **2008**, *50*, 362–390. [CrossRef]
17. Eremin, R.A.; Osipov, A.P. On genesis of the Natalkinskoe gold ore deposit. *Kolyma* **1974**, *6*, 41–43. (In Russian)
18. Kalinin, A.I.; Kanishchev, V.K.; Orlov, A.G.; Gashtol'd, V.V. Structure of the Natalkinskoe ore field. *Kolyma* **1992**, *10–11*, 10–14. (In Russian)

19. Voroshin, S.V.; Shakhtyrov, V.G.; Tyukova, E.E.; Gashtold, V.V. Geology and genesis of the Natakinskoe gold ore deposit. *Kolyma* **2000**, *2*, 22–23. (In Russian)
20. Goryachev, N.A.; Sidorov, V.A.; Litvinenko, I.S.; Mikhalitsyna, T.I. Mineral composition and petrogeochemical peculiarities of ore zones of the Natalkinskoe deposit deep horizons. *Kolyma* **2000**, *2*, 38–49. (In Russian)
21. Mezhov, S.V. Geological structure of the Natalkinskoe gold ore deposit. *Kolymskie Vesti* **2000**, *9*, 8–17. (In Russian)
22. Pristavko, V.A.; Sidorov, V.A.; Mikhalitsyna, T.I.; Burova, A.S.; Krasnaya, E.N. Geological-geochemical model the Natalkinskoe gold ore deposit. *Kolymskie Vesti* **2000**, *9*, 18–24. (In Russian)
23. Grigorov, S.A. Genesis and formation dynamics of the Natalkinskoe gold ore deposit according to the system analysis of the geochemical field. *Rudy Met.* **2006**, *3*, 44–48. (In Russian)
24. Struzhkov, S.F.; Natalenko, M.N.; Chekvaidze, V.B.; Isaakovich, I.Z.; Golubev, S.Y.; Danil'chenko, V.A.; Obuskhov, A.V.; Zaitsev, M.A.; Kryazhev, S.G. Multi-factor model of the Natalkinskoe gold ore deposit. *Rudy Met.* **2006**, *3*, 34–44. (In Russian)
25. Volkov, A.V.; Murashov, K.Y.; Sidorov, A.A. Geochemical peculiarities of ores from the largest Natalka gold deposit in Northeastern Russia. *Dokl. Earth Sci.* **2016**, *466*, 161–164. [CrossRef]
26. Grigorov, S.A.; Vorozhbenko, V.D.; Kushnarev, P.I.; Markevich, V.Y.; Tokarev, V.N.; Chichev, V.I.; Yagubov, N.P.; Mikhailov, B.K. Geology and key signatures of the Natalkinskoe gold deposit. *Otechestvennaya Geol.* **2007**, *3*, 43–50. (In Russian)
27. Sharafutdinov, V.M.; Khasanov, I.M.; Mikhalitsyna, T.I. Petrophysical zoning of the Natalka ore field. *Russ. J. Pac. Geol.* **2008**, *2*, 441–453. [CrossRef]
28. Sotskaya, O.; Goryachev, N.; Goryacheva, E.; Nikitenko, E. Micromineralogy of "Black Shale" Disseminated-Sulphide Gold Ore Deposits of the Ayan-Yuryakh Anticlinorium (North–East of Russia). *J. Earth Sci. Eng.* **2012**, *2*, 744–753.
29. Bunakova, N.Y.; Zakharenko, V.M. *Gold, Platinum, Palladium, Rhodium, Iridium and Ruthenium Detection in Rocks of Various Composition by Atomic Absorption with Electrothermal Atomization Following Concentration*; NSAM No 430-Kh Methods; VIMS: Moscow, Russia, 2005; p. 24. (In Russian)
30. Lebedeva, M.I.; Rogozhin, A.A. *Gold Determination in Rocks, Ores and Their Processed Products by Extraction–Atomic–Absorption Method with Organic Sulfides*; NSAM No 237-S Methods; VIMS: Moscow, Russia, 2016; p. 18. (In Russian)
31. Men'shikov, V.I.; Vlasova, V.N.; Lozhkin, V.I.; Sokol'nikova, I.V. Determination of platinum-group elements in rocks by ICP-MS with external calibration after cation exchange separation of matrix elements by KU-2-8 resin. *Anal. Kontrol* **2016**, *20*, 190–201. (In Russian)
32. Tauson, V.L.; Bessarabova, O.I.; Kravtsova, R.G.; Pastushkova, T.M.; Smagunov, N.V. Separation of forms of gold occurrence in pyrites by studying statistic samples of analytical data. *Russ. Geol. Geophys.* **2002**, *43*, 57–67.
33. Tauson, V.L.; Lustenberg, E.K. Quantitative determination of modes of gold occurrence in minerals by the statistical analysis of analytical data samplings. *Geochem. Int.* **2008**, *46*, 423–428. [CrossRef]
34. Ripley, E.M.; Chryssoulis, S.L. Ion microprobe analysis of platinum-group elements in sulfide and arsenide minerals from the Babbitt Cu–Ni deposit, Duluth Complex, Minnesota. *Econ. Geol.* **1994**, *89*, 201–210. [CrossRef]
35. Trigub, A.L.; Tagirov, B.R.; Kvashnina, K.O.; Chareev, D.A.; Nickolsky, M.S.; Shiryaev, A.A.; Baranova, N.N.; Kovalchuk, E.V.; Mokhov, A.V. X-ray spectroscopy study of the chemical state of "invisible" Au in synthetic minerals in the Fe–As–S system. *Am. Mineral.* **2017**, *102*, 1057–1065. [CrossRef]
36. Parviainen, A.; Gervilla, F.; Melgarejo, J.-C.; Johanson, B. Low-temperature, platinum-group elements-bearing Ni arsenide assemblages from the Atrevida mine (Catalonian Coastal Ranges, NE Spain). *Neues Jahrb. Mineral. Abh.* **2008**, *185*, 33–49. [CrossRef]
37. Tauson, V.L.; Kravtsova, R.G.; Smagunov, N.V.; Spiridonov, A.M.; Grebenshchikova, V.I.; Budyak, A.E. Structurally and superficially bound gold in pyrite from deposits of different genetic types. *Russ. Geol. Geophys.* **2014**, *55*, 273–289. [CrossRef]
38. Tauson, V.L.; Lipko, S.V.; Smagunov, N.V.; Kravtsova, R.G. Trace element partitioning dualism under mineral-fluid interaction: Origin and geochemical significance. *Minerals* **2018**, *8*, 282. [CrossRef]

39. Tauson, V.L.; Lipko, S.V.; Smagunov, N.V.; Kravtsova, R.G.; Arsent'ev, K.Y. Distribution and segregation of trace elements during the growth of ore mineral crystals in hydrothermal systems: Geochemical and mineralogical implications. *Russ. Geol. Geophys.* **2018**, *59*, 1718–1732. [CrossRef]

40. Tauson, V.L.; Babkin, D.N.; Akimov, V.V.; Lipko, S.V.; Smagunov, N.V.; Parkhomenko, I.Y. Trace elements as indicators of the physicochemical conditions of mineral formation in hydrothermal sulfide systems. *Russ. Geol. Geophys.* **2013**, *54*, 526–543. [CrossRef]

41. Tauson, V.L.; Lipko, S.V.; Arsent'ev, K.Y.; Mikhlin, Y.L.; Babkin, D.N.; Smagunov, N.V.; Pastushkova, T.M.; Voronova, I.Y.; Belozerova, O.Y. Dualistic distribution coefficients of trace elements in the system mineral–hydrothermal solution. IV. Platinum and silver in pyrite. *Geochem. Int.* **2017**, *55*, 753–774. [CrossRef]

42. Tauson, V.; Lipko, S.; Kravtsova, R.; Smagunov, N.; Belozerova, O.; Voronova, I. Distribution of "invisible" noble metals between pyrite and arsenopyrite exemplified by minerals coexisting in orogenic Au deposits of North–Eastern Russia. *Minerals* **2019**, *9*, 660. [CrossRef]

43. Tauson, V.L.; Kravtsova, R.G. Evaluation of the gold admixture in structure of pyrite of epithermal gold-silver ore deposits (the North–East Russia). *Zap. Ross. Mineral. Obshch.* **2002**, *131*, 1–11. (In Russian)

44. Tauson, V.L.; Kravtsova, R.G. Chemical typomorphism of mineral surfaces: Surface composition specifics (by the example of gold-bearing pyrite from epithermal deposit). *Russ. Geol. Geophys.* **2004**, *45*, 204–209.

45. Tauson, V.L.; Kravtsova, R.G.; Grebenshchikova, V.I. Chemical typomorphism of the surface of pyrite crystals of gold ore deposits. *Dokl. Earth Sci.* **2004**, *399*, 1291–1295.

46. Tauson, V.L.; Babkin, D.N.; Pastushkova, T.M.; Krasnoshchekova, T.S.; Lustenberg, E.E.; Belozerova, O.Y. Dualistic distribution coefficients of elements in the system mineral-hydrothermal solution. I. Gold accumulation in pyrite. *Geochem. Int.* **2011**, *49*, 568–577. [CrossRef]

47. Zhitova, L.M.; Kinnaird, J.A.; Gora, M.P.; Shevko, E.P. Magmatogene fluids of metal-bearing reefs in the Bushveld Complex, South Africa: Based on research data on fluid inclusions in quartz. *Geol. Ore Depos.* **2016**, *58*, 58–81. [CrossRef]

48. Tauson, V.L.; Kravtsova, R.G.; Grebenshchikova, V.I.; Lustenberg, E.E.; Lipko, S.V. Surface typochemistry of hydrothermal pyrite: Electron spectroscopic and scanning probe microscopic data. II. Natural pyrite. *Geochem. Int.* **2009**, *47*, 231–243. [CrossRef]

49. Mitkin, V.N.; Galizky, A.A.; Korda, T.M. Some observations on the determination of gold and the platinum-group elements in black shales. *Geostandard Newslett.* **2000**, *24*, 227–240. [CrossRef]

50. Oberthür, T.; Cabri, L.J.; Weiser, T.W.; McMahon, G.; Müller, P. Pt, Pd and other trace elements in sulfides of the main sulfide zone, Great Dyke, Zimbabwe: A reconnaissance study. *Can. Mineral.* **1997**, *35*, 597–609.

51. Öhlander, B.; Müller, B.; Axelsson, M.; Alakangas, L. An attempt to use LA-ICP-SMS to quantify enrichment of trace elements on pyrite surfaces in oxidizing mine tailings. *J. Geochem. Explor.* **2007**, *92*, 1–12. [CrossRef]

52. Kovalenko, A.N.; Tugova, E.A. Thermodynamics and kinetics of non-autonomous phase formation in nanostructured materials with variable functional properties. *Nanosyst. Phys. Chem. Math.* **2018**, *9*, 641–662. [CrossRef]

53. Tauson, V.L. The principle of continuity of phase formation at mineral surfaces. *Dokl. Earth Sci.* **2009**, *425*, 471–475. [CrossRef]

54. Palenik, C.S.; Utsunomiya, S.; Reich, M.; Kesler, S.E.; Wang, L.; Ewing, R.C. "Invisible" gold revealed: Direct imaging of gold nanoparticules in a Carlin-type deposit. *Am. Mineral.* **2004**, *89*, 1359–1366. [CrossRef]

55. Gonzalez-Jimenez, J.M.; Reich, M. An overview of the platinum-group element nanoparticles in mantle-hosted chromite deposits. *Ore Geol. Rev.* **2017**, *81*, 1236–1248. [CrossRef]

56. Gonzalez-Jimenez, J.M.; Deditius, A.; Gervilla, F.; Reich, M.; Suvorova, A.; Roberts, M.P.; Roque, J.; Proenza, J.A. Nanoscale partitioning of Ru, Ir, and Pt in base-metal sulfides from the Caridad chromite deposit, Cuba. *Am. Mineral.* **2018**, *103*, 1208–1220. [CrossRef]

57. Filimonova, O.N.; Nickolsky, M.S.; Trigub, A.L.; Chareev, D.A.; Kvashnina, K.O.; Kovalchuk, E.V.; Vikentyev, I.V.; Tagirov, B.R. The state of platinum in pyrite studied by X-ray absorption spectroscopy of synthetic crystals. *Econ. Geol.* **2019**, *114*, 1649–1663. [CrossRef]

 *minerals*

Article

Zoned Laurite from the Merensky Reef, Bushveld Complex, South Africa: "Hydrothermal" in Origin?

Federica Zaccarini * and Giorgio Garuti

Department of Applied Geological Sciences and Geophysics, University of Leoben, A-8700 Leoben, Austria; giorgio.garuti1945@gmail.com
* Correspondence: federica.zaccarini@unileoben.ac.at; Tel.: +43-(0)3842-402-6218

Received: 21 March 2020; Accepted: 19 April 2020; Published: 21 April 2020

Abstract: Laurite, ideally $(Ru,Os)S_2$, is a common accessory mineral in podiform and stratiform chromitites and, to a lesser extent, it also occurs in placer deposits and is associated with Ni-Cu magmatic sulfides. In this paper, we report on the occurrence of zoned laurite found in the Merensky Reef of the Bushveld layered intrusion, South Africa. The zoned laurite forms relatively large crystals of up to more than 100 μm, and occurs in contact between serpentine and sulfides, such as pyrrhotite, chalcopyrite, and pentlandite, that contain small phases containing Pb and Cl. Some zoned crystals of laurite show a slight enrichment in Os in the rim, as typical of laurite that crystallized at magmatic stage, under decreasing temperature and increasing sulfur fugacity, in a thermal range of about 1300–1000 °C. However, most of the laurite from the Merensky Reef are characterized by an unusual zoning that involves local enrichment of As, Pt, Ir, and Fe. Comparison in terms of Ru-Os-Ir of the Merensky Reef zoned laurite with those found in the layered chromitites of the Bushveld and podiform chromitites reveals that they are enriched in Ir. The Merensky Reef zoned laurite also contain high amount of As (up to 9.72 wt%), Pt (up to 9.72 wt%) and Fe (up to 14.19 wt%). On the basis of its textural position, composition, and zoning, we can suggest that the zoned laurite of the Merensky Reef is "hydrothermal" in origin, having crystallized in the presence of a Cl- and As-rich hydrous solution, at temperatures much lower than those typical of the precipitation of magmatic laurite. Although, it remains to be seen whether the "hydrothermal" laurite precipitated directly from the hydrothermal fluid, or it represents the alteration product of a pre-existing laurite reacting with the hydrothermal solution.

Keywords: laurite; sulfides; fluids; platinum group elements (PGE); platinum group minerals (PGM); Merensky Reef; Bushveld Complex; South Africa

1. Introduction

Minerals of ruthenium are very rare and only five of them, namely anduoite $(Ru,Os)As_2$, laurite $(Ru,Os)S_2$, ruarsite RuAsS, ruthenarsenite $(Ru,Ni)As$, and ruthenium (Ru,Ir,Os), have been approved by the International Mineralogical Association (IMA). They occur as accessory minerals associated with mafic–ultramafic rocks, especially with chromitite, and as nuggets in placer deposits. Among the minerals of ruthenium, laurite is the most common. It was discovered in 1866 in a placer from Laut, Banjar, South Kalimantan Province, Borneo, Indonesia [1]. Laurite is a common constituent of the suite of platinum group minerals (PGM) inclusions (usually less than 20 μm) in podiform and stratiform chromitites [2–4]. Less frequently, laurite has been reported from placers and Ni-Cu magmatic sulfide deposits [5–7]. Laurite forms a complete solid solution with erlichmanite (OsS_2) [5], and their typical mode of occurrence, i.e., included in chromite grains, indicate that they crystallized at high temperatures, in a thermal range of about 1300–1000 °C prior to, or coeval with, the precipitation of the host chromitite [2–4]. The reciprocal stability of laurite and erlichmanite is strongly controlled by sulfur fugacity and temperature. In particular, laurite precipitates at a higher temperature and lower sulfur fugacity, compared to erlichmanite. This order of crystallization can be observed in the zoning of the small crystals of laurite and erlichmanite enclosed in fresh chromite grains that, typically, show an Os-poor core, grading into a high-Os rim [2–4]. This magmatic zoning can be obliterated by low temperature processes such as serpentinization and weathering, as documented in laurite associated with podiform chromitites [8–10]. During alteration processes at low temperature, laurite and erlichmanite lose their original S and release part of Os and Ir to form secondary Ru-Os-Ir alloys, in which the lost S may be replaced by Fe-oxide [8–10]. The occurrence of laurite in the Bushveld Complex of South Africa has been documented by several authors [2,7,11–17]. Most commonly, the mineral occurs as small polygonal grains enclosed in chromite grains of the Critical Zone chromitite layers and has only occasionally been found as part of the sulfide ore of the Merensky Reef. In this contribution, we have investigated in detail the mineral chemistry of the laurite associated with the sulfide-rich zone of the Merensky Reef. The grains are characterized by an unusual zoning and composition compared with laurite inclusions in the Bushveld chromitites, suggesting that the mineral was generated under different thermodynamic conditions in the two cases.

2. Sample Provenance and Petrographic Description

The Bushveld layered intrusion is located in the central part of the Transvaal province, north of Pretoria, South Africa (Figure 1A), and it is divided into Eastern, Western, and Northern limbs (Figure 1B). The Bushveld intrusion is well known among economic geologists because it contains the world's largest deposits of platinum group elements (PGE), namely: the UG-2 chromitite and the Merensky Reef [18]. The noun Merensky Reef refers to a sulfide-bearing pegmatoidal feldspathic pyroxenite enriched in PGE, marked at the bottom and top, by two centimetric layers of chromitite. The Reef can be traced for a total strike of about 280 km, marking the limit between the Critical and Main Zone [18,19]. The investigated samples were collected by one of the authors (G.G.) in the Rustenburg underground mine, during the third International Platinum Symposium held in Pretoria from 6 to 10 July 1981 [20,21]. The Rustenburg mine is located in the Western limb of the Bushveld Complex, about 100 km west of Pretoria (Figure 1B,C). Here, four different zones of Bushveld (undifferentiated in Figure 1C) are intruded by the Pilanesberg Alkaline Complex [22]. Four square polished blocks, about 2.5 × 2.5 cm (Figure 2), were prepared for petrographic and mineralogical investigation. The blocks consist of a thin layer of chromitite, about 0.2 cm thick, in contact with pegmatoidal feldspathic pyroxenite and large blebs of sulfide. In agreement with observations made by several authors [21,23,24], the sulfide-rich zone contains accessory actinolite, micas, talc, chlorite, and a serpentine subgroup mineral.

3. Methods

The polished blocks were previously studied by reflected-light microscope. Quantitative chemical analyses of laurite were performed with a JEOL JXA-8200 electron microprobe (JEOL, Tokyo, Japan), installed in the E. F. Stumpfl laboratory, Leoben University, Leoben, Austria, operated in WDS (wavelength dispersive spectrometry) mode. Major and minor elements were determined at 20 kV accelerating voltage and 10 nA beam current, with 20 s as counting time for the peak and 10 s for the backgrounds. The beam diameter was about 1 μm in size. The Kα lines were used for S, As, Fe and Ni; Lα for Ir, Ru, Rh, Pd, and Pt, and Mα for Os. The reference materials were pure metals for the six PGE (Ru, Rh, Pd, Os, Ir and Pt), synthetic NiS, natural pyrite and niccolite for Ni, Fe, S and As. The following diffracting crystals were selected: PETJ for S; PETH for Ru, Os, Pd and Rh; LIFH for Fe, Ni, Ir and Pt; and TAP for As. Automatic correction was performed for the Ru-Rh and Rh-Pd interferences. The detection limits were calculated by the software and are: Os (0.07 wt%), Ir (0.06 wt%), Ru, Pd, and Pt (0.04 wt%), Rh (0.03 wt%), Fe, Ni, As and S (0.02 wt%). The grains smaller than 10 μm were analyzed by EDS. The same instrument was used to obtain back-scattered electron images (BSE) and X-ray elemental distribution maps.

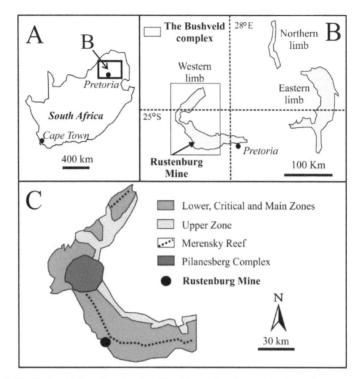

Figure 1. The Bushveld Complex, South Africa (**A**) and locations of the Rustenburg mine and the Merensky Reef (**B,C**) in the Western limb (modified after [15,18]).

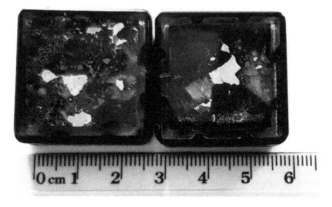

Figure 2. Example of the studied polished blocks from the Merensky Reef, showing the sulfide blebs (creamy–yellow) and the cumulitic chromitite (small polygonal dark grey grains) in the pegmatoidal feldspathic pyroxenite.

4. Laurite: Morphology, Texture, and Composition

The investigated samples contain laurite in two different textural positions, either included in fresh chromite of the thin chromitite layer (Figure 3A), or at the contact between sulfide patches (pyrrhotite, chalcopyrite, pentlandite) and serpentine (Figure 3B–D).

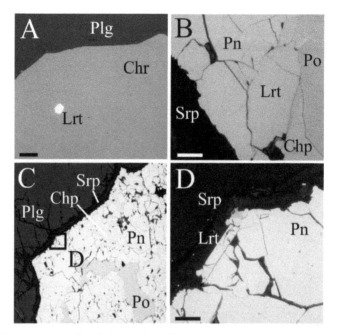

Figure 3. Digital image in reflected plane polarized light showing laurite from the Merensky Reef. (**A**) Laurite enclosed in fresh chromite. (**B,C**) Zoned laurite and (**D**) enlargement of (**C**). Abbreviations: Lrt = laurite, Plg = plagioclase, Chr = chromite, Srp = serpentine, Pn = pentlandite, Po = pyrrhotite, Chp = chalcopyrite. Scale bar = 20 μm.

Quantitative analyses of laurite enclosed in chromite and associated with sulfides are listed in Tables 1 and 2, respectively. Laurite included in chromite forms tiny crystals, usually not exceeding 10 μm in size, characterized by euhedral to subeuhedral morphology and homogenous composition. Laurite associated with sulfides and serpentine is bigger, up to more than 100 μm, and may occur as single crystals or clusters of grains (Figure 3B–D), characterized by subeuhedral to anhedral shape. The BSE images of large laurite display remarkable zoning emphasized by marked contrast in the electronic reflectivity (Figure 4A–D). Laurite in the sulfide assemblage is accompanied by a complex association of precious minerals comprising: cooperite (PtS), moncheite (PtTe$_2$), platarsite (PtAsS), rustenburgite (Pt$_3$Sn), Pt-Fe alloy, undetermined Pt-Te-Bi and Pd-Te-Bi compounds, Au-Ag alloy, and the recently discovered PGM bowlesite PtSnS [21].

Table 1. Selected wavelength dispersive spectrometry (WDS) electron microprobe analyses of Merensky Reef laurite enclosed in chromite.

Sample	As	S	Ru	Os	Ir	Rh	Pt	Pd	Ni	Fe	Total
					wt%						
mr18a	1.69	36.81	50.36	5.42	2.85	1.24	0.00	1.79	0.03	1.00	101.18
mr8a	1.85	36.93	49.99	4.97	3.03	1.06	0.00	1.68	0.08	0.90	100.48
Sample	As	S	Ru	Os	Ir	Rh	Pt	Pd	Ni	Fe	
					at%						
mr18a	1.28	65.25	28.32	1.62	0.84	0.69	0.00	0.96	0.03	1.02	
mr8a	1.40	65.58	28.16	1.49	0.90	0.59	0.00	0.90	0.08	0.91	

Table 2. WDS electron microprobe analyses of Merensky Reef zoned laurite.

Sample	As	S	Ru	Os	Ir	Rh	Pt	Pd	Ni	Fe	Total
					wt%						
MR2a1	1.08	37.50	39.32	4.79	7.23	1.09	0.00	1.32	0.26	6.54	99.13
MR2a2	1.28	36.97	37.20	7.97	9.49	0.90	0.00	1.18	0.06	5.76	100.81
MR2a3	1.63	37.63	43.23	2.53	6.22	1.38	0.00	1.35	0.42	5.87	100.27
MR2a4	1.51	37.58	45.24	3.15	5.75	1.32	0.00	1.55	0.81	4.23	101.15
MR2a5	1.38	37.68	43.61	3.59	6.02	1.35	0.00	1.47	0.90	4.73	100.73
MR2a6	1.74	37.32	42.99	2.78	6.46	1.37	0.00	1.41	0.65	5.22	99.94
MR2a7	1.44	38.69	44.54	1.88	5.27	1.19	0.00	1.47	0.49	5.24	100.21
MR2a8	1.42	39.13	44.65	2.00	4.50	1.63	0.00	1.58	0.39	5.44	100.73
MR2a9	1.50	38.46	43.86	1.86	6.06	1.05	0.10	1.42	0.53	5.28	100.13
MR2a10	1.54	38.98	43.03	2.77	6.25	1.16	0.00	1.45	0.31	6.17	101.66
MR2a11	2.22	38.17	44.63	1.57	5.35	1.53	0.00	1.63	0.57	4.57	100.23
MR2a12	9.72	28.45	28.24	5.90	13.43	1.75	7.20	1.05	0.80	3.05	99.57
MR2a13	3.66	36.78	42.64	1.90	5.85	2.02	1.78	1.58	0.63	4.35	101.20
MR2a14	2.90	37.25	43.43	1.64	5.83	1.98	1.04	1.55	0.54	4.75	100.91
MR2a15	2.74	36.97	41.61	2.55	6.73	1.77	0.85	1.48	0.76	5.57	101.02
MR2a16	1.39	39.25	44.22	2.38	5.75	1.08	0.00	1.39	0.42	5.26	101.14
MR2a17	1.41	39.21	44.41	1.92	5.11	1.12	0.00	1.37	0.34	5.74	100.63
MR2a18	1.49	39.68	44.34	1.70	4.52	1.18	0.00	1.75	0.31	6.26	101.24
MR2a19	1.40	39.70	43.20	1.73	5.22	1.38	0.00	1.42	0.35	6.38	100.80
MR2a20	1.42	38.14	41.29	3.31	6.79	1.36	0.00	1.49	0.27	5.78	99.84
MR2a21	1.42	39.38	44.15	2.30	5.31	1.11	0.00	1.38	0.41	5.51	100.97
MR2a22	1.49	39.29	45.22	1.47	4.26	1.23	0.00	1.44	0.22	5.84	100.46
MR2a23	1.50	39.63	43.90	1.38	4.63	1.32	0.00	1.48	0.39	6.40	100.63
MR2a24	1.54	37.92	43.50	3.35	7.10	0.82	0.00	1.55	0.82	4.61	101.20
MR4a1	2.06	41.46	33.93	0.94	2.68	3.03	0.57	1.49	0.23	14.19	100.58
MR4a2	3.90	38.45	35.37	1.51	2.98	3.50	2.53	1.61	0.30	10.13	100.27
MR4a3	8.19	32.52	33.93	3.15	2.83	2.69	8.69	1.37	0.36	5.66	99.38
MR4a4	4.32	37.85	37.63	2.52	2.25	2.75	3.79	1.58	0.35	7.79	100.83
MR4a5	5.78	34.99	34.15	3.55	3.08	2.80	5.61	1.50	0.31	7.78	99.56
MR4a6	6.88	34.58	36.72	2.47	2.64	2.68	7.33	1.36	0.35	5.90	100.90
MR4a7	7.94	32.54	33.50	3.92	4.15	2.51	8.20	1.41	0.31	5.97	100.44
MR4a8	4.26	37.60	36.69	3.21	3.26	2.54	2.95	1.62	0.27	8.77	101.17
MR1a1	1.38	37.10	42.11	4.44	9.00	1.18	0.00	1.40	0.56	4.03	101.19
MR1a2	1.36	37.54	42.06	4.85	8.61	1.22	0.00	1.35	0.56	3.82	101.37
MR1a3	1.44	36.34	42.08	4.46	8.24	1.09	0.00	1.45	0.71	3.70	99.51
MR1a4	1.43	37.43	41.91	4.47	8.97	1.10	0.00	1.43	0.52	3.94	101.21
MR1a5	1.45	37.60	39.58	6.03	7.48	1.14	0.00	1.22	0.12	5.98	100.59
MR1a6	1.46	38.23	37.79	5.25	6.84	1.31	0.00	1.24	0.13	8.68	100.92
MR1a7	1.45	36.94	41.04	7.93	7.14	1.09	0.00	1.31	0.73	3.75	101.38
MR1a8	2.03	36.96	41.34	2.77	9.50	1.56	0.00	1.45	0.67	4.13	100.41
MR1a9	1.51	37.75	36.32	7.72	8.20	1.21	0.00	1.10	1.39	4.95	100.16
MR1a10	1.69	36.81	50.36	5.42	2.85	1.24	0.00	1.79	0.03	1.00	101.18
MR1a11	1.85	36.93	49.99	4.97	3.03	1.06	0.00	1.68	0.08	0.90	100.48

Table 2. *Cont.*

Sample	As	S	Ru	Os	Ir	Rh	Pt	Pd	Ni	Fe
					at%					
MR2a1	0.81	65.69	21.85	1.41	2.11	0.59	0.00	0.70	0.25	6.58
MR2a2	0.97	65.76	20.99	2.39	2.82	0.50	0.00	0.63	0.05	5.88
MR2a3	1.20	64.94	23.67	0.74	1.79	0.74	0.00	0.70	0.39	5.82
MR2a4	1.12	64.99	24.82	0.92	1.66	0.71	0.00	0.81	0.77	4.20
MR2a5	1.02	65.21	23.94	1.05	1.74	0.73	0.00	0.76	0.85	4.70
MR2a6	1.29	64.96	23.74	0.82	1.88	0.74	0.00	0.74	0.62	5.21
MR2a7	1.05	65.89	24.06	0.54	1.50	0.63	0.00	0.75	0.45	5.13
MR2a8	1.02	65.97	23.89	0.57	1.26	0.86	0.00	0.80	0.36	5.26
MR2a9	1.10	65.81	23.81	0.54	1.73	0.56	0.03	0.73	0.50	5.19
MR2a10	1.11	65.72	23.02	0.79	1.76	0.61	0.00	0.73	0.28	5.98
MR2a11	1.63	65.43	24.27	0.45	1.53	0.82	0.00	0.84	0.53	4.50
MR2a12	8.48	58.03	18.27	2.03	4.57	1.11	2.41	0.64	0.89	3.57
MR2a13	2.73	64.07	23.56	0.56	1.70	1.10	0.51	0.83	0.60	4.35
MR2a14	2.15	64.45	23.84	0.48	1.68	1.07	0.30	0.81	0.51	4.72
MR2a15	2.03	64.14	22.90	0.75	1.95	0.96	0.24	0.77	0.72	5.55
MR2a16	1.01	66.26	23.68	0.68	1.62	0.57	0.00	0.71	0.39	5.10
MR2a17	1.02	66.10	23.75	0.55	1.44	0.59	0.00	0.69	0.31	5.56
MR2a18	1.06	66.04	23.41	0.48	1.25	0.61	0.00	0.88	0.28	5.98
MR2a19	1.00	66.30	22.89	0.49	1.45	0.72	0.00	0.72	0.32	6.11
MR2a20	1.05	65.91	22.63	0.96	1.96	0.73	0.00	0.78	0.25	5.73
MR2a21	1.03	66.27	23.58	0.65	1.49	0.58	0.00	0.70	0.38	5.32
MR2a22	1.07	66.01	24.10	0.42	1.19	0.64	0.00	0.73	0.20	5.63
MR2a23	1.07	66.11	23.23	0.39	1.29	0.68	0.00	0.74	0.35	6.13
MR2a24	1.14	65.45	23.82	0.97	2.04	0.44	0.00	0.80	0.78	4.56
MR4a1	1.39	65.32	16.96	0.25	0.71	1.49	0.15	0.71	0.20	12.84
MR4a2	2.78	64.02	18.68	0.42	0.83	1.82	0.69	0.81	0.28	9.68
MR4a3	6.50	60.32	19.97	0.98	0.88	1.55	2.65	0.77	0.37	6.02
MR4a4	3.13	64.09	20.21	0.72	0.63	1.45	1.06	0.81	0.32	7.58
MR4a5	4.40	62.15	19.24	1.06	0.91	1.55	1.64	0.80	0.30	7.94
MR4a6	5.25	61.69	20.78	0.74	0.78	1.49	2.15	0.73	0.34	6.04
MR4a7	6.29	60.18	19.66	1.22	1.28	1.45	2.49	0.79	0.31	6.34
MR4a8	3.09	63.62	19.69	0.91	0.92	1.34	0.82	0.82	0.25	8.52
MR1a1	1.04	65.42	23.56	1.32	2.65	0.65	0.00	0.74	0.54	4.08
MR1a2	1.02	65.85	23.41	1.43	2.52	0.66	0.00	0.72	0.54	3.85
MR1a3	1.11	65.22	23.96	1.35	2.47	0.61	0.00	0.78	0.70	3.81
MR1a4	1.08	65.77	23.36	1.32	2.63	0.60	0.00	0.76	0.50	3.98
MR1a5	1.08	65.67	21.93	1.77	2.18	0.62	0.00	0.64	0.11	6.00
MR1a6	1.07	65.12	20.43	1.51	1.94	0.70	0.00	0.63	0.12	8.49
MR1a7	1.10	65.50	23.09	2.37	2.11	0.60	0.00	0.70	0.71	3.81
MR1a8	1.53	65.24	23.15	0.82	2.80	0.86	0.00	0.77	0.65	4.19
MR1a9	1.14	66.34	20.25	2.29	2.41	0.66	0.00	0.58	1.33	5.00
MR1a10	1.28	65.25	28.32	1.62	0.84	0.69	0.00	0.96	0.03	1.02
MR1a11	1.40	65.58	28.16	1.49	0.90	0.59	0.00	0.90	0.08	0.91

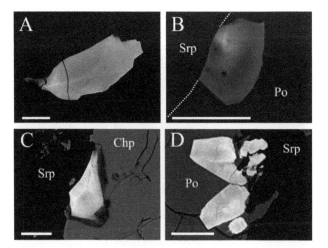

Figure 4. Back-scattered electron (BSE) images of zoned laurite. (**A**) See reflected-light image of Figure 3B for the mineralogical assemblage of the grain. (**B**) Laurite in contact with pyrrhotite and serpentine, (**C**) laurite in contact with chalcopyrite and serpentine and (**D**) laurite grains in contact with pyrrhotite and serpentine Abbreviations as in Figure 3. Scale bar = 20 μm.

As previously reported by [14], abundant Pb-Cl minerals, less than 10 μm in size, were also observed enclosed in the sulfides (Figure 5A), and qualitatively identified by EDS (Figure 5B). The EDS overlap between Pb and S was checked by a WDS semi-quantitative analysis that gave a composition (wt%) of 77.8 for Pb and 18.9 for Cl, very similar to the mineral analyzed by [14].

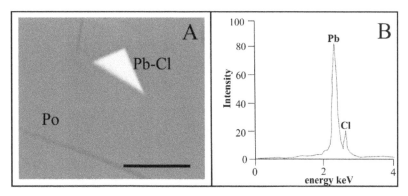

Figure 5. (**A**) BSE image of a Pb-Cl mineral (Pb-Cl) enclosed in pyrrhotite (Po), scale bar = 10 μm, and (**B**) its EDS spectrum (see the text for the Pb-S overlap).

Electron microprobe analyses of the zoned laurite (Table 2) and elemental distribution maps (Figures 6–8) showed unusual enrichments in As, Ir, Os, Pt, and Fe. Distribution of Rh, Pd, and Ni was not visible in the X-ray maps because of the low concentrations, while Cu (not analyzed, but visible in Figure 7) was due to a Cu-phase filling fissures in laurite. Substitution of As for S occurs systematically from a homogeneous background of about 1.00–1.50 wt% (Figure 6) up to patchy enrichment of 3.66–9.7 wt% (Figures 7 and 8). The enrichments of Os and Ir are closely related and may occur either at the rim of grains as described by [14] (Figure 6), or as irregular patches (Figures 7 and 8). The Pt appears to be particularly concentrated, up to 8.69 wt%, in the As-rich zones (Figures 7 and 8).

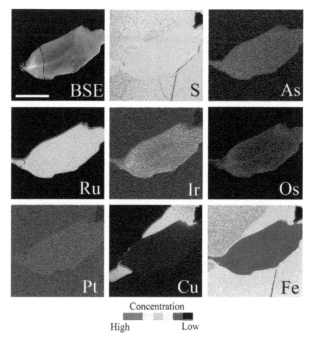

Figure 6. BSE image and X-ray element-distribution maps of S, As, Ru, Ir, Os, Pt, Cu, and Fe in zoned laurite from the Merensky Reef. See Figures 3B and 4A for the paragenetic assemblage. Scale bar = 20 μm.

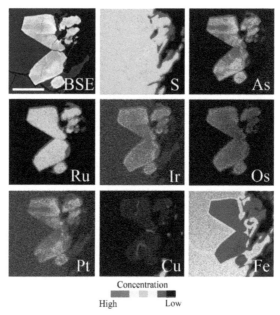

Figure 7. BSE image and X-ray element-distribution maps of S, As, Ru, Ir, Os, Pt, Cu, and Fe, showing the zoning of the laurite from the Merensky Reef, see reflected-light image of Figure 4D for the mineralogical assemblage. Scale bar is 20 μm.

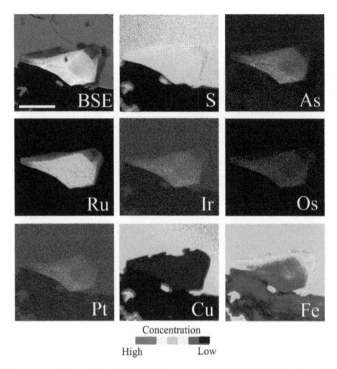

Figure 8. BSE image and X-ray element-distribution maps of S, As, Ru, Ir, Os, Pt, Cu, and Fe, showing the zoning of the laurite from the Merensky Reef, see reflected-light image of Figure 4C for the mineralogical assemblage. Scale bar is 20 µm.

The Ru-Os-Ir (wt%) ternary diagram (Figure 9) shows that the zoned laurite of the Merensky Reef are significantly enriched in Ir, compared with laurite enclosed in the chromitite of the same Reef, and other chromitite layers of the Bushveld. They also do not display the Ru-Os negative correlation inferred by the Ru-Os substitution trend due to the laurite-erlichmanite solid solution trend (Figure 8). Based on more than 1000 published analyses, and unpublished data of the authors, laurite associated with ophiolitic, stratiform, and Alaskan-type magmatic chromitites exhibit a pronounced negative correlation between Ru and Os (R = −0.97). In contrast, the correlation matrix calculated from our electron microprobe analyses (Table 3) indicates that zoned laurite of the Merensky Reef are characterized by the absence of Ru-Os correlation (R = −0.07).

Table 3. Element correlation in the zoned laurite for the Merensky Reef.

at%	As	S	Ru	Os	Ir	Rh	Pt	Pz	Ni	Fe
As	1.00									
S	**−0.98**	1.00								
Ru	−0.56	0.53	1.00							
Os	0.10	−0.14	−0.07	1.00						
Ir	0.04	−0.09	−0.04	0.56	1.00					
Rh	0.72	−0.70	−0.68	−0.25	−0.42	1.00				
Pt	**0.97**	**−0.95**	−0.58	0.06	−0.10	0.75	1.00			
Pd	0.03	−0.05	0.45	−0.37	−0.50	0.20	0.03	1.00		
Ni	0.07	−0.09	−0.02	0.20	0.50	−0.12	−0.01	−0.21	1.00	
Fe	0.11	−0.10	**−0.78**	−0.35	−0.37	0.56	0.18	−0.27	−0.29	1.00

Note: the relevant correlations are highlighted in bold.

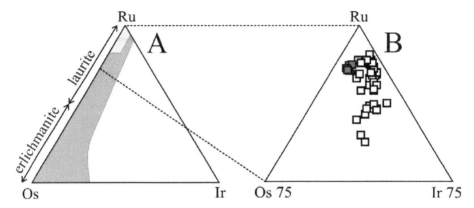

Figure 9. (**A**) Ru-Os-Ir ternary diagram (wt%) for magmatic laurite enclosed in the Lower-, Middle-, and Upper-group chromitite layers: from the Bushveld Complex (yellow field) and podiform chromitite (blue field). Compositional fields after [11,12] and unpublished data of the authors. (**B**) Merensky Reef zoned laurite (open square) and laurite enclosed in the Merensky Reef chromite (red square).

In addition, the high concentrations of As (up to 9.72 wt%), Pt (up to 9.72 wt%), and Fe (up to 14.19%) (Table 2), distinguish the zoned laurite of the Merensky Reef from the laurite inclusions in different types of chromitite (ophiolitic, stratiform, Alaskan-type).

The correlation matrix (Table 3) and distribution X-ray maps (Figure 7) clearly support a positive correlation between Pt-As (R = +0.97), and a negative correlation of both elements with S (R = −0.95 and −0.98, respectively). The possible existence of submicroscopic inclusions of sperrylite ($PtAs_2$) or platarsite (PtAsS) in laurite was carefully checked, and discarded.

The fact that the sum of S+As atoms is consistently close to stoichiometry (S + As = 2.00) supports that Pt and As are parts of the laurite structure. Notably also the high concentrations of Fe are not due to exotic inclusions, but Fe appears to be the major substitute for Ru, showing a negative correlation of R = −0.78 (Table 3), that is a clear discrepancy with common magmatic laurite in general.

5. Origin of the Zoned Laurite in the Merensky Reef

Several theories for the origin of the Merensky Reef have been proposed, and they have been recently summarized by [25]. The proposed genetic models include: (i) gravitational settling of crystals that precipitated in the magma chamber during the orthomagmatic stage; (ii) hydrodynamic sorting of a mobilized cumulate slurry in a large magma chamber, under slow cooling; (iii) crystallization at the crystal mush-magma interface caused by a replenishment event; (iv) interaction of a hydrous melt with a partially molten cumulate assemblage; (v) intrusion of magmas as sills into undifferentiated norite; and (vi) intrusion of magma into a pre-existing cumulate pile ([25] and references therein).

To explain the formation of the coarse-grained feldspathic orthopyroxenite enriched in PGE, and associated chromitite in the Merensky Reef, several authors have invoked the reaction between a late-stage hydrous melt with an unconsolidated cumulate assemblage [23–27]. On the basis of natural observations supported by experimental results, formation of tiny Os-Ir-Ru PGM inclusions in chromitite can be modeled by a sequence of crystallization events controlled by sulfur fugacity and temperature (T) [2–4]. The sulfur fugacity is expected to increase with decreasing T in magmatic systems between about 1300–1000 °C, and a consistent order of crystallization can be observed. Alloys in the system Os-Ir-Ru are the first to precipitate, followed by laurite, usually characterized by a core-to-rim increment of Os content, and finally, erlichmanite. Incorporation of IrS_2 molecules in the laurite structure is generally low, controlled by Ir activity in the system. However, the systematic substitution of Os for Ru can be remarkable, and the composition can enter the field of erlichmanite if sulfur fugacity increases sufficiently during magmatic crystallization.

At a first instance, the Os enrichment observed at the rim of some zoned laurite of the Merensky Reef may correspond to such a magmatic crystallization trend. However, other characteristics of the Merensky Reef zoned laurite, such as the unusual zoning that involves local enrichment of As, Pt, Ir, and Fe (Figures 7 and 8), and the absence of Ru-Os substitution, are in apparent contrast with this conclusion. The observed zoning also does not reflect a fluctuation of the sulfur fugacity, suggested to explain the oscillatory zoning of Ru and Os, described in the laurite from the Penikat Layered Complex of Finland [28]. The presence of abundant minerals containing Pb and Cl and occurring enclosed in the sulfides associated with the zoned laurite indicates the presence of Cl in the system [14]. According to [14], the Cl-rich phase precipitated from a late-stage solution or formed as a result of replacement of a precursor galena by an aqueous hydrochloric solution in the final stage of hydrothermal alteration, at low-temperature. Theoretical and experimental work coupled with natural observations suggest that both Cl and As may be important for the transport and mobilization of the PGE during metasomatic and hydrothermal events [23,28–32].

The textural position and the coarse grain size of laurite crystals, as well as their paragenesis including hydrous silicates, suggest crystallization at a late stage from a volatile-rich melt enriched in As and Cl, after coalescence of an immiscible sulfide liquid. The close stabilization of serpentine points to relatively low temperatures for the precipitation of zoned laurite, certainly much lower than those required for the crystallization of tiny laurite included in the chromite seams of the Merensky Reef. Although it is not possible to provide a precise temperature for the genesis of the zoned laurite in the Merensky Reef, we suggest they were in the range of 400–200 °C, similar to temperatures calculated for PGM deposition in the hydrothermal Waterberg platinum deposit of Mookgophong, South Africa [31]. Therefore, we can suggest that the zoned laurite of the Merensky Reef is "hydrothermal" in origin, having crystallized in the presence of a Cl- and As-rich hydrous solution, at temperatures much lower than those typical of the precipitation of magmatic laurite. Although, it remains to be seen whether the "hydrothermal" laurite precipitated directly from the hydrothermal fluid, or it represents the alteration product of a pre-existing laurite reacting with the hydrothermal solution.

Author Contributions: F.Z. and G.G. wrote the manuscript and provided support in the data interpretation. G.G. collected the studied sample, and F.Z. performed the chemical analyses. All authors have read and agreed to the published version of the manuscript.

Funding: The authors are grateful to the University Centrum for Applied Geosciences (UCAG) for the access to the E. F. Stumpfl electron microprobe laboratory.

Acknowledgments: Many thanks are due to the editorial staff of Minerals, to the guest editor Maria Economou-Eliopoulos, and two referees for their useful comments. We are honored to dedicate this manuscript to the memory of our friend and colleague, Demetrios G. Eliopoulos.

Conflicts of Interest: The authors declare no conflicts of interest.

References

1. Blackburn, W.H.; Dennen, W.H. *Encyclopedia of Minerals Names, The Canadian Mineralogist Special Publication*; Robert, M., Ed.; Mineralogical Society of Canada: Ottawa, ON, Canada, 1997; Volume 1, p. 360. ISBN 0-921294-45-x.

2. Prichard, H.M.; Barnes, S.J.; Fisher, P.C.; Zientek, M.L. Laurite and associated PGM in the Stillwater chromitites: Implications for processes on formation, and comparisons with laurite in the Bushveld and ophiolitic chromitites. *Can. Mineral.* **2017**, *55*, 121–144. [CrossRef]

3. Zaccarini, F.; Garuti, G.; Pushkarev, E.; Thalhammer, O. Origin of Platinum Group Minerals (PGM) Inclusions in Chromite Deposits of the Ural. *Minerals* **2018**, *8*, 379. [CrossRef]

4. Garuti, G.; Proenza, J.; Zaccarini, F. Distribution and mineralogy of platinum-group elements in altered chromitites of the Campo Formoso layered intrusion (Bahia State, Brazil): Control by magmatic and hydrothermal processes. *Mineral. Pet.* **2007**, *89*, 159–188. [CrossRef]

5. Bowles, J.F.W.; Tkin, D.A.; Lambert, J.L.M.; Deans, T.; Phillips, R. The chemistry, reflectance, and cell size of the erlichmanite (OsS$_2$)-laurite (RuS$_2$) series. *Mineral. Mag.* **1983**, *47*, 465–471. [CrossRef]

6.   Bowles, J.F.W.; Suárez, S.; Prichard, H.M.; Fisher, P.C. The mineralogy, geochemistry and genesis of the alluvial platinum-group minerals of the Freetown Layered Complex, Sierra Leone. *Mineral. Mag.* **2018**, *82*, 223–246. [CrossRef]

7.   Oberthür, T. The Fate of Platinum-Group Minerals in the Exogenic Environment—From Sulfide Ores via Oxidized Ores into Placers: Case Studies Bushveld Complex, South Africa, and Great Dyke, Zimbabwe. *Minerals* **2019**, *9*, 581. [CrossRef]

8.   Garuti, G.; Zaccarini, F. In-situ alteration of platinum-group minerals at low temperature: Evidence from chromitites of the Vourinos complex (Greece). *Can. Mineral.* **1997**, *35*, 611–626.

9.   Zaccarini, F.; Proenza, J.A.; Ortega-Gutierrez, F.; Garuti, G. Platinum Group Minerals in ophiolitic chromitites from Tehuitzingo (Acatlan Complex, Southern Mexico): Implications for postmagmatic modification. *Mineral. Petrol.* **2005**, *84*, 147–168. [CrossRef]

10. Zaccarini, F.; Bindi, L.; Garuti, G.; Proenza, J. Ruthenium and magnetite intergrowths from the Loma Peguera chromitite, Dominican Republic, and relevance to the debate over the existence of platinum-group element oxides and hydroxides. *Can. Mineral.* **2015**, *52*, 617–624. [CrossRef]

11. Maier, W.D.; Prichard, H.M.; Fisher, P.C.; Barnes, S.J. Compositional variation of laurite at Union Section in the Western Bushveld Complex. *S. Afr. J. Geol.* **1999**, *102*, 286–292.

12. Zaccarini, F.; Garuti, G.; Cawthorn, G. Platinum group minerals in chromitites xenoliths from the ultramafic pipes of Onverwacht and Tweefontein (Bushveld Complex). *Can. Mineral.* **2002**, *40*, 481–497. [CrossRef]

13. Kaufmann, F.E.D.; Hoffmann, M.C.; Bachmann, K.; Veksler, I.V.; Trumbull, R.B.; Hecht, L. Variations in Composition, Texture, and Platinum Group Element Mineralization in the Lower Group and Middle Group Chromitites of the Northwestern Bushveld Complex, South Africa. *Econ. Geol.* **2019**, *14*, 569–590. [CrossRef]

14. Barkov, A.Y.; Martin, R.F.; Kaukonen, R.J.; Alapieti, T.T. The occurrence of Pb–Cl–(OH) and Pt–Sn–S compounds in the Merensky Reef, Bushveld layered complex, South Africa. *Can. Mineral.* **2001**, *39*, 1397–1403. [CrossRef]

15. Prichard, H.M.; Barnes, S.J.; Maier, W.D.; Fisher, P.C. Variations in the nature of the platinum-group minerals in a cross-section through the Merensky Reef at Impala platinum: Implications for the mode of formation of the Reef. *Can. Mineral.* **2004**, *42*, 423–437. [CrossRef]

16. Hutchinson, D.; Foster, J.; Prichard, H.; Gilbertm, S. Concentration of particulate Platinum-Group Minerals during magma emplacement; a case study from the Merensky Reef, Bushveld Complex. *J. Petrol.* **2015**, *56*, 113–159. [CrossRef]

17. Junge, M.; Oberthür, T.; Melcher, F. Cryptic variation of chromite chemistry, platinum group element and platinum group mineral distribution in the UG-2 chromitite: An example from the Karee mine, Western Bushveld complex, South Africa. *Econ. Geol.* **2014**, *109*, 795–810. [CrossRef]

18. Cawthorn, R.G. The Platinum Group Element Deposits of the Bushveld Complex in South Africa. *Plat. Met. Rev.* **2010**, *54*, 205–215. [CrossRef]

19. Chistyakova, S.; Latypov, R.; Youlton, K. Multiple Merensky Reef of the Bushveld Complex, South Africa. *Contrib. Mineral. Petrol.* **2019**, *174*, 26. [CrossRef]

20. Vermaak, C.F.; Von Gruenewaldt, G. Third international platinum symposium, excursion guidebook. *Geol. Soc. S. Afr.* **1981**, *62*, 5.

21. Vymazalová, A.; Zaccarini, F.; Garuti, G.; Laufek, F.; Mauro, D.; Stanley, C.J.; Biagioni, C. Bowlesite, IMA 2019-079. CNMNC Newsletter No. 52. *Mineral. Mag.* **2019**, *83*. [CrossRef]

22. Cawthorn, R.G. The geometry and emplacement of the Pilanesberg Complex, South Africa. *Geol. Mag.* **2015**, *152*, 1–11. [CrossRef]

23. Ballhaus, C.G.; Stumpfl, E.F. Sulfide and platinum mineralization in the Merensky Reef: Evidence from hydrous silicates and fluid inclusions. *Contrib. Mineral. Petrol.* **1986**, *94*, 193–204. [CrossRef]

24. Nicholson, D.M.; Mathez, E.A. Petrogenesis of the Merensky Reef in the Rustenburg section of the Bushveld Complex. *Contrib. Mineral. Petrol.* **1991**, *107*, 293–309. [CrossRef]

25. Hunt, E.J.; Latypov, R.; Horváth, P. The Merensky Cyclic Unit, Bushveld Complex, South Africa: Reality or Myth? *Minerals* **2018**, *8*, 144. [CrossRef]

26. Mathez, E.A. Magmatic metasomatism and formation of the Merensky Reef, Bushveld Complex. *Contrib. Mineral. Petrol.* **1995**, *119*, 277–286. [CrossRef]

27. Boudreau, A.E. Modeling the Merensky Reef, Busvheld Complex, Republic of South Africa. *Contrib. Mineral. Petrol.* **2008**, *156*, 431–437. [CrossRef]

28. Barkov, A.; Fleet, M.E.; Martin, R.F.; Alapieti, T.T. Zoned sulfides and sulfarsenides of the platinum-group elements from the Penikat layered complex, Finland. *Can. Mineral.* **2004**, *42*, 515–537. [CrossRef]

29. Boudreau, A.E. Chlorine as an exploration guide for the platinum-group elements in layered intrusions. *J. Geochem. Explor.* **1993**, *48*, 21–37. [CrossRef]

30. Kislov, E.V.; Konnikov, E.G.; Orsoev, D.; Pushkarev, E.; Voronina, L.K. Chlorine in the Genesis of the Low-Sulfide PGE Mineralization in the Ooko-Dovyrenskii Layered Massif. *Geokhimiya* **1997**, *5*, 521–528.

31. Oberthür, T.; Melcher, F.; Fusswinkel, T.; van den Kerkhof, A.M.; Sosa, G.M. The hydrothermal Waterberg platinum deposit, Mookgophong (Naboomspruit), South Africa. Part 1: Geochemistry and ore mineralogy. *Mineral. Mag.* **2018**, *82*, 725–749. [CrossRef]

32. Le Vaillant, M.; Barnes, S.J.; Fiorentini, M.; Miller, J.; Mccuaig, C.; Mucilli, P. A hydrothermal Ni-As-PGE geochemical halo around the Miitel komatiite-hosted nickel sulfide deposit, Yilgarn craton, Western Australia. *Econ. Geol.* **2015**, *110*, 505–530. [CrossRef]

Article

Geochemistry of Rare Earth Elements in Bedrock and Till, Applied in the Context of Mineral Potential in Sweden

Martiya Sadeghi *, Nikolaos Arvanitidis and Anna Ladenberger

Department of Mineral Resources, Geological Survey of Sweden, Box 670, 751 28 Uppsala, Sweden;
Nikolaos.arvanitidis@sgu.se (N.A.); Anna.Ladenberger@sgu.se (A.L.)
* Correspondence: martiya.sadeghi@sgu.se; Tel.: +46-1817-9232

Received: 15 March 2020; Accepted: 15 April 2020; Published: 18 April 2020

Abstract: The Rare Earth Element (REE) mineralizations are not so "rare" in Sweden. They normally occur associated and hosted within granitic crystalline bedrock, and in mineral deposits together with other base and trace metals. Major REE-bearing mineral deposit types are the apatite-iron oxide mineralizations in Norrbotten (e.g., Kiruna) and Bergslagen (e.g., Grängesberg) ore regions, the various skarn deposits in Bergslagen (e.g., Riddarhyttan-Norberg belt), hydrothermal deposits (e.g., Olserum, Bastnäs) and alkaline-carbonatite intrusions such as the Norra Kärr complex and Alnö. In this study, analytical data of samples collected from REE mineralizations during the EURARE project are compared with bedrock and till REE geochemistry, both sourced from databases available at the Geological Survey of Sweden. The positive correlation between REE composition in the three geochemical data groups allows better understanding of REE distribution in Sweden, their regional discrimination, and genetic classification. Data provides complementary information about correlation of LREE and HREE in till with REE content in bedrock and mineralization. Application of principal component analysis enables classification of REE mineralizations in relation to their host. These results are useful in the assessment of REE mineral potential in areas where REE mineralizations are poorly explored or even undiscovered.

Keywords: rare earth elements; Sweden; lithogeochemistry; till geochemistry

1. Introduction

The Rare Earth Element (REE) occurrences in Sweden are widely distributed all over the country but some regions are more endowed than others. Major REE-bearing mineral deposit types, found mainly in Norrbotten and Bergslagen ore regions, are the apatite-iron oxide mineralizations in Kiruna and Grängesberg, the skarn mineralizations of the Riddarhyttan-Norberg belt, the alkaline igneous rock-associated in Norra Kärr, and the hydrothermal mineralizations in Olserum and Alnö deposit areas [1] (see Figure 1). Distribution of REEs in Swedish soil and bedrock has previously been investigated by Sadeghi and Andersson [2] with the aim to identify the main changes in REE geochemistry related to geology and weathering, taking also into account the current baseline level for REE in soil and bedrock over Sweden. The Geochemical Atlas of Sweden provides a harmonized, countrywide database with modern baseline geochemical data from C horizon (element concentration in the weathered parent rock horizon of soil profile) in till [3] and REEs in different solid media (topsoil, subsoil, and stream sediments) have been investigated using the Forum of European Geological Surveys (FOREGS) database in order to identify the REEs regional background values [4]. In Sadeghi et al. [5], the Geochemical Mapping of Agricultural and Grazing Land Soil (GEMAS) project data were investigated with focus on REEs in two solid media (topsoil from agricultural (Ap) and grazing land (Gr) soil) to identify the background values of REEs both in Sweden and in Europe. The Ap samples were collected from regularly plowed fields at a depth interval of 0 to 20 cm and the Gr samples were collected from soil under permanent grass cover at a 0 to 10 cm depth range.

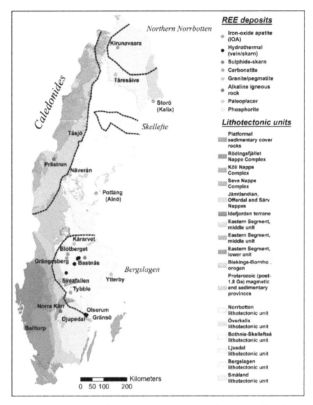

Figure 1. Simplified map of major lithotectonic units and ore districts in Sweden, with a selection of Rare Earth Elements (REEs) deposits, prospects, and occurrences (modified from [1] and bedrock database at Geological Survey of Sweden)

REE deposits can be referred to regolith, basinal, metamorphic and magmatic associations based on a mineral-systems approach [6]. Various deposit types form either directly from the crystallization of the melt and/or fluids predominantly derived from the melt. The magmatic deposits can be divided into orthomagmatic and hydrothermal types. Basinal deposit types are inferred to be formed through mechanical (e.g., placer) and chemical (e.g., phosphorite) sedimentary processes, and from diagenetic fluids generated in sedimentary basins. Deposit types of the regolith association require an REE-bearing rock source to form feasible secondary contents of REE. The REE deposits are formed either due to enrichment of REE in the residual material and/or from local remobilization of REE. Deposit types of the metamorphic association are generated during regional and/or contact metamorphism and involve related metamorphic fluids.

This study analyzes, explores and evaluates the composition of REE mineralized samples collected during the EURARE project (2013–2017) enabling their genetic classification into specific deposit types. This information is integrated with the Geological Survey of Sweden (SGU)'s lithogeochemistry and till geochemistry databases to better approach and interpret the distribution of REE anomalous provinces at the scale of the country.

2. Overview of REE Mineralizations in Sweden

2.1. REE Mineralizations in Granitic Pegmatites and Granitoids

Partial melting of crustal material produces felsic melts that are enriched in REEs. Elements in pegmatites that can be enriched up to minable (ore) grades include Li, Cs, Be, Sn, Nb, Ta, U, Y, Zr, and REEs. This long and diverse list of elements, to which a significantly high content of B, P, F, Rb, Bi, Hf, etc., could be added, make fractionated pegmatites among the most mineralogically complex deposit types on the Earth [7]. Pegmatitic facies associated with syenite-alkaline granite complexes constitute important deposits of REEs, U, Nb and Zr, and less commonly, Be and P [7].

In Sweden, granitic pegmatites, typically of a moderately to highly fractionated character, often host variable amounts of minerals rich in trace elements, including Be, Li, Nb, Ta, Sn, U, Th, as well as REEs. [8,9]. One notable granitic pegmatite, that shows locally relatively high REE content, (a porphyritic granite to gneiss with age 1300 Ga so-called "RA-granite") is located in Balltorp (Eastern Segment lithotectonic unit) (Figure 1) in southwestern Sweden [10]. The RA-granite is a Be-F-Nb-REE-Sn-Ta-Th-U-Zr-anomalous gneissic granite that does show some potential but has so far not been systematically explored (cf. [10] and references therein). Moreover, several other granitic to syenitic rocks in southern Sweden and elsewhere exhibit increased REE contents. The most important granitic pegmatites and granitoids with potential for REE mineralization are, Tåresåive (highest contents of REE_{tot} = 94,900 ppm) in northern Sweden (Norrbotten lithotectonic unit), Näverån in Central Sweden (close to the Caledonides), as well as granitoids in Bergslagen (e.g., Ytterby; highest contents of REE_{tot} = 30,624 ppm), and granite in the southwestern part of Sweden, (e.g., Balltorp; highest contents of REE_{tot} = 8800 ppm), shown in Figure 1. A contrasting case to these primary enrichments is the granitoid-hosted but epigenetic, shear zone-related mineralization at Näverån (highest contents of REE_{tot} = 6929 ppm) in central Sweden [1].

2.2. REE Mineralization in Alkaline Intrusive Rocks and Carbonatites

Several REE deposits associated with peralkaline complexes (syenitic rocks) are usually enriched in U and Th and have a relatively high Heavy Rare Earth Element (HREE) content when compared with carbonatites [11]. In Sweden, Norra Kärr deposit represents one of the most advanced and promising REE mining projects in northern Europe. This concentrically zoned alkaline intrusion, with nepheline-syenite, contains eudialyte-group minerals highly enriched in REEs with up to 8.5 wt% (La + Ca + Nd + Y) [1], showing also a strong resource potential in heavy REEs.

Carbonatites are common hosts of orthomagmatic-type REE mineralizations that tend to be variably enriched in REEs, Sr, Ba, U, Th, Nb, Ta, P, and F. Although the overall REE content of carbonatites may vary, their shape on the chondrite-normalized plot almost invariably displays high Light Rare Earth Element (LREE) content and no negative Eu anomalies [12]. Barium and Sr are generally abundant, whereas U, Th, Nb, and P, are variably abundant in the mineralized carbonatite. REE-bearing mineral phases have the highest REE content but major carbonate minerals, including calcite and dolomite, may also contain substantial contents of REEs [13]. Sövitic carbonatitic melts tend to show the highest levels of REE enrichment. They also show significant enrichment in LREE (LREE/HREE ratio of ~40, compared to ~7 of alkaline and felsic melts).

In Sweden, the Alnö complex consists of alkaline and carbonatite intrusive rocks located on the island of Alnö, in the Bothnian Bay and on the mainland north to northwest of Alnö, all along the coast of east-central Sweden [14–16]. The carbonatite, and specifically the calcite-dominated søvite, have significantly elevated total REE contents, between ca 500 and 1500 ppm [17]. Alnö intrusive complex and associated co-magmatic rocks on the mainland have also been the focus of exploration for REEs, including that carried out by the Boliden company in the 1970s. The main lithology ijolite contains clinopyroxene with LREEs up to 100 times chondritic values [18,19]. The Söråker intrusion located on the mainland north of Alnö comprises Ca- and Mg-carbonatite dikes and melilitolites. Another small carbonatite occurrence is known from the Middle Allochthon in the Caledonides where calc-silicate rocks at Prästrun in west-central Sweden host a REE-Nb-(U-Th) mineralization (highest contents of REE$_{tot}$ = 2300 ppm). This intrusive-hosted REE mineralization was first described at Prästrun.

Alkaline lamprophyre is another lithology that tends to have elevated REE contents. Chemically, potassic lamprophyre dikes are often characterized by relatively low SiO$_2$ and TiO$_2$, and high MgO and K$_2$O contents. In addition, they show high content of large-ion lithophile trace elements (e.g., Rb, Sr, and Ba) and LREE, but low content of high-field-strength elements (e.g., Nb, Ta, Zr, Hf, and Ti) [20]. An example of the REE-enriched lamprophyre is the 1.1 Ga alkaline ultramafic dykes up to 1-metre-wide, outcropping along the northernmost part of the coast in the Gulf of Bothnia, (Kalix) known as Storön deposit (Figure 1; [20,21]). The dikes represent olivine- and mica-rich lamprophyres and silico-carbonatites [21].

2.3. REE Mineralization in Apatite-Iron Oxide Deposits

At present, over forty iron oxide apatite deposits are known in northern Sweden [22–24]. The main (at over 2000 Mt) deposit in the Norrbotten district is the famous Kiruna (Kiirunavaara) deposit and active mine (cf. [24]). A hydrothermal (including replacement) and (ortho-) magmatic origin of iron oxide apatite deposits in Norrbotten have been suggested in numerous studies (e.g., [25–31] and references therein). The main REE host minerals in this deposit are apatite, typically a fluorapatite, and monazite, occurring often as inclusions in apatite [32–35]. Apatite abundance and composition in the ore varies, e.g., at Kiruna, where the phosphorus content in the upper parts of the ore is ca. 2%, equivalent to almost 11% apatite. REE enrichment is represented mainly by LREEs, Ce and Y anomalies observed as early as the 1930s by Geijer [34]. REE geochemistry of the host rocks in the northern district shows two patterns, one with a negative Eu anomaly and strong enrichment in LREEs at Kiruna and area close to this mine and the second one without a well pronounced Eu anomaly e.g., at the Malmberget and the areas close to this mine (Figure 2) [36].

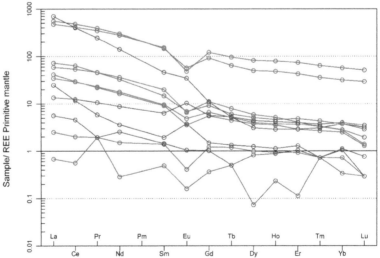

Figure 2. REE spider plot [37] of host rocks from the northern districts; Kiruna mine and occurrences close to this area (red) display a distinct Eu trough and a relative enrichment in LREEs compared to the HREEs, and while the samples from the Malmberget mine and occurrences nearby show extensive metamorphism and deformation overprint compared to the Kiruna area (Blue). Data sources: SGU and [36].

In central Sweden, (the Bergslagen district), a number of iron oxide apatite deposits are known with the biggest being Grängesberg and Blötberget, which were mined in historical times (Grängesberg mine closed in 1989) and both represent Kiruna-type deposits. The previous studies reveal that they still constitute a rather important potential reserve of REEs and phosphorus, both from the main ore and from mining waste [38]. The deposits are predominantly hosted by 1.9 Ga volcanosedimentary rocks variably metamorphosed and deformed. The ore is usually enriched in LREEs and to a minor degree in U, Th, Sm, Tb, Yb, and Y. The majority of REE enrichment is located in phosphate minerals such as apatite, xenotime, and monazite and to a lesser degree in silicates such as allanite and epidote [1,30,33,35].

2.4. Fe-REE Hydrothermal (Bastnäs-Type) Mineralizations

Evaluation of the REE geochemistry of hydrothermal mineral deposits, and the processes by which they are concentrated, is a complex task [39]. However, these mineralizations can be grouped on the basis of their geochemical and mineralogical characteristics. The degree of REE enrichment in a deposit is a function of the concentration of REEs in the fluid, the water-rock ratio, the efficiency of the precipitation process and the nature and amount of co-precipitated phases [39]. Oreskes and Einaudi [40] suggested that the high abundance of fluorocarbonates and the lack of Ca minerals in REE deposits (e.g., Olympic Dam) indicated that the fluids behind hydrothermal REE mineralizations may have been F- and CO_2-rich and that the REEs were transported as Cl and/or F complexes. Lottermoser [41] suggested that the association of REE with U minerals indicated that U and REE were complexed by the same ligands and inferred that those were CO_3, F, or SO_4.

The Bastnäs-type Fe-REE deposits in Sweden are early Proterozoic, skarn-hosted iron oxide (magnetite-dominated), locally polymetallic (±Cu, Au, Co, Bi, Mo) mineralizations that in part carry very REE-rich mineral assemblages (cf. [42]). They are located in the Bergslagen district and are characterized by the occurrence of locally abundant REE-rich silicate minerals such as cerite-(Ce) and allanite (*sensu lato*) but also include REE fluorocarbonates such as bastnäsite-(Ce).

3. Materials and Methods

In this study the lithogeochemical and till geochemistry databases have been used for re-interpretation of the REE baseline in different rock types and till in Sweden, and for the evaluation of REE mineral potential. More than 20,000 rock samples were collected and analyzed as part of the bedrock mapping program at the Geological Survey of Sweden (SGU; www.sgu.se). Within the EURARE project (www.eurare.org), field work activities were carried out in several REE-mineralized areas and were followed up by detailed investigations on selected mineralizations. The modal mineral composition, petrographical features and chemical compositions carried out for more than 200 rock samples [1]. The lithogeochemical analyses were carried out at the ACME Lab (Vancouver, BC, Canada) and ALS Scandinavia AB (Luleå, Sweden). At the ACME Lab the samples were analyzed for major, minor, and trace elements by ICP-emission spectrometry following a lithium metaborate/tetraborate fusion and dilute nitric digestion. At the ALS, trace elements including the full rare earth element suites were obtained from fused beads followed by acid digestion and measured by either ICP-AES or an inductively coupled plasma sector field mass spectrometer (ICP-SFMS).

The till database used in this study contains over 2500 samples analyzed by aqua regia digestion followed by ICP-MS. These national-wide results have been published as the Geochemical Atlas of Sweden [3] and are available for the public online (https://www.sgu.se/mineralnaring/geokemisk-kartlaggning/geokemisk-atlas/).

The open-source software Geochemical Data Toolkit in R (GCDKIT-version 4.1) [43] which is built using the freeware R language, has been applied for the data processing and biplot diagrams in this paper (http://www.gcdkit.org).

Principal component (PC) analysis is a conventional multivariate technique that is often used for studying geochemical data [4,44–47]. PC analysis reduces a large number of variables to a smaller number, allowing the user to determine the components (groups of variables) that account for variation in multivariate data [47]. PC analysis has often been used to process and interpret geochemical and other types of spatial data e.g., [4,5,47,48]. PC analysis builds on the correlation (covariance) matrix, which measures the interrelationships among multiple variables. The first PC (PC1) explains most of the variance within the original data, and each subsequent PC (PC2–n) explains progressively less of the variance. A multivariate dataset can usually be reduced to two or three PCs that account for the majority of the variance within the dataset. PC analysis was performed on average values of the different mineralization types as discussed above. Thirty-seven elements were chosen for PC analysis based on correlation coefficients. The Statistica software (version 13.2, Dell Software, Aliso Viejo, CA, USA) has been used for the principal component analysis (https://statistica.software.informer.com/).

The Geochemical Atlas of Sweden [3] is a country-wide harmonized database with modern baseline geochemical data from the C horizon in till. The database encompasses 2578 till samples from SGU archive as well as new sampled till collected mainly in the mountainous areas of western Sweden. Since the C horizon is considered as an anthropogenically undisturbed layer, the major geochemical signature in till should originate from underlying bedrock, its lithology, mineralogy, and potential mineralization. Secondary processes such as ice transport, leaching, and biological activity have minor impact on till chemical composition. During weathering, REEs are generally not very mobile, but this varies depending on the host mineral and local pH. REE mobility is further controlled by adsorption onto iron oxides, phosphates, and clay minerals [2,49]. In general, till can be used as proxy for the underlying bedrock in areas where outcrops are not available. Till can be treated then as composite rock material and its composition can be often interpolated to the average parent rock.

4. Results and Discussion

4.1. REE Content in Bedrock

Lithogeochemistry data from the SGU database have been used to examine links between REE contents in bedrock with known REE mineralizations. The main intention was to establish a tool for future exploration and predictive mapping surveys. Generally, primary REEs are associated with magmatic (granite and pegmatite) and alkaline rocks. Rocks enriched in monazite and zircon often contain higher contents of rare earth elements. In sedimentary rocks, higher contents of rare earth elements occur in shale and greywacke. There are no obvious differences in the geochemical distribution between the LREE group (La, Ce, Pr, Nd, and Sm) and the HREE group (Eu-Lu + Y) in the lithogeochemical data. However, their relative contents may differ for each element, probably due to the mineral composition and type of the host rock. For example, Eu is often found in rock-forming minerals such as plagioclase and in accessory minerals, mainly in allanite, bastnäsite, monazite, apatite, zircon, and fluorite. Typically, Eu can replace Sr and Ca in rock-forming minerals, therefore, its enrichment in plagioclase is common. This results in Eu enrichment in plagioclase-rich rocks in northernmost Sweden, such as gabbro and granodiorite.

The data with a higher content of the total REE and Y (above 500 ppm) were extracted from the database and compared with the extracted data with a content of Y more than 100 ppm (Figure 3). Our results show a linear correlation and match between the samples with a higher content of REE + Y in total with Y content, and more importantly, there is a direct link between Y content in samples with existing known mineralizations in Sweden (Figure 3).

4.2. Geochemistry of REE Mineralization in Sweden Inferred from the EURARE Project

During the EURARE field campaign, approximately 200 samples were taken for petrography and mineralogy studies and for geochemical analysis. The location of the samples and type of mineralization are shown in Figure 4, and the geochemical results are provided in Table 1.

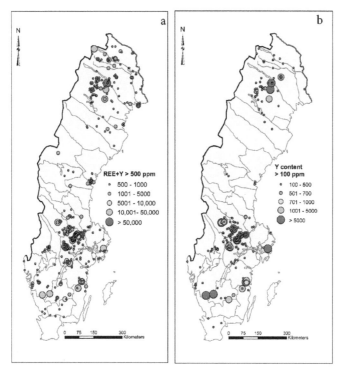

Figure 3. Simplified map of lithogeochemical data in Sweden: (**a**) total REE + Y > 500 ppm and (**b**) Y content >100 ppm (source of data: lithogeochemical database at the Geological survey of Sweden).

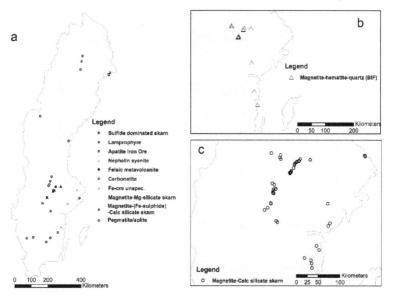

Figure 4. Simplified location map of samples in the EURARE project: (**a**) an overview of sample locations and types, excluding magnetite-calc silicate skarn and BIF ones; (**b**) magnetite-hematite-quartz samples; (**c**) magnetite-calc silicate skarn samples.

Table 1. Summary of samples analyzed with a total REE (TotREE) content above 1% (10,000 ppm). N and E are SwerefTM coordinates.

Occurrence/Deposit	Sample	N	E	Type	TotREE (ppm)	LREE (ppm)	HREE (ppm)	Th (ppm)	U (ppm)	P (ppm)
Ingridstorp	KES155002A	6,392,246	559,703	Pegmatite/aplite	47,172	43,007	4165	>1000	266	70
Stålklockan	PNY130005	6,635,349	532,852	Magnetite-Calc silicate skarn	45,512	34,001	11,511	2.01	58.5	90
Tåresåive	ING090076A	7,421,196	709,317	Pegmatite/aplite	42,651	33,056	9596	>1000	1505	140
Bastnäs	BAST140001	6,634,470	532,926	Magnetite-Calc silicate skarn	38,078	33,149	4929	20.7	66.4	70
Bastnäs	PNY130001	6,634,425	533,003	Magnetite-Calc silicate skarn	37,907	33,134	4773	9.64	92	60
Gyttorp	Gyttorp	6,597,427	497,660	Magnetite-Calc silicate skarn	37,069	33,233	3836	50.4	1120	<10
Mörkens	MÖRK2	6,662,160	577,465	Magnetite-Calc silicate skarn	36,921	33,225	3697	3.36	18.85	40
Djupedalsgruvan	JSM150007A	6,425,442	578,440	Fe-ore unspec.	34,643	22,746	11,898	158	9	2640
Ingridstorp	KES155003A	6,358,023	372,718	Pegmatite/aplite	33,895	19,577	14,318	>1000	1330	1260
Ytterby	YBY140002	6,592,332	690,226	Pegmatite/aplite	30,624	13,624	17,000	>1000	1185	6200
Rödbergsgruvan	Rödbergsgruvan A	6,597,194	494,237	Magnetite-Calc silicate skarn	30,536	29,379	1158	1.53	4.15	80
Östanmossa	PNY130015	6,660,543	551,764	Magnetite-Calc silicate skarn	30,319	28,850	1469	1.21	3.7	40
Djupedalsgruvan	JSM150008A	6,425,442	578,440	Fe-ore unspec.	29,990	16,626	13,365	214	8.39	3360
Holmtjärn	HOLM140001	6,691,407	514,383	Pegmatite/aplite	29,550	12,550	17,000	>1000	>10,000	450
Ytterby	YBY140003	6,592,332	690,226	Pegmatite/aplite	26,419	10,239	16,180	>1000	1270	1470
Djupedalsgruvan	JSM150006A	6,425,442	578,440	Fe-ore unspec.	25,904	11,133	14,771	134	21.5	690
Mörkens	MÖRK3	6,662,160	577,465	Magnetite-(Fe-sulphide)-Calc silicate skarn	24,971	21,636	3335	0.43	2.67	30
Gruvhagen	GRUVHAG1	6,662,377	577,026	Magnetite-Calc silicate skarn	24,574	23,618	956	3.34	0.77	10
Ingelsbo	KES155006A	6,364,317	411,300	Pegmatite/aplite	22,192	5192	17,000	>1000	>10,000	<10
Ytterby	YBY140001	6,592,332	690,226	Pegmatite/aplite	18,695	3801	14,894	>1000	2360	6280
Reunavaare	REU140001	7,342,033	697,981	Pegmatite/aplite	18,207	13,158	5049	>1000	438	30
Tybble	SGUR10007	6,517,096	516,016	Iron oxide apatite	17,428	17,274	154	6.41	457	1400
Ingridstorp	KES155001A	6,392,246	559,703	Pegmatite/aplite	16,320	16,026	294	257	53.9	90
Flakaberget	FLA140001	7,389,427	703,658	Pegmatite/aplite	15,556	1627	13,929	>1000	1600	520
Johanna	PNY130013	6,658,660	550,883	Magnetite-Calc silicate skarn	14,008	12,966	1042	0.72	8.68	30
Sveafallen	SVEAFALLEN	6,563,279	467,518	Pegmatite/aplite	13,220	12,579	641	>1000	16	1460
Mörkens	MÖRK1	6,662,160	577,465	Magnetite-(Fe-sulphide)-Calc silicate skarn	11,077	9203	1874	0.09	2.47	40

The geochemical analyses of the samples collected during the EURARE project show that 68 samples have a total REE content above 1000 ppm and of those 27 have more than 1% total REE.

The samples with more than 1% REEs represent mineralizations classified as: iron oxide apatite (1 sample), Fe-ore unspecified (3 samples), magnetite-calc-silicate skarn (9 samples), magnetite-(Fe-sulphide)-Calc silicate skarn (2 samples), and granitic pegmatite (12 samples). These results are summarised in Table 1.

A bivariate plot of total REE content vs. LREE (Figure 5a) and vs. HREE (Figure 5b) show that there is a general bimodal linear correlation between total REE and LREE (Figure 5a), one trend is characterized by LREE enrichment (for magnetite-skarn mineralization) and a second one by LREE depletion, especially in some late magmatic pegmatite REE mineralizations and some related to iron mineralizations. The comparison of total REEs with HREEs also shows clear decoupling between two groups, the first composed of low HREE mineralization types (e.g., various iron ores, skarn mineralizations, carbonatite, and sulphide-dominated skarn ores) and these with elevated HREE content are mainly represented by late magmatic pegmatites and unspecified iron ore types. It is unclear how the magmatic rocks differ in primary composition, so they host both low- and high-level HREE mineralizations. BIF deposits (magnetite-hematite-quartz) do not show a very high content of REE_{tot} and generally are most likely to have elevated LREE rather than HREE. The number of two samples of sulfide dominated skarn is limited and, therefore, it is excluded from interpretation of the results to avoid any over interpretation.

A bivariate plot of La vs. La/Yb (Figure 6) shows a weak but overall positive correlation between the two variables, except for a few samples, which show higher HREE contents. This suggests that those samples with a high LREE content also have relatively high LREE/HREE ratios. They mostly represent REE mineralizations associated with granitic pegmatites. One sample from iron oxide-apatite deposits contains REE-bearing minerals such as xenotime and monazite and shows higher contents of HREEs (Figure 6).

Thorium and uranium content and their ratio are useful for recognizing geochemical facies and also the content of radioactive elements in different types of mineralization which may have an impact on future exploitation. Figure 7a shows Th content versus total REEs. In general, skarn and iron oxide-apatite mineralizations with higher contents of total REEs (REE_{tot} > 1%) show a positive correlation with low contents of Th (Th < 50ppm). A few samples from iron oxide mineralizations show higher content of Th up to 200 ppm and total REE contents of up to 3.5%. Two samples of granitic pegmatites show higher contents of Th, whereas those of total REEs are low, which means that the content of Th in those samples is not directly related to REE-bearing minerals and may also correspond to the primary origin of the host rocks and its later deformation. For example, rocks with sedimentary origin have usually higher Th/U ratios than magmatic rocks, and these ratios can be further modified during regional metamorphism [11,12,49].

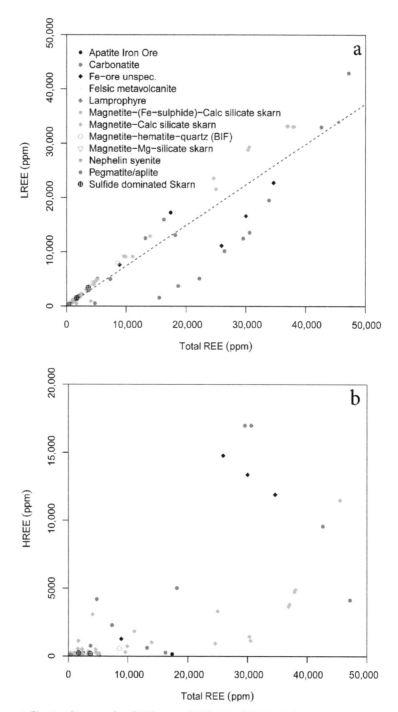

Figure 5. Bivariate diagrams of total REEs versus LREEs (**a**) and HREEs (**b**) for EURARE project samples.

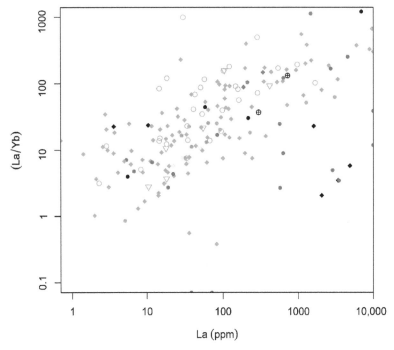

Figure 6. Bivariate plot of La (ppm) versus La/Yb for the EURARE samples (see the legend in Figure 5).

The U/total REE (Figure 7b) of the EURARE samples shows that high U contents (U > 400 ppm) are typical of REE mineralization hosted in granitic pegmatites. An additional few samples collected from magnetite-calc silicate skarn type deposits show higher contents of U (>1000 ppm). In the sample from magnetite-calc silicate skarn, one sample shows a high content of total REE correlates with high U content which may indicate the U enrichment in REE-bearing minerals.

Comparison of total REE with Th/U ratios based on two analyzed samples from lamprophyres (Figure 7c) shows that the samples with low contents of total REE (REE$_{tot}$ < 400) typically have higher Th than U contents (Th/U > 20). The Th/U ratio is generally low in the samples from skarn and iron oxide-apatite REE mineralizations (Th/U < 10) with the exception of four samples from iron oxide-apatite REE mineralizations (at the Djupedalsgruvan) that show high contents of both Th and REE (and contain xenotime and allanite minerals) (Figure 7c).

Plots of quantitative values of selected REEs (Nd, Pr, and Y) and phosphorus compared to binary plots of LREE versus HREE show at least two contradicting trends. For example, in Figure 8a samples with high Nd contents show two opposite trends; one positive with well correlated LREE and Nd contents accompanied by low HREE and the second negative trend where samples with decreasing Nd contents plot towards a lower LREE content and a higher HREE. Few exceptions of samples with high content of HREE and Nd content (>4000 ppm) can also be observed (Figure 8a).

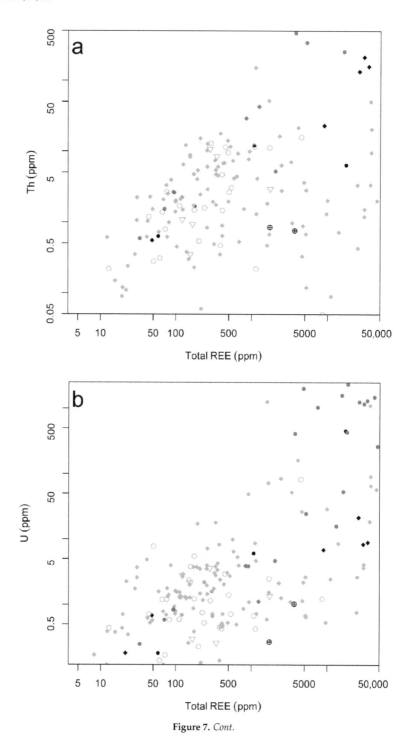

Figure 7. *Cont.*

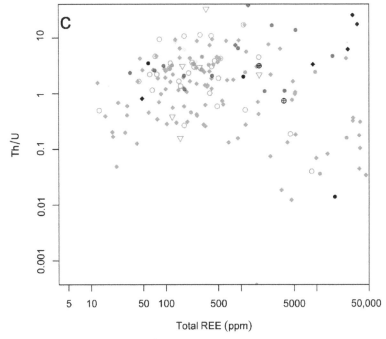

Figure 7. Binary plots of (**a**) total REE vs. Th (ppm); (**b**) total REE vs. U (ppm); (**c**) total REE vs. Th/U; for the EURARE samples (see the legend in Figure 5).

In a similar type of plot, quantitative content of Pr in the samples shows a semi-concordant trend with Nd, and this may reflect that the minerals bearing Pr and Nd are the ones containing HREE, such as xenotime-(HREE) and fergusonite-(HREE) as well as gadolinite-(REE).

Figure 8c shows linear positive correlation of high Y samples with a higher content of HREE. Since Y has a similar geochemical behavior as HREE the low HREE contents follow low Y contents in EURARE samples. High LREE samples are less affected by Y content with weak negative correlation. Samples with low and high Y content when compared to LREE might reflect various degrees of host magma differentiation or secondary remobilization during regional metamorphism (competitive re-crystallization and crystallization of metamorphic minerals) [11,12,49].

Comparison of phosphorus content, indicative of the presence of phosphate minerals shows two different clusters (Figure 8d). The first cluster shows that some high P samples have low contents of both LREEs and HREEs, which can be interpreted as apatite mineralization in iron ore with no REE enrichment. The second cluster shows that samples with high P content correlate with high HREE content, which may indicate the presence of minerals other than apatite as a host of HREEs, such as xenotime.

Within the sample set collected during the EURARE project, there are several iron ore samples (Fe-ore unspec.) that cannot be easily classified (more mineralogical and petrological investigations are needed). In an attempt to classify them, the classical PM-normalized REE diagram has been used [37] (McDonough and Sun 1995) (Figure 9). At a first glance, their REE patterns suggest the presence of two genetic groups. The first group with REE enrichment (>100 ppm) shows a similar trend to the iron oxide-apatite- and skarn-type REE mineralizations, with a negative Eu anomaly and relative HREE enrichment. The second group shows a weak positive Eu anomaly and steep decreasing trends from LREEs to HREEs accompanied by a rather low total of REEs (<100 ppm). These two samples may have an alkaline or peralkaline origin but a more detailed study is needed (Figure 9).

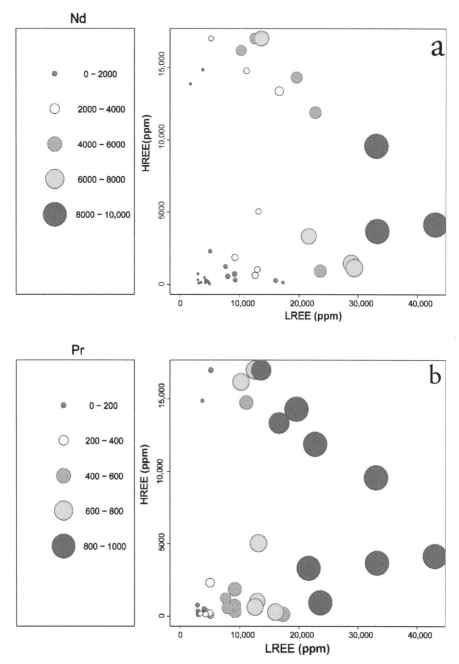

Figure 8. *Cont.*

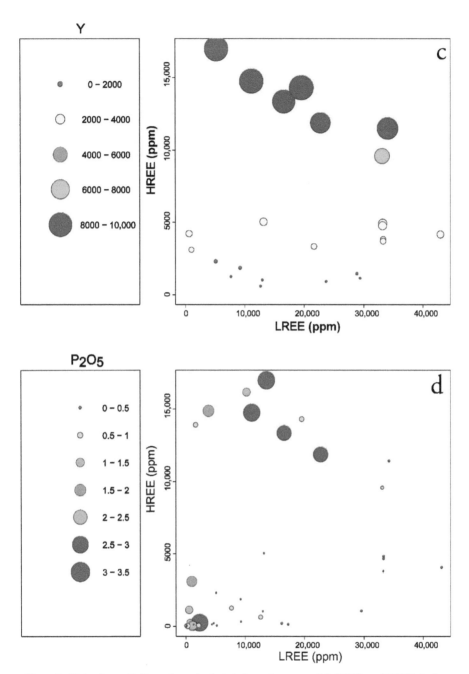

Figure 8. Plots of quantitative values of selected elements compared to LREEs and HREEs in the EURARE samples: (**a**) Nd (ppm); (**b**) Pr (ppm); (**c**) Y (ppm); (**d**) P$_2$O$_5$ (wt%). Samples with contents less than background values are excluded.

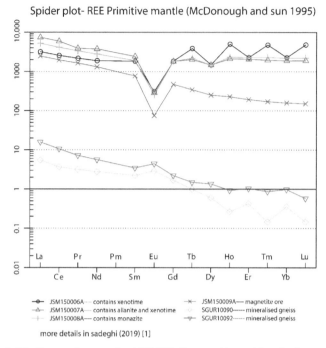

Figure 9. Primitive mantle-normalized [37] REE diagram of iron oxides of unknown origin.

4.3. Principal Component (PC) Analysis

The results of PC analysis are reported in Table 2 for the PCs with eigenvalue >1 and show that PC1 and PC2 explain 30% and 17%, respectively, of the total variance among the REE-mineralized samples. It implies that in PC1 with an association of REEs, and Co, Fe, and Mn most likely represent magnetite calc-silicate skarn samples, while an association of Cs-Hf-Nb-Sr-Ta-Th-Zr-Al-Ba in PC2 likely indicates REE mineralizations related to granitic pegmatite bearing Nb-Ta mineralization.

Table 2. Principal component loading, eigenvalues, % variance explained, and cumulative % variance for PC1 to PC8.

Value Number	Eigenvalues of Correlation Matrix and Related Statistics			
	Eigenvalue	% Total (Variance)	Cumulative % (Eigenvalue)	Cumulative % Variance
PC1	17.56	29.77	17.56	29.77
PC2	9.99	16.93	27.55	46.69
PC3	4.65	7.88	32.20	54.57
PC4	3.41	5.78	35.61	60.35
PC5	2.89	4.90	38.50	65.25
PC6	2.20	3.74	40.70	68.98
PC7	1.94	3.30	42.64	72.27
PC8	1.86	3.16	44.51	75.43

The PC3 defines an association of Cs-Rb-K-Li, which probably represents alkaline and/or felsic rocks including granitic pegmatite and related polymetallic mineralization (Table 3).

Table 3. Explanation of the five principal components for elements analyzed, accounting for 65% of the total variance.

Component	Centered Log-Ratio (clr)-Transformed Data		
	% of Variance Explained	Association	Interpretation
PC1	29.77	(i) REEs (ii) Fe-Mn	(i) REEs associations (ii) Magnetite calc-silicate skarn
PC2	16.93	(i) REEs (ii) Cs-Hf-Nb-Sr-Ta-Th-Zr-Al-Ba	(i) REEs associations (ii) Granitic pegmatit-bearing Nb-Ta mineralization
PC3	7.88	(i) Cs-Rb-K-Li (ii) Ag-Cd-Bi-Co-Cu-Zn-C	(i) alkaline/granitic rocks (ii) Base metals component
PC4	5.78	(i) C-Hf-Sr-Ba-Ca (ii) Fe- -Mg-Tl	(i) Carbonatite/lamprophyr? (ii) Magnetite calc-silicate skarn?
PC5	4.9	(i) Hf-Zr-In (ii) V-P-Ni-Cr	(i) Felsic component (ii) Mafic component

Plots of pairs of PCs show that it is likely possible to distinguish between some different types of REE mineralization or to distinguish between samples with a higher content of REEs and those with low one (Figure 10). For example, in the plot of PC1 versus PC2 (Figure 10) there are several clusters representing an association of REEs and granitic pegmatite bearing Nb-Ta mineralization. A plot of PC3 versus PC4 distinguishes between REE mineralization associated with magnetite calc-silicate skarn, alkaline, and/or granitic rocks and polymetallic mineralization (Figure 10).

4.4. Comparison of REE Till Geochemistry with Lithogeochemistry of Underlying Bedrock

Table 4 shows simple statistics of REE distribution in till. The median values are similar to upper crust values by Rudnick and Gao [49]. Due to the fact that Rudnick and Gao's values are calculated for total contents and till results originate from aqua regia digestion, one of the conclusions is that the Swedish till is enriched in REEs in comparison to the global values.

Table 4. Percentiles, minimum, and maximum values for the REE + Y + Sc from the till dataset. P50 is a median value. * abundance in the upper continental crust [49].

Element	n = 2578	UCC *	Min	P25	P50	P75	P90	P99	Max
La	ppm	31	8.1	25.6	31.6	39.6	49.9	85.5	198.9
Ce	ppm	63	16.3	60.1	74.8	96.0	123.7	211.8	388.1
Pr	ppm	7.1	1.9	6.3	7.7	9.4	11.8	20.6	38.1
Nd	ppm	27	7.0	23.3	28.5	34.9	43.4	73.1	129.3
Sm	ppm	4.7	1.4	4.4	5.4	6.7	8.3	13.3	24.2
Eu	ppm	1.0	0.1	0.6	0.8	1.0	1.2	1.9	5.5
Gd	ppm	4.0	1.1	3.6	4.4	5.4	6.6	10.4	22.3
Tb	ppm	0.7	0.2	0.5	0.6	0.8	1.0	1.5	3.6
Dy	ppm	3.9	0.9	2.9	3.5	4.3	5.3	8.5	22.6
Ho	ppm	0.8	0.1	0.6	0.7	0.9	1.1	1.7	4.8
Er	ppm	2.3	0.4	1.5	1.9	2.3	3.0	4.6	14.3
Tm	ppm	0.3	0.1	0.2	0.3	0.3	0.4	0.6	2.0
Yb	ppm	2.0	0.3	1.3	1.7	2.1	2.7	4.2	14.6
Lu	ppm	0.3	0.0	0.2	0.2	0.3	0.4	0.6	2.6
Y	ppm	21	3.7	14.9	18.3	22.6	27.9	47.1	163.2
Sc	ppm	14	0.6	3.5	4.5	5.6	6.8	9.6	21.2

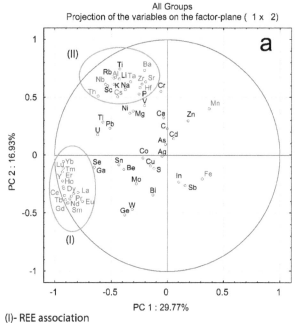

(I)- REE association
(II)- Granitic pegmatite bearing Nb-Ta mineralisation

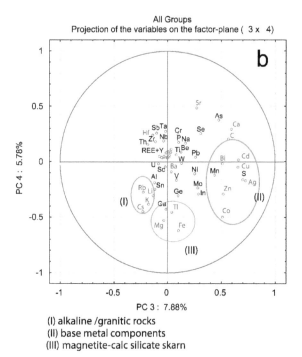

(I) alkaline /granitic rocks
(II) base metal components
(III) magnetite-calc silicate skarn

Figure 10. Biplots of (**a**) PC1 vs. PC2; (**b**) PC3 vs. PC4.

Maps of LREE and HREE contents in till (see [3]) show both similar and distinct distribution patterns. Generally, LREE (La-Sm) and HREE groups (Gd-Lu, Y) have similar distribution within the group, therefore, we use here the La till map as a representative for all LREEs and the Yttrium map as a representative for HREEs (Figure 11). Raster La and Y maps with regional anomalies show large-scale REE trends, while plotted lithogeochemistry results (circle symbols) underlie local spots and extreme values.

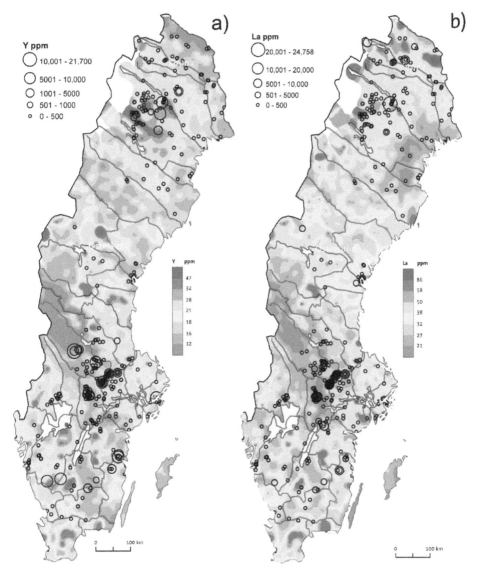

Figure 11. Geochemical maps comparing till geochemistry (raster, after Andersson et al., 2014 [3]) with lithogeochemistry (circle symbols) in Sweden: (**a**) Y in till and bedrock; (**b**) La in till and bedrock.

LREEs and Eu have major anomalies in northernmost Sweden (outlying Archean lithologies of magmatic origin) while these anomalies are absent in HREEs. Minor enrichment in LREEs visible in lithogeochemistry data reflects possible LREE enrichment in granite-pegmatite bedrock and apatite-iron deposits near Kiruna.

Further south, in central Norrbotten (Figures 1 and 11), the bedrock (granite-pegmatite and felsic metavolcanics rocks) and overlying till are highly enriched both in LREE and HREE and correlation with many REE (+W, Mn, and U) mineralizations, e.g., Tåresåive (REEs, Mo, U, Th, Nb, and Ta).

In central Sweden, LREE and HREE enrichment in till occurs in the Caledonides within the tectonic windows where many sulfide mineralizations and U deposits have been mapped. LREE enrichment at the Caledonian front and within the Lower Allochthon can be attributed to the REE enrichment in the fine-grained sedimentary rocks (black shale and phosphorite), e.g., Tåsjö deposit (Figures 1 and 11). On the Fennoscandian Shield side, few REE mineralizations are known and they possibly may be reflected in till with its prominent anomaly south of Östersund, possibly in relation to the so-called Rätan granite. NE from Östersund, in central Sweden, -there is a prominent HREE anomaly and it is probably related to the outcrops of younger granite (so-called Revsund granite). The lithogeochemistry database has poor coverage in central Sweden, however, Alnö carbonatitic intrusion by the coast is both visible in till and few collected rock samples with elevated LREE contents. In the Ljusdal (Ljusdal lithotectonic unit, Figure 1), high HREE (and elevated LREE) contents can be noticed in till and several small REE (W, Mn, U, and Fe-Mn) mineralizations are known in direct relation to the anomaly extension. Lithogeochemistry data are missing in this region.

In the Bergslagen ore district, there is a general decoupling between REE lithogeochemistry and REE content in till. The known REE mineralization line in western Bergslagen is not visible in till geochemistry. Point REE till anomalies can be observed in the NW part of Bergslagen (often related to iron mineralizations) and LREE high contents in till occur in southern Bergslagen by the coast.

In southern Sweden, LREE anomalies in the bedrock are often related to mineralizations and granitic lithologies. Younger granites of the Blekinge lithotectonic unit (Figure 1) and fluorite mineralization in southern Sweden also seem to be well reflected in till geochemistry. High contents of Y in bedrock and till in southern Sweden, south and south-west of the lake Vättern (Figure 11) seem to correlate with occurrences of highly metamorphosed rocks including occurrences of eclogites with Y-enriched garnet and pyroxene. Minor REE (W, Mn, and U) occurrences are known north of the Y till anomaly.

The comparison of till geochemical maps with lithogeochemistry and mineralizations shows clear correlation between REE anomalies in surficial deposits and the underlying bedrock and the presence of mineralization. Therefore, till can be effectively used for REE mineral exploration and mineral potential studies in areas where there is no sufficient information about the bedrock and mineralizations.

5. Conclusions

The analysis of REE geochemistry in mineralized samples compared to bedrock lithogeochemistry and till geochemistry reveals several regional trends in REE distribution in Sweden, which may have practical implications for mineral exploration:

1. Lithogeochemistry of the bedrock correlates with geochemical REE provinces and even single REE ore deposits;
2. REE geochemistry of mineralized EURARE samples allows geochemical classification of various REE occurrences in Sweden;
3. Till geochemistry presents the complex picture of REE distribution at the country scale. Many known REE locations are underlain by till anomalies. On the other hand, there are many high REE spots in till, which cannot be easily explained by known REE deposits. They may present interesting exploration targets;
4. Taking into account favorable geological and metallogenetic conditions in Sweden, there is a good potential for discovering new REE deposits;

5. Applied geochemical methods should be integrated in a multidisciplinary way with complementary data and methods, such as mineralogy, geophysics, and numerical modeling and statistics, to improve understanding of REE ore genesis and related geo-modeling. This holistic approach, also including ore geochemistry and mineralogy, will provide a well-documented REE mineral knowledge base, intelligence, and exploration methodology.

Author Contributions: Conceptualization, M.S. and N.A.; methodology, M.S. and A.L.; software, M.S. and A.L.; validation, M.S., A.L. and N.A. and.; formal analysis, M.S. and N.A.; investigation, M.S.; resources, M.S.; data curation, M.S.; writing—original draft preparation, M.S and A.L.; writing—review and editing, N.A. and A.L.; visualization, M.S and A.L. All authors have read and agreed to the published version of the manuscript.

Funding: The EURARE project was funded by the European Community Seventh Framework Programme (FP7/2007-2013) under grant agreement no. 309373. The project, number 35242, was co-funded by the Geological Survey of Sweden (SGU).

Acknowledgments: We thank Erik Jonsson, Torbjörn Bergman, Magnus Ripa, Per Nysten, Johan Söderhielm, and Dick Claeson for their contributions to the EURARE project. The EURARE project was funded by the European Community Seventh Framework Programme (FP7/2007–2013) under grant agreement no. 309373. The project, number 35242, was co-funded by the Geological Survey of Sweden (SGU). The authors thank two anonymous reviewers for their constructive and valuable comments and suggestions.

Conflicts of Interest: The authors declare no conflict of interest.

References

1. Sadeghi, M. (Ed.) Rare earth elements distribution, mineralisation and exploration potential in Sweden. In *Sveriges Geologiska Undersökning, Rapporter Och Meddelanden*; Geological Survey of Sweden: Uppsala, Sweden, 2019; Volume 146, p. 184.

2. Sadeghi, M.; Andersson, M. *Sällsynta Jordartsmetaller i Sverige, Förekomst och Utbredning i berg och jord "the Rare Earth Element Distribution over Sweden" (in Swedish and English)*; Report No 2015-21; Geological Survey of Sweden: Uppsala, Sweden, 2015; p. 98.

3. Andersson, M.; Carlsson, M.; Ladenberger, A.; Morris, G.; Sadeghi, M.; Uhlbäck, J. *Geochemical Atlas of Sweden (Geokemisk Atlas over Sverige)*; Sveriges Geologiska Undersökning: Uppsala, Sweden, 2014; p. 210. ISBN 978-91-7403-258-1.

4. Sadeghi, M.; Morris, G.A.; Carranza, E.J.M.; Ladenberger, A.; Andersson, M. Rare earth element distribution and mineralisation in Sweden: An application of principal component analysis to FOREGS soil geochemistry. *J. Geochem. Exp.* **2013**, *133*, 160–175. [CrossRef]

5. Sadeghi, M.; Petrosino, P.; Ladenberger, A.; Albanese, S.; Andersson, M.; Morris, G.; Lima, A.; De Vivo, B. Ce, La and Y concentration in agricultural and grazing land soils of Europe—An exercise for mapping and exploration. *J. Geochem. Exp.* **2013**, *133*, 202–213. [CrossRef]

6. Hudson, M. *2007: Drilling Expands Near-Surface Uranium at Tåsjö, Sweden*; Mawson Resources Ltd.: Vancouver, BC, Canada, 2007.

7. London, D. Rare-element granitic pegmatites. In *Reviews in Economic Geology*; Society of Economic Geologists, Inc.: Littleton, CO, USA, 2016; Volume 18, pp. 165–193.

8. Sundius, N. Kvarts, fältspat och glimmer samt förekomster där av i Sverige. In *Sveriges Geologiska Undersökning C*; Geological Survey of Sweden: Uppsala, Sweden, 1952; Volume 520, p. 231.

9. Lundegårdh, P.H. *Nyttosten i Sverige*; Almqvist & Wiksell: Stockholm, Sweden, 1971; p. 271.

10. Holmqvist, A. Be-F-Nb-REE-Sn-Ta-Th-U-Zr-mineraliserad granit i Västsverige. In *Sveriges Geologiska AB, Division Prospektering, PRAP-Rapport 89026*; Geological Survey of Sweden: Uppsala, Sweden, 1989; p. 14 + appendices.

11. Dostal, J. Rare metal deposits associated with alkaline/peralkaline igneous rocks. In *Reviews in Economic Geology*; Society of Economic Geologists, Inc.: Littleton, CO, USA, 2016; Volume 18, pp. 33–54.

12. Verplanck, P.L.; Mariano, A.N.; Mariano, A., Jr. Rare earth element ore geology of carbonatites. In *Reviews in Economic Geology*; Society of Economic Geologists, Inc.: Littleton, CO, USA, 2016; Volume 18, pp. 5–32.

13. Mariano, A.N. Nature of economic mineralization in carbonatites and related rocks. In *Carbonatites: Genesis and Evolution*; Bell, K., Ed.; Unwin Hyman: London, UK, 1989; pp. 149–176.

14. Kresten, P. A magnetometric survey of the Alnö complex. *Geol. Fören. Stockh. Förh.* **1976**, *98*, 364–365. [CrossRef]

15. Kresten, P. The Alnö complex: Discussion of the main features, bibliography and excursion guide. In Proceedings of the Nordic Carbonatite Symposium, Sundsvall, Sweden, 20–27 May 1979; p. 67.

16. Kresten, P. The Alnö area (Alnöområdet). In *Beskrivning till Berggrundskartan över Västernorrlands län*; Lundqvist, T., Gee, D., Kumpulainen, R., Karis, L., Kresten, P., Eds.; Sveriges Geologiska Undersökning ser Ba nr 31: Uppsala, Sweden, 1990; pp. 238–278.

17. Hornig-Kjarsgaard, I. Rare earth elements in sövitic carbonatites and their mineral phases. *J. Petrol.* **1998**, *39*, 2105–2121. [CrossRef]

18. Svensson, U. *Sällsynta Jordartsmetaller på Alnön. Boliden Mineral AB, Årsrapport 1971, Bilaga 36*; Boliden Mineral AB: Boliden, Sweden, 1972; p. 5 + appendix (unpublished report).

19. Chai, F.; Zhang, Z.; Mao, J.; Parat, A.; Wang, L.; Dong, L.; Ye, H.; Chen, l.; Zheng, R. Lamprophyre or Lamproite Dyke in the SW Tarim Block?—Discussion on the Petrogenesis of These Rocks and Their Source Region. *J. China Univ. Geosci.* **2006**, *17*, 13–24. [CrossRef]

20. Kresten, P.; Åhman, E.; Brunfelt, A.O. Alkaline ultramafic lamprophyres and associated carbonatite dykes from the Kalix area, northern Sweden. *Geol. Rundsch.* **1981**, *70*, 1215–1231. [CrossRef]

21. Kresten, P.; Rex, D.C.; Guise, P.G. 40Ar-39Ar ages of ultramafic lamprophyres from the Kalix area, northern Sweden. In *Radiometric Dating Results 3. SGU C830*; Lundqvist, T., Ed.; Geological Survey of Sweden: Uppsala, Sweden, 1997.

22. Jonsson, E. Epigenetic REE-U-Th-anomalous Fe oxide mineralisation in the Narken area, NE Sweden. In Proceedings of the 33rd International Geological Congress, Oslo, Norway, 6–14 August 2008. Abstracts, MRD-11.

23. Geijer, P. *The Iron Ores of the Kiruna Type: Geographical Distribution, Geological Characters, and Origin*; Sveriges Geologiska Undersökning C: Uppsala, Sweden, 1931; Volume 367, p. 39.

24. Geijer, P. *The Rektorn Ore Body at Kiruna*; Sveriges Geologiska Undersökning C: Uppsala, Sweden, 1950; Volume 514, p. 18.

25. Parák, T. Kiruna iron ores are not "intrusive-magmatic ores of the Kiruna type". *Econ. Geol.* **1975**, *70*, 1242–1258. [CrossRef]

26. Frietsch, R. On the magmatic origin of iron ores of the Kiruna type. *Econ. Geol.* **1978**, *73*, 478–485. [CrossRef]

27. Hitzman, M.W.; Oreskes, N.; Einaudi, M.T. Geologic characteristics and tectonic setting of Proterozoic iron oxide (Cu-U-Au-REE) deposits. *Precambrian Res.* **1992**, *58*, 241–287. [CrossRef]

28. Nyström, J.O.; Henriquez, F. Magmatic features of iron ores of the Kiruna type in Chile and Sweden: Ore textures and magnetite geochemistry. *Econ. Geol.* **1994**, *89*, 820–839. [CrossRef]

29. Williams, P.J.; Barton, M.D.; Johnson, D.A.; Fontboté, L.; de Haller, A.; Mark, G.; Oliver, N.H.S.; Marschik, R. Iron oxide copper-gold deposits: Geology, space-time distribution, and possible modes of origin. In *Economic Geology, 100th Anniversary Volume*; Society of Economic Geologists: Littleton, CO, USA, 2005; pp. 371–405.

30. Jonsson, E.; Troll, V.R.; Högdahl, K.; Harris, C.; Weis, F.; Nilsson, K.P.; Skelton, A. Magmatic origin of giant central Swedish "Kiruna-type" apatite-iron oxide ores. *Sci. Rep.* **2013**, *3*, 1–8. [CrossRef] [PubMed]

31. Bergman, S.; Kübler, L.; Martinsson, O. *Description of Regional Geological and Geophysical Maps of Northern Norrbotten County (East of the Caledonian Orogen)*; Sveriges Geologiska Undersökning Ba: Uppsala, Sweden, 2001; Volume 56, p. 110.

32. Harlov, D.E.; Andersson, U.B.; Förster, H.-J.; Nyström, J.O.; Dulski, P.; Broman, C. Apatite-monazite relations in the Kiirunavaara magnetite-apatite ore, northern Sweden. *Chem. Geol.* **2002**, *191*, 47–72. [CrossRef]

33. Jonsson, E.; Harlov, D.; Majka, J.; Högdahl, K.; Persson-Nilsson, K. Fluorapatite-monazite-allanite relations in the Grängesberg apatite-iron oxide ore district, Bergslagen, Sweden. *Am. Mineral.* **2016**, *101*, 1769–1782. [CrossRef]

34. Pan, Y.; Fleet, M.E. Composition of the fluorapatite-group minerals: Substitution mechanisms and controlling factors. *Rev. Mineral. Geochem.* **2002**, *48*, 13–49. [CrossRef]

35. Jonsson, E.; Persson Nilsson, K.; Hallberg, A.; Högdahl, K. The Palaeoproterozoic apatite-iron oxide deposits of the Grängesberg area: Kiruna-type deposits in central Sweden. In *NGF Abstracts and Proceedings, 29th Nordic Geological Winter Meeting, Oslo, Norway*; Nakrem, H.A., Harstad, A.O., Haukdal, G., Eds.; Norsk Geologisk Forening: Trondheim, Norway, 2010; Volume 1, pp. 88–89.

36. Lundh, J. A Lithogeochemical Study of Northern Sweden and the Kiruna and Malmberget Iron-Apatite Ore Deposits (Dissertation). 2014. Available online: http://urn.kb.se/resolve?urn=urn:nbn:se:uu:diva-227039 (accessed on 24 June 2014).

37. McDonough, W.F.; Sun, S.S. The composition of the Earth. *Chem. Geol.* **1995**, *120*, 223–253. [CrossRef]
38. Hallberg, A.; Albrecht, L.; Jonsson, E.; Olsson, A. The Grängesberg apatite-iron deposit—biggest in Bergslagen, Sweden. In Proceedings of the 27th Nordic Geological Winter Meeting, Bulletin of the Geological Society of Finland, Oulu, Finland, 9–12 January 2006; Special issue 1. p. 45.
39. Samson, I.M.; Wood, S.A. The rare earth elements: Behavior in hydrothermal fluids and concentration in hydrothermal mineral deposits, exclusive of alkaline setting. In *Rare elements Geochemistry and Mineral Deposits, Geological Association of Canada short course Notes*; Linnen, R.I., Samson, I.M., Eds.; Society of Economic Geologists: Littleton, CO, USA, 2005; Volume 17, pp. 269–297.
40. Oreskes, N.; Einaudi, M.T. Origin of rare earth element-enriched hematite breccias at the Olympic Dam Cu-U-Au-Ag deposit, Roxby Downs, South Australia. *Econ. Geol.* **1990**, *85*, 1–28. [CrossRef]
41. Lottermoser, B.G. Rare earth elements in Australian uranium deposits. In *Uranium–Past and Future Challenges, Proceeding of the 7th International Conference on Uranium Mining and Hydrogeology*; Merkel, B.J., Arab, A., Eds.; Springer: Berlin, Germany, 1995; pp. 25–30.
42. Holtstam, D.; Andersson, U.B. The REE minerals of the Bastnäs-type deposits, south-central Sweden. *Can. Mineral.* **2007**, *45*, 1073–1114. [CrossRef]
43. Janoušek, V.; Farrow, C.M.; Erban, V. Interpretation of whole-rock geochemical data in igneous geochemistry: Introducing Geochemical Data Toolkit (GCDkit). *J. Petrol.* **2006**, *47*, 1255–1259. [CrossRef]
44. Carranza, E.J.M. Geochemical Anomaly and Mineral Prospectivity Mapping in GIS. In *Handbook of Exploration and Environmental Geochemistry*; Elsevier: Amsterdam, The Netherlands, 2008; Volume 11.
45. Grunsky, E.C. The interpretation of geochemical survey data. *Geochem. Exp. Environ. Anal.* **2010**, *10*, 27–74. [CrossRef]
46. Güller, C.; Thyne, G.; McGray, J.E.; Turner, A.K. Evaluation of graphical and multivariate statistical methods for classification of water chemistry data. *Hydrogeol. J.* **2002**, *10*, 455–474. [CrossRef]
47. Harris, J.R.; Grunsky, E.C.; Wilkinson, L. Developments in the effective use of lithogeochemistry in regional exploration programs: Application of GIS technology. In *Proceedings of the Exploration 97, Fourth Decennial International Conference on Mineral Exploration*; Gubins, A.G., Ed.; Prospectors and Developers Association of Canada: Toronto, ON, Canada, 1997; pp. 285–292.
48. Hode Vuorinen, J.; Hålenius, U.; Whitehouse, M.J.; Mansfeld, J.; Skelton, A.D.L. Compositional variations (major and trace elements) of clinopyroxene and Ti-andradite from pyroxenite, ijolite and nepheline syenite, Alnö Island, Sweden. *Lithos* **2005**, *81*, 55–77. [CrossRef]
49. Rudnick, R.L.; Gao, S. Composition of the Continental Crust. In *The Crust*; Rudnick, R.L., Ed.; Elsevier-Pergamon: Oxford, UK, 2003; pp. 1–64. [CrossRef]

Article

Effect of Mineralogy on the Beneficiation of REE from Heavy Mineral Sands: The Case of Nea Peramos, Kavala, Northern Greece

Christina Stouraiti [1,*], **Vassiliki Angelatou** [2], **Sofia Petushok** [1], **Konstantinos Soukis** [1] **and Demetrios Eliopoulos** [2]

[1] Faculty of Geology and Geoenvironment, National and Kapodistrian University of Athens, 15784 Athens, Greece; sofpetushok@geol.uoa.gr (S.P.); soukis@geol.uoa.gr (K.S.)

[2] Department of Mineral Processing, Institute of Geology and Mineral Exploration, 13677 Acharnes, Greece; vasaggelatou@igme.gr (V.A.); deliopoulos1000@gmail.com (D.E.)

* Correspondence: chstouraiti@geol.uoa.gr; Tel.: +30-210-727-4941

Received: 24 February 2020; Accepted: 22 April 2020; Published: 26 April 2020

Abstract: Beneficiation of a rare earth element (REE) ore from heavy mineral (HM) sands by particle size classification in conjunction with high-intensity magnetic separation (HIMS) was investigated. The HM sands of Nea Peramos, Kavala, Northern Greece, contain high concentrations of REE accommodated mainly in silicate minerals, such as allanite. However, the potential of the Northern Greek placer for REE exploitation has not been fully evaluated due to limited on-shore and off-shore exploration drilling data. Characterization of the magnetic separation fractions using XRD and bulk ICP-MS chemical analysis showed that the magnetic products at high intensities were strongly enriched in the light REE (LREE), relative to the non-magnetic fraction. Allanite and titanite are the major host mineral for REE in the magnetic products but mainly allanite controls the REE budget due its high concentration in LREE. SEM/EDS and ICP-MS analysis of the different particle size fractions showed LREE enrichment in the fractions −0.425 + 0.212 mm, and a maximum enrichment in the −0.425 + 0.300 mm. The maximum enrichment is achieved after magnetic separation of the particle size fractions. Mass balance calculations showed that the maximum REE recovery is achieved after magnetic separation of each particle size fraction separately, i.e., 92 wt.% La, 91 wt.% Ce, and 87 wt.% Nd. This new information can contribute to the optimization of beneficiation process to be applied for REE recovery from HM black sands.

Keywords: rare earth elements (REE); heavy mineral sands; EURARE; allanite; monazite; HIMS; mineralogical characterization; geochemical characterization; magnetic separation; particle size fractions

1. Introduction

Heavy mineral sands (or black sands) are coastal deposits of resistant dense minerals that locally form economic concentrations of the heavy minerals. They serve as a major source of titanium worldwide with main minerals rutile and ilmenite and, in some cases, show high accumulation in rare earth elements (REE) and Th [1]. The rare earth elements (REE) are a group of 17 chemically similar elements, the lanthanides, scandium (Sc), and yttrium (Y) which behave similarly in most environments in the Earth's crust. REE are considered as "critical metals" for the European Union economy due to the vast application in a variety of sectors, a complicated production process as well as political issues associated with the monopoly in supply from China, especially the supply of heavy rare earths [2–4]. According to the recent EU Joint Research Center (JRC) report among the REE, six are identified as more critical, because their combined importance for strategic sectors of the economy such as high-efficiency electronics and energy technologies with risks of supply shortage. These are Dy, Eu, Tb, Y, Pr, and Nd [5].

Currently REE are not exploited in Europe, however, due to the current situation several exploration projects have been assessed in the course of the recently ended EURARE and ASTER European projects, which showed that some of them are in an advanced stage of exploration and development [6,7]. The most promising cases are the alkaline igneous rock-hosted deposits in South Greenland, the Norra Kärr deposit in Sweden and Fen Complex in Norway (Goodenough et al. [6]) and the alkaline volcanic-derived placers of Aksu Diamas in Turkey [8]. In Greece, the most significant REE concentrations are associated with heavy mineral sands on the coast of Nea Peramos and Strymonikos Gulf. Moreover, the EURARE project highlighted the significance of the secondary REE deposits, such as the bauxite residue (red mud) from the processing of Greek bauxites [4,6,9].

Mudd and Jowitt [1] stretched the economic potential of heavy mineral sands as an important underestimated REE resource especially for monazite and xenotime minerals. HM sands remain excluded from mineral resource considerations mostly due to the environmental problems that are associated with the radioactivity of tailings and the reagents used. However, there has been limited research work on the quantification of these impacts [3].

Geological prospecting by the Institute of Geology and Mineral Exploration (I.G.M.E.) of Greece on the black sands in the broader area of Strymon bay started in 1980's and focused primarily on the natural enrichment in actinides (U-Th) and associated radioactivity in the on-shore and offshore zones of Loutra Eleftheron to Nea Peramos regions [10–13] (Figure 1). In the last decade, there is an increasing number of geochemical and mineralogical studies that have been carried out on the coastal areas of Kavala [14–19], Sithonia Peninsula of Chalkidiki [20], Touzla Cape [21], and the area of Maronia, Samothrace [22]. Most of these studies focused on the characterization of the placers and their natural radioactivity ([23] and references therein). Previous studies have demonstrated that the heavy minerals, monazite, allanite, titanite, uraninite, zircon, and apatite, are traced in the Kavala black sands, derive from the Symvolon/Kavala pluton, a deformed granodioritic complex of Miocene age (Table 1) [6,9,14,24]. Despite the low grade of the Greek placers at Nea Peramos, as emphasized in the EURARE project, there is a good potential of beneficiation due to the coarse particle size and the liberation of REE minerals. The Northern Greece heavy mineral sands potential was not feasible to be fully evaluated as a potential REE resource in the course of EURARE project due to limited exploration data.

Worldwide, the commonly exploited rare earth-bearing minerals in industrial scale are bastnäsite, monazite, and xenotime [25]. Other REE-bearing minerals such as eudialyte, synchysite, samarskite, allanite, zircon, steenstrupine, cheralite, rhabdophane, apatite, florencite, fergusonite, loparite, perovskite, cerianite, and pyrochlore are rarely found in deposits of economic significance [26]. However, there are new deposits being under development containing many new REE minerals that seek further understanding, such as zircon, allanite, and fergusonite [27].

Beneficiation of the three commercially extracted heavy minerals, bastnäsite, monazite, and xenotime involves gravity, magnetic, electrostatic, and flotation separation methods with froth flotation being the most commonly applied REE mineral separation operation ([27–29] and references therein). There are numerous research articles on REE mineralogy and hydrometallurgical processing but there is still a lack of comprehensive descriptions of the beneficiation methods necessary to concentrate REE minerals. A main reason for this lack is the fact that concentrates of monazite and xenotime worldwide are produced from heavy mineral sands, therefore, comminution is scarcely required [26].

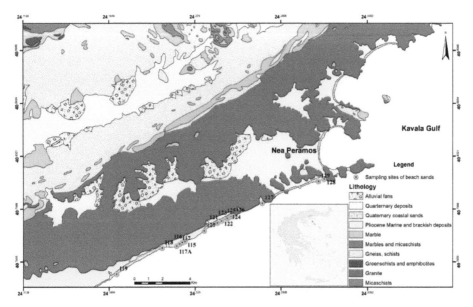

Figure 1. Geological map of the sampling area of Nea Peramos Loutra Eleftheron coast, Kavala region [12,13]. The large igneous body behind the coastline is Symvolon granite.

Table 1. Common silicate, phosphate, and carbonate rare earth element (REE) bearing minerals in heavy mineral (HM) sands (bold, this study) and bauxite residues in Greece [6,9,14,24]. Data for mineral properties from [26] (and references therein) and apatite data from [30].

REE-Mineral	Chemical Formula	Density (g/cm³)	Magnetic Properties	Weight % REO	ThO₂	UO₂
Silicates						
Allanite (Ce)	$(Ce,Ca,Y)_2(Al,Fe^{2+},Fe^{3+})3(SiO_4)_3(OH)$	3.50–4.20	paramagnetic	3–51	0–3	-
Allanite (Y)	$(Y,Ce,Ca)_2(Al,Fe^{3+})_3(SiO_4)_3(OH)$	n/a	paramagnetic	3–51	0–3	-
Cheralite (Ce)	$(Ca,Ce,Th)(P,Si)O_4$	5.28	n/a	-	<30	-
Sphene (titanite)	$(Ca,REE)TiSiO_5$	3.48–3.60	paramagnetic	<3	-	-
Thorite	$(Th,U)SiO_4$	6.63–7.20	paramagnetic	<3	70–80	10–16
Zircon	$(Zr,REE)SiO_4$	4.60–4.70	diamagnetic	-	0.1–0.8	-
Phosphates						
Apatite	$Ca_5(PO_4)_3(F,Cl,OH)$	3.17	n/a	~19	-	-
Fluorapatite	$(Ca,Ce)_5(PO_4)$	3.10–3.25	n/a	-	-	-
Monazite (Ce)	$(Ce,La,Nd,Th)PO_4$	4.98–5.43	paramagnetic	35–71	0–20	0–16
Monazite (La)	$(La,Ce,Nd,Th)PO_4$	5.17–5.27	paramagnetic	35–71	0–20	0–16
Monazite (Nd)	$(Nd,Ce,La,Th)PO_4$	5.43	paramagnetic	35–71	0–20	0–16
Rhabdophane (Ce)	$(Ce,La)PO_4.H_2O$	3.77–4.01	n/a	-	-	-
Xenotime (Y)	YPO_4	4.40–5.10	paramagnetic	52–67	-	0–5
Carbonates						
Bastnäsite (Ce)	$(Ce,La)(CO_3)F$	4.9–5.2	paramagnetic	70–74	0–0.3	0.09
Bastnäsite (La)	$(La,Ce)(CO_3)F$	n/a	paramagnetic	70–74	0–0.3	0.09
Bastnäsite (Y)	$Y(CO_3)F$	3.90–4.00	paramagnetic	70–74	0–0.3	0.09
Synchysite (Nd)	$Ca(Nd,La)(CO_3)_2F$	4.11 (calc)	n/a	-	-	-

n/a: not available; (-): this information is not known.

This paper presents results of an ongoing beneficiation study of REE-rich HM sands form the Nea Peramos, Kavala (Northern Greece) at laboratory scale. Two process schemes were applied and tested at laboratory scale in this stage in order to improve understanding of mineral separation by different physical processes including, particle size classification through wet sieving as a preconcentration process followed by high-intensity magnetic separation (HIMS) [26,27]. Scanning electron microscopy (SEM), X-ray diffraction analysis, and inductively coupled plasma–mass spectrometry (ICP-MS) were used in combination, for the characterization of feed material as well as beneficiation products.

Geological Setting

The studied area is situated in Northern Greece, in the internal domain of the Aegean arc (Figure 1). It is part of the Rhodope massif, a polymetamorphosed nappe stack, which comprises high-grade Pre-Alpine felsic to intermediate orthogneisses (mostly Variscan age), schists, and marble interlayered or tectonically overlain by basic to ultrabasic rocks, that were exhumed in the Cenozoic [31–38]. The lowermost Pangaion-Pirin or Lower Unit is exposed in several domes in the Rhodope area and as the footwall to the top to SW Strymon detachment [37,39]. The Pangaion-Pirin Unit comprises a thick marble sequence alternating with schists, both underlain by Variscan orthogneiss and schists [37,40,41]. The Pangaion-Pirin Unit is intruded by several Oligocene to early Miocene granitoids [37,42]. The syn-tectonic early Miocene Symvolon (or Kavala) granodiorite, which occupies a large part of the studied area, has intruded and deformed along the southeastern end of the Strymon detachment [39,41]. Marine deposits of Pliocene age, mainly sandstones and marls, are observed along the coast, covering the granodiorite.

2. Samples and Methods

2.1. Sampling

Fifteen samples of coastal sands were collected alongside a coastal line of 10 km from Loutra Eleftheron to Nea Peramos areas, Kavala region, Northern Greece (Figure 1). Sample campaign was carried out in the course of EURARE project during the period of 2013–2015. The sampling technique involved opening holes of 40 cm deep. According to field observations, locally, there were black sand concentrations in layers of few cm thick in alternation with typical light-colored sands. Due to sample inhomogeneity a large amount of ~15 kg of bulk sand from each sampling location was taken for securing a representative sample. A 20 kg sample of placer sand with an initial top size of 1.7 mm was prepared by mixing equal weights of individual collected samples (composite A-mixed sample) and divided by splitting into samples of 2.5 kg.

2.2. Sample Characterization

2.2.1. XRD Analysis

The major mineralogy was determined by powder X-ray diffraction using a Siemens D-5005 diffractometer with Cu K radiation, at the Geology Department of NKUA. Intensities were recorded at 0.02° 2θ step intervals from 3° to 70°, with a 2 s counting time per step. The resultant diffraction patterns were processed using EVA software by Bruker AXS Inc (Madison, WI, USA), in order first to identify peaks and then relate them to selected mineral phases that are present in the Kavala black sands samples (Figure 2). Detection limits are of the order of 1 wt.% approximately, but this is mineral and sample dependent.

The magnetic and non-magnetic fractions and all the particle size fractions were finely ground in the appropriate size, placed in a sample holder, and smeared uniformly onto a glass slide, assuring a flat upper surface.

Semiquantitative analysis was performed using the Reference Intensity Ratio method (RIR) of I/Ic [43]. For the semiquantitative analysis, the Diffract Plus software by Bruker-AXS was used. The minerals that did not have I/Ic ratios, the ratios from similar minerals of the same mineral group were used. For allanite, intensity ratio of epidote (0.9) was used, and for magnesiohornblende, an average ratio of hornblende (0.6) was used.

2.2.2. SEM/EDS Analysis

Sand particles were mounted within a plug of epoxy resin and then polished particle-mount sections were prepared. Before SEM analysis, the thin section was covered with a thin veneer of carbon using a vacuum carbon coater. Textural analysis and semiquantitative elemental analysis of heavy minerals was undertaken using a JEOL JSM-5600 Scanning Electron Microscope (SEM) coupled to an energy dispersive X-ray spectrometer (EDS) of OXFORD LINK ISIS 300 (OXFORD INTRUMENTS), with the use of software for ZAF correction, at the Faculty of Geology and Geoenvironment, NKUA, using secondary electron (SE) and backscatter electron (BSE) modes.

2.2.3. Bulk Chemical Analysis

REE ore sample preparation included initial digestion with Aqua Regia and HF (in order to solve dissolution issues in silicate samples) in Teflon® containers. Subsequently, the residue was chemically attacked with HCl and H_2O_2. Finally, the samples are preserved in 5% concentrated HCl. REE were analyzed by inductively coupled plasma-mass spectrometry (ICP-MS) at I.G.M.E. (Supplementary Table S1). External calibration solutions with matrix correction were used to measure the instrumental sensitivity of ICP-MS.

2.3. Beneficiation Tests

The beneficiation tests were conducted at the Laboratory of Beneficiation and Metallurgy, I.G.M.E.

2.3.1. Particle Size Analysis

From the composite sample A-mixed, a representative sample (of 2327 g initial weight) was prepared properly in order to be submitted to particle size analysis. Samples were submitted to wet sieving in order to provide data on particle size distribution, using sieves with aperture of 1.70, 0.850, 0.500, 0.425, 0.355, 0.300, 0.212, and 0.150 mm. All the samples were placed in a dryer for 24 h at 90 °C.

The information acquired by particle size distribution contributes to the study of the effects of particle size, mineral liberation, and association characteristics aiming at the selection of appropriate process schemes to be tested and used. In the course of wet sieving process, a change in the color was observed as we moved onto smaller particle size fractions.

Preconcentration step helps in rejecting early gangue materials thus achieving benefits such as higher feed grades, lower waste stream production in further processes, and finally, low operating costs.

2.3.2. Magnetic Separation

Magnetic separation was selected as a possible concentration technique as the gangue minerals are known to have lower magnetic susceptibility than value minerals which contain rare earth elements. Based on strong or weak magnetic properties, iron-bearing minerals are characterized as ferromagnetic or paramagnetic, respectively. Ferromagnetic refers to minerals strongly attracted to a magnet, like a piece of iron. Magnetite, maghemite, and pyrrhotite are the most common ferromagnetic minerals [20].

A laboratory-scale High Intensity Magnetic Separator (HIMS; Model 10/1, ERIEZ EUROPE, Caerphilly, UK) with standard intensity was used to recover REE-bearing minerals (Figure 2). More than 20 tests were carried out during magnetic separation process, applying different conditions and parameters (vibration, inclination, and speed). Two process routes were followed:

(i)   The whole sample passing through −0.500 mm was driven to magnetic separation.
(ii)   Each size fraction separately (down to +0.212 mm) was tested for magnetic separation.

The sample was fed by a vibrating feeder. The moving velocity of the feed carrying conveyor was about 40 str/s, tilt was set initially to 72°. The applied magnetic intensity was 2 T. After recovering magnetic products at 72°, the magnetic sample was fed again to the magnetic separator adjusted to a 74° and repeated to 76° inclination.

Wet high-intensity magnetic separation (WHIMS) was also conducted on the composite A-mixed using an ERIEZ separator by applying different voltages, i.e., at 30 V (0.48 T) and 150 V (2.4 T). The potential difference was selected to correspond to extreme conditions of lower and higher magnetic field strength in order to check the effect of the intensity of the magnetic field on the efficiency of the magnetic separation process. The results after magnetic separation were evaluated on the basis of semiquantitative mineralogical analysis using the Reference Intensity Ratio method (RIR) of I/Ic.

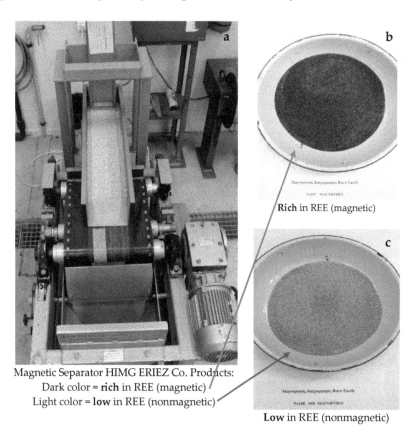

Magnetic Separator HIMG ERIEZ Co. Products:
Dark color = **rich** in REE (magnetic)
Light color = **low** in REE (nonmagnetic)

Rich in REE (magnetic)

Low in REE (nonmagnetic)

Figure 2. (**a**) Magnetic Separator HIMG ERIEZ Co. and products of separation, (**b**) dark colored magnetic product and, (**c**) non-magnetic product.

2.4. Evaluation of Beneficiation Tests

All the separated fractions from the beneficiation tests were tested for mineralogical and chemical composition, except for the fraction +1.7 mm due to its low quantity and low content in REE. The SEM/EDS analysis of the particle size fractions and the magnetic separates provided a rough evaluation of the test performance. The degree of liberation of the REE-bearing minerals from the gangue at various particle sizes was also established by SEM/EDS analysis. The fractions from particle size analysis and the final magnetic separation of the undersize 0.500 mm fractions were evaluated on the basis of mass balance calculations.

3. Results

3.1. Mineralogy of the Sands

Table 2 and Figure S1 shows the results of mineral identification and semiquantitative modal composition of the composite sample "A-mixed" obtained by XRD analysis and RIR method. The studied HM sands consist mainly of silicate minerals, Fe-oxides (magnetite, hematite), titanite and minor amounts (<3%) of Ti oxides (ilmenite and rutile), and phosphate phases (apatite, monazite, and xenotime) (Table 2). The major heavy minerals contained in the studied sands are amphibole (Mg-hornblende and pargasite), magnetite, titanite, allanite-epidote, and hematite (Table 2). Zircon, monazite, cheralite, ilmenite, rutile, thorite, apatite, xenotime, baryte, and sulfides were identified as minor constituents by SEM/EDS analysis. Allanite is the major host mineral of REE in relative abundance of 3% in the composite sample (Table 2). Based on SEM/EDS analysis and published EPMS data, the studied monazite is classified as a cerian type (monazite-Ce). However, due to the low abundance of the mineral as well as overlapping of monazite peak intensities with those of magnesiohornblende and allanite, this mineral was not clearly identified by the XRD spectra (Figure S1).

Table 2. Semiquantitative mineral contents of the composite sample (A-mixed), based on Reference Intensity Ratio (RIR) method. Specific gravity and magnetic property data from [44–46].

Mineral	Weight %	Nominal Specific Gravity	Magnetic Property
Albite	39	2.68	Diamagnetic
Quartz	31	2.63	Diamagnetic
K-feldspar	11	2.57	Diamagnetic
Titanite	8	3.4–3.6	Paramagnetic
Mg-hornblende	6	3.24	Paramagnetic
Allanite	3	3.75	Paramagnetic
Hematite	2	5.30	Ferromagnetic
Magnetite	1	5.20	Ferromagnetic
Trace minerals	<0.5	-	-
Total	100		

(-): not applicable.

SEM Analysis of REE Minerals

(a) Allanite-(Ce)

SEM/EDS semiquantitative analysis of allanite indicated that total concentration of REE oxides (TREO) range ca. from 10.7 to 16.8 wt.%, which is in agreement with the published data [18]. The stoichiometry of allanite indicates that Ce-allanite is the dominant type (Supplementary Table S2). The backscattered election (BSE) images suggest that allanite is liberated in the particle size −0.500 mm. The surface of the allanite is weathered with several cracks. Backscattered electron images showed that most allanite exhibits pronounced zoning toward the marginal areas of the grain (Figure 3a,b,d). The chemical composition of the peripheral zones indicates replacement of the allanite with epidote. Zoned allanite contains significant Th in the central part of the grain, ca. up to 2.5 wt.%, and they show metamict texture. This is documented by the abundant radiation lines in the structure of the grain displayed by this type of allanite, which results from the destructive effects of its own radiation on the crystal lattice (Figure 3).

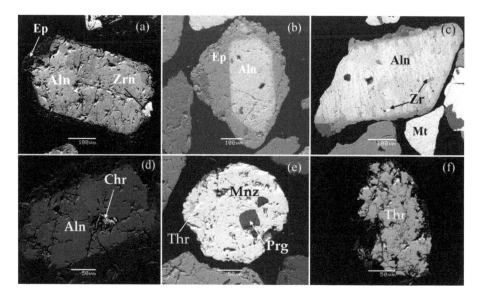

Figure 3. Backscattered electron images from SEM of various rare earth element (REE) minerals and other heavy minerals of different particle size fractions. (**a**,**b**) zoned allanite–(Ce) showing metamict structure and epidote peripheral zone (−0.425 mm to +0.212 mm), (**c**) allanite containing zircon inclusions from the magnetic fraction, (**d**) zoned allanite with weathering features and cracks containing inclusion of Ca-rich monazite (cheralite) (**e**) monazite containing pargasite and thorite inclusions, from the magnetic fraction, (**f**) thorite-weathered grain, darker zones are U depleted); abbreviations: Aln—allanite, Ep—epidote, Zr-zircon, Mnz—monazite, Thr—thorite, Mt—magnetite, Prg—pargasite, and Chr—cheralite.

(b) Monazite (-Ce)

Monazite occurs as very fine grains, i.e., <10–20 μm in diameter, disseminated in allanite, and only rare liberated grains up to 200 μm are found (Figure 3b,e). In most of the cases, monazite is a replacement phase of allanite. Monazite shows higher content in Ce than Nd or La, which agrees for a cerian-type monazite and is consistent with reported data from the same area [18] (Supplementary Table S2). ThO_2 content of the monazite is generally high, i.e., 17 wt.%. The low totals reported in the analyses are probably related to undetected MREE and HREE by SEM/EDS.

(c) Titanite

Titanite is more abundant than allanite (>5%). SEM/EDS analysis of titanites from this study showed that the REE content is below the detection limit of the method (<0.1–0.2 wt.%). The results are consistent with published EPMA analysis for titanites from the same area, which showed that La, Ce, or Nd content is ≤1000 mg/kg [18].

(d) Other Heavy REE Minerals

Zircon is another REE-hosting phase, with relatively higher concentration in heavy REE (HREE) (253–890 mg/kg) relative to light REE (LREE) (8–44 mg/kg) [18,23]. Apatite is a common fine-grained accessory phase disseminated in silicate minerals. Neither apatite nor zircon or thorite was feasible to be analyzed by SEM/EDS for REE due to the low REE content and/or the small particle size of the minerals for this method (see Supplementary Table S2).

3.2. REE Geochemistry

The chondrite-normalized REE contents of the Nea Peramos HM sand samples display similar patterns but a large compositional range (Figure 4). The total REE (TREE) abundances are generally higher than that of the average upper continental crust, i.e., 183 mg/kg [47], except for one sample (#NP120) (Table 3; detailed data in Supplementary Table S1). A comparison of average REE concentrations of the HM sands of Greece based on reported analysis [23] and HM sands from this study shows that the Nea Peramos samples display the highest total REE enrichment. The chondrite-normalized REE patterns of the studied samples show a pronounced enrichment of light REE, e.g., ranging from 70 to 7000 times chondrite values, and flat heavy REE (HREE) with a small negative Eu anomaly (Figure 4). Allanite controls the LREE budget of the studied sands but the flat HREE characteristic is controlled by another major REE mineral, possibly titanite and/or epidote. This characteristic is typically shown by titanite which is the major heavy mineral phase that holds HREE (see Discussion). These observations are in agreement with previous mineralogical studies of the studied sands [18,23]. The average content of Sc is 7.5 mg/kg, which is lower compared to the respective value in upper continental crust, i.e., 14 mg/kg [47]. Average Y concentration in the studied samples is 67 mg/kg and is enriched by a factor of 3 relatively to the upper continental crust, i.e., 21 mg/kg [47]. Yttrium enrichment is associated with xenotime abundance.

Our samples present a large compositional variation in total REE (Figure 4). On the other hand, all samples show a systematic enrichment in light REE. Therefore, the preliminary beneficiation tests were performed on a composite sample (A-mixed) produced by mixing of all the samples.

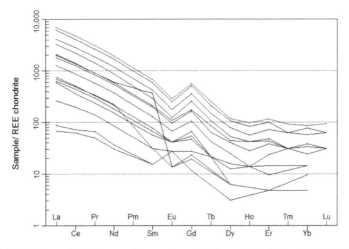

Figure 4. REE patterns of Nea Peramos heavy mineral (HM) sands. Red-colored line highlights the most REE-enriched sample #123. Normalization values of chondrite from [48].

Table 3. REE (Lanthanides, Sc, Y) contents of the Nea Peramos (NP) HM sands (in mg/kg). Light rare earth elements (LREE) group includes the elements from La to Sm, while heavy rare earth elements (HREE) group includes the elements from Eu to Lu; Y and Sc are included in total REE content, according to European Union (EU) definition [2].

Sample	NP 115	NP 116	NP 117	NP 117A	NP 118	NP 119	NP 120	NP 121	NP 122	NP 123	NP 124	NP 125	NP 126	NP 127	NP 128
LREE	600	826	1407	315	707	3504	118	2024	2225	7645	6646	4552	98	2297	787
HREE	47	61	98	27	51	204	18	147	160	465	409	295	15	76	47
TREE	654	894	1509	344	762	3716	143	2181	2393	8124	7068	4857	116	2381	842

3.3. Particle Size Analysis and REE Distribution

The distribution of REE and Th in the particle size fractions of the composite sample A-mixed is determined by chemical assays (Table 3).

A combination of SEM/EDS and ICP-MS chemical analysis of the different particle size fractions showed a LREE enrichment in the fractions −0.500 to +0.212 mm and under 0.150 mm (Figure 5a). In Table 4, mass balance calculations show that the −0.150 mm fraction, despite its composition, is insignificant in terms of beneficiation due to its very low mass, i.e., 1.5 wt.% Moreover, the oversize +1.70 mm particle size fraction has a very low REE content and hence is also considered as insignificant for further beneficiation (Figure 5b).

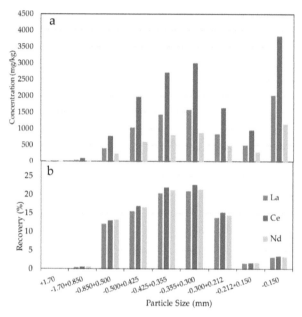

Figure 5. (**a**) Particle size analysis and light rare earth element (LREE) distribution (mg/kg) in sample A-mixed and (**b**) metal recovery in each particle size fraction, based on mass balance calculations (see text).

Table 4. Particle size analysis and REE distribution in each particle size fraction.

Particle Size (mm)	Mass (g)	Weight (%)	La	Ce	Nd	Th	La	Ce	Nd	Th
			Concentration (mg/kg)				Mass (mg)			
+1.70	55.6	2.42	16	25	10	7	0.89	1.39	0.56	0.39
−1.70 + 0.850	230.7	10.02	54	106	38	25	12.46	24.45	8.77	5.77
−0.850 + 0.500	640.48	27.82	416	777	247	190	266.44	497.65	158.20	121.69
−0.500 + 0.425	324.62	14.10	1044	1983	607	455	338.90	643.72	197.04	147.70
−0.425 + 0.355	307.32	13.35	1445	2719	821	641	444.08	835.60	252.31	196.99
−0.355 + 0.300	286.26	12.43	1598	3018	891	723	457.44	863.93	255.06	206.97
−0.300 + 0.212	353.02	15.33	852	1643	487	390	300.77	580.01	171.92	137.68
−0.212 + 0.150	69.12	3.00	507	968	305	240	35.04	66.91	21.08	16.59
−0.150	35.05	1.52	2031	3840	1162	945	71.19	134.59	40.73	33.12
Total	2302.17						1927.21	3648.27	1105.66	866.90

The maximum effective recovery is achieved in the fractions −0.425 + 0.300 mm; this size range corresponds to the "liberation" size of allanite. A similar trend of enrichment is observed in the sample #123 [17]. Thorium follows the enrichment trend of LREE and its concentration in the undersize

0.500 mm ranges from 240 to 945 mg/kg. Thorium concentration is associated with the abundance of allanite and monazite in the HM sands, which contain 2.3 and 17–35 wt.% ThO_2, respectively, according to EDS/SEM analysis (Supplementary Table S2).

3.4. REE Distribution in the Magnetic Fractions

High-intensity magnetic separation (HIMS) is a common separation step in REE containing beach sands in order to concentrate the targeted paramagnetic REE-bearing part and is usually applied for monazite or xenotime [26,27]. The LREE tend to concentrate in the magnetic fraction because of their magnetic properties. Dry HIMS was applied initially in the most REE-enriched sample #123. Mineralogical evaluation based on XRD spectra shows that from the paramagnetic minerals allanite, titanite and magnesiohornblende tend to concentrate in the magnetic fraction under high intensities. Qualitative evaluation of the mineralogy of magnetic and nonmagnetic separates by XRD spectra is shown in Figure 6. The peak intensities of allanite can be clearly distinguished in the magnetic separate of the sample #123 but are absent in the nonmagnetic part. Following that observation, magnetic separation of particle size fractions from the composite A-mixed sample was carried out. The chemical assays of the magnetic fractions are shown in Table 5 and Figure 7.

The same REE-enrichment trend, as shown in the particle size analysis, is also recorded in the HIMS test (Figure 7). The maximum LREE concentration was achieved in the particle size: −0.355 + 0.212 and a second concentration maximum in the −0.425 + 0.355 mm fraction. Two test of HIMS were conducted: in the first test, all particle sizes (Feed 1 in Figure 8) were processed by magnetic separation, whereas in the second test, only the undersize 0.500 mm were tested after removing the REE-poor +0.500 mm fraction from the initial composite sample A-mixed (preconcentration stage) (Table 5). The results of mass balance calculations and REE distribution in every product are presented in Table 5. Similar recoveries were achieved in the two feeds, i.e., from 92% to 87.6% and 95.9% to 78.87% for the particle size fractions and the undersize 0.500 mm feed, respectively. The maximum LREE enrichment is attained in the magnetic products of the different particle sizes. It is clearly shown in the mass balance calculations in Table 5 that magnetic separation for each particle size separately improves the recovery of REE and reaches recoveries of 75–90% in just the 22% of feed material (fractions −0.500 to +0.212 mm).

Table 5. REE distribution and recovery of REE and Th in the magnetic products of each particle size of the A-mixed sample and the undersize 0.500 mm.

Feed: Sample A (All Fractions)		La	Ce	Nd	Th	La	Ce	Nd	Th
Mass (g)		Concentration (mg/kg)				Mass (g)			
	1995	942	1647	512	365	1.87	3.29	1.021	0.728
Particle size		magnetic fractions (mg/kg)				Mass (g)			
−1.70 + 0.850	42.55	261	447	137	93	0.0111	0.0190	0.0058	0.0040
−0.850 + 0.500	162.23	1301	2309	717	392	0.2111	0.3746	0.1163	0.0636
−0.500 + 0.425	101.65	2764	4802	1444	1099	0.2810	0.4881	0.1468	0.1117
−0.425 + 0.355	109.81	3627	6321	1865	1472	0.3983	0.6941	0.2048	0.1616
−0.355 + 0.300	100.6	3838	6649	1968	1595	0.3861	0.6689	0.1980	0.1605
−0.300 + 0.212	142.83	3044	5280	1563	1302	0.4348	0.7541	0.2232	0.1860
Total	659.67					1.7223	2.9989	0.8950	0.6873
Recovery (wt.%)						92.1	91.1	87.6	94.4
Feed: −0.500 mm		La	Ce	Nd	Th	La	Ce	Nd	Th
Mass (g)		Concentration (mg/kg)				Mass (g)			
	530	1260	2223	670	460	1.9278	3.4012	1.0251	0.7038
Magnetic fraction		490	3774	6089	1650	1300	1.8493	2.9836	0.8085
Recovery (wt.%)						95.9	89.7	78.87	90.5

Notably, Th enrichment follows the LREE enrichment trend, and its recovery is high in the either tested feed. The steps of magnetic separation process and the REE grade of concentrates are summarized in the flowsheet of Figure 8.

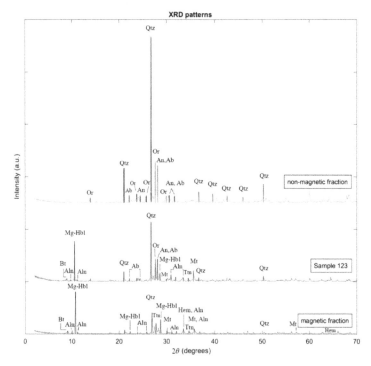

Figure 6. Comparison of XRD patterns of magnetic and nonmagnetic separates after high-intensity magnetic separation (HIMS) from sample #123. Note that the clear peaks of allanite are absent in the nonmagnetic fraction.

Further testing on the magnetic susceptibility of the paramagnetic minerals in the composite sample was conducted under extreme conditions of the magnetic field, i.e., low (0.48 T) and high (2.4 T) strength, by wet HIMS. The results are evaluated by semiquantitative analysis of the XRD spectra of the magnetic and nonmagnetic products. The maximum content of allanite, i.e., 8 wt.%, is achieved in the magnetic fraction at magnetic field strength of 0.48 T (30 V) and remains constant at 2.4 T (150 V) (Supplementary Table S3).

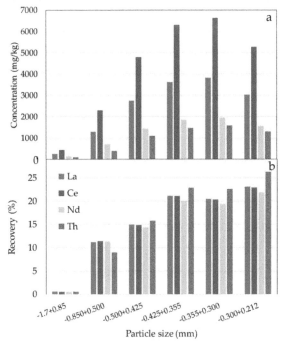

Figure 7. (**a**) LREE and Th distribution in the magnetic products of the particle size fractions and (**b**) metal recovery in each particle size fraction, based on mass balance calculations.

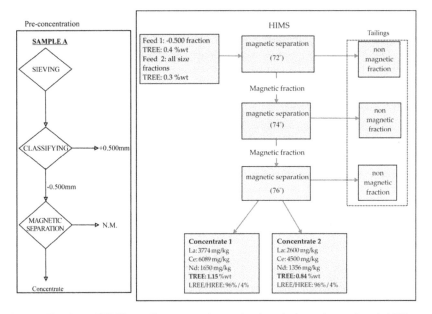

Figure 8. Flowsheet of HIMS tests. Preconcentration step involves sieving and removing of +0.500 mm (Feed 1). Feed 2 corresponds to all the particle size fractions (−1.70 to +0.212 mm) of the initial composite (sample A-mixed).

4. Discussion

The beneficiation of REE ores is challenging from the technological point of view due to complexity in the raw material processing [49–52]. Such processing involves three main stages: an initial beneficiation step, where REE minerals are concentrated from the ore; a second step, where rare earth oxides are extracted from their host minerals producing a mixed rare earth concentrate; and final step, where individual rare earth metals are obtained through metallurgical separation [6].

Mineral characterization is essential for potential process development impacts. Setting the criteria for initial target grind sizes to liberate the economic minerals and determining the possible mineral associations among REE minerals and various gangue phases are both strongly depended on the mineral characterization study [53]. Therefore, the information obtained herein can be used, apart from the prediction of the concentrate quality, in the associated concentration of deleterious elements.

X-ray diffraction (XRD) analysis allows the identification of minerals. However, when mineral modal abundance is less than ~1 wt.% considerable uncertainty and error can be introduced. It is noteworthy that the ore grade in most REE placer deposits is commonly below 0.1% for the REE-host mineral, e.g., monazite [54]. The evaluation of WHIMS magnetic separation products by XRD analysis agrees with the enrichment of allanite in the final cumulative magnetic concentrate where the content of allanite is increased by a factor of 2.5, i.e., from 3 wt.% in the initial composite feed to final 8 wt.% in the magnetic concentrate (Supplementary Table S3). Correspondingly, REE contents of the final magnetic concentrates increased by a factor of 2.8 compared to the initial feed (Table 4, Figure 8). This case study revealed a close association of REE-silicate minerals of allanite with monazite inclusions, as well as a high Th content in the magnetic separate associated with the REE-host minerals, e.g., allanite-monazite/cheralite and thorite. Therefore, the information obtained herein can be used, apart from the prediction of the concentrate quality, in the associated concentration of deleterious elements. One of the most serious issues associated with REE processing is the radioactive wastes produced as by-product of the extraction stage.

The results from the present beneficiation tests of Nea Peramos HM sands in the course of EURARE project also showed the high content of radionuclides, specifically Th. Thorium distribution follows the enrichment trend of LREE in the concentrates and this poses limitations in the concentration process (e.g., [10,11,23]). There are still many questions and knowledge gaps regarding the beneficiation processes of the REE-bearing minerals which require a great deal of investigation [26,28]. Previous research papers described a series of physical separation processes applied to preconcentrated rare earth minerals (REM) and discard iron oxide minerals from the magnetic fractions. These include, gravity, magnetic, electrostatic, and flotation separation techniques ([52] and references therein).

Effects of Mineral Magnetic Susceptibility on the HIMS Treatment

Magnetic separation process of minerals is based on different behavior (magnetic susceptibility) of mineral particles when exposed to an applied magnetic field. The magnetization of a material is a measure of the density of magnetic dipoles induced in the material [44].

The magnetic susceptibility of minerals is mainly controlled by chemical composition. REE minerals such as allanite, monazite, bastnäsite, and xenotime exhibit moderate paramagnetic property [44,48]. Previous studies on the beneficiation applicable for silicate REE minerals, including allanite and cerite from complex ores, have shown that wet high-intensity or high-gradient magnetic separation is more effective than flotation [29,55]. Yang et al. (2015) and Jordens et al. (2014) [55,56] showed that allanite have normally poor floatability using conventional reagents.

The main host mineral for LREE in Nea Peramos HM sands is allanite-(Ce), commonly containing inclusions of other REE-bearing phases such as thorite, zircon, monazite, and cheralite. Therefore, REE concentration in the magnetic fraction is attributed largely to allanite which was efficiently concentrated in the magnetic fraction under high-intensity magnetic field at 0.48 T. The results of magnetic separation from this study are contradictory to a recent study [18]. In the previous study, allanite was recorded to be rather separated in the nonmagnetic part but the results are not comparable since the conditions of the applied magnetic separation test are not described by the authors.

Dry HIMS (DHIMS), applied to each particle size separately, improved significantly the recovery of REE with achieved recoveries of 92–88% (Table 5). Similar REE recoveries were achieved by processing the undersize 0.500 mm, i.e., 96–79%. REE grade of the magnetic concentrate starting from the undersize 0.500 mm is better than the magnetic product of the particle size fractions. In general, the actual TREE concentration in the magnetic products was 2.8 times higher compared to the initial feed contents, in the two feeds tested. The optimum magnetic separation was effected after repetitive passages of the magnetic fraction at increasing belt inclination from 72° to 76°. Moreover, the preliminary results of WHIMS corroborated for an effective recovery of allanite in the magnetic fraction, and the enrichment achieved was in the order of 2.5 times in the magnetic product (Supplementary Table S3).

Additional laboratory tests for REE dissolution and extraction from concentrate was tried by acid treatment, either leaching or acid baking followed by water leaching conducted by I.G.M.E. during EURARE project [57]. The acids used were HCl, H_2SO_4, or both. The results showed that rather low recoveries were achieved. In direct acid leaching, decreasing pulp density leads to increasing rare earth recovery, while the opposite is observed in the recovery procedure with acid baking; an increasing pulp density at acid baking step at about 15% leads to an increasing recovery tendency of about 15–20% (unpublished data, Angelatou pers. comm.). The duration of acid baking test seems to have no effect on the recovery of rare earth elements.

5. Conclusions

HM sands from the shoreline of Nea Peramos, Kavala, Northern Greece was studied in order to be able to determine the best process to concentrate the ore by testing simple screening as a prebenefication method and high-intensity magnetic separation as the main beneficiation method. The results of the test work lead to the following conclusions:

- Allanite-(Ce) is the major host mineral for light REE (LREE), whereas monazite, zircon, and thorite constitute trace amounts. Titanite displays low concentration in LREE <0.1–0.2%. Metamict allanite is common and is thorium-enriched relative to the nonmetamict allanite.
- A simple screening can achieve a satisfactory prebenefication.
- A stepwise magnetic separation improves the recovery of REE.
- Magnetic separation for each particle size fraction separately improves the recovery of REE and reaches recoveries of 75–90% in just the 20% of feed material.
- The grades of magnetic concentrate for the two processes (all particle size fractions and undersize 0.500 mm) were 0.84% and 1.15% TREE, respectively, at the recoveries of 87–92% and 79–96%.
- The increase in REE content is associated with the increase of thorium content in concentrates regarding placer sands from Nea Peramos, Greece.
- The use of gravimetric methods (such as Wilfley shaking table) did not contribute much to the beneficiation of the ores.
- Radioactive wastes from a potential REE processing operation of Nea Peramos HM sands should be carefully encountered for the risk to the environment and humans.

Supplementary Materials: The following are available online at http://www.mdpi.com/2075-163X/10/5/387/s1, Table S1: REE contents of the Nea Peramos HM sands (in mg/kg), Table S2: SEM/EDS spot analysis of REE-bearing minerals, and Table S3: Semiquantitative mineralogical composition of the wet HIMS-treated samples and the composition of the initial A-mixed composite sample for comparison. Applied voltage values of 30 and 150 V correspond to 0.48 and 150 T magnetic field strength, respectively, Figure S1: X-ray diffraction patterns of composite A-mixed and sample #123, in the range of 20–43° showing in magnification the identified heavy minerals. Quartz appears as the stronger peak. The major peaks of intensities for the heavy minerals only are marked. Monazite (-Ce) peaks shown are indicative (database for RRUFF™ Project [43]). Ttn = titanite, Mz = monazite, Mg-Hbl = magnesiohornblende, Prg = pargasite, Aln = allanite, Mt = magnetite, Hem = hematite).

Minerals **2020**, *10*, 387

Author Contributions: V.A. and D.E. collected the samples and provided the field information; V.A. conceived and designed the experiments undertaken by I.G.M.E. in the framework of EURARE project; V.A. and S.P. performed the experiments; C.S. supervised the mineralogical analysis; and C.S., S.P., V.A. and K.S. wrote the paper. All authors have read and agreed to the published version of the manuscript.

Funding: Part of this research was supported by EURARE research project, funded from the European Community's Seventh Framework Programme ([FP7/2007–2013]) under grant agreement no 309373. Project web site: www. eurare.eu.

Acknowledgments: The authors would like to thank N. Xirokostas for elaborating the chemical analysis of REE ore at the ICP-MS facility of I.G.M.E., V. Skounakis for assistance in performing the SEM/EDS analysis at NKUA, Geology and Geoenvironment, Athens, and I. Marantos from I.G.M.E. for assistance in the quantitative evaluation of the XRD patterns by RIR method. This publication reflects only the authors' view, exempting the Community from any liability.

Conflicts of Interest: The authors declare no conflict of interest.

References

1. Mudd, G.M.; Jowitt, S.M. Rare earth elements from heavy mineral sands: Assessing the potential of a forgotten resource. *Appl. Earth Sci. (Trans. Inst. Min. Metall. B)* **2016**, *125*, 107–113. [CrossRef]
2. EU Commission. *Study on the Review of the List of Critical Raw Materials*; European Commission: Brussels, Belgium, 2017.
3. McLennan, B.; Gorder, G.D.; Ali, S.H. Sustainability of rare earths—An overview of the state of knowledge. *Minerals* **2013**, *3*, 304–317. [CrossRef]
4. Balomenos, E.; Davris, P.; Deady, E.; Yang, J.; Panias, D.; Friedrich, B.; Binnemans, K.; Seisenbaeva, G.; Dittrich, C.; Kalvig, P.; et al. The EURARE Project: Development of a sustainable exploitation scheme for Europe's Rare Earth Ore deposits. *Johns. Matthey Technol. Rev.* **2017**, *61*, 142–153. [CrossRef]
5. Moss, R.; Tzimas, E.; Willis, P.; Arendorf, J.; Thompson, P.; Chapman, A.; Morley, N.; Sims, E.; Bryson, R.; Peason, J.; et al. Critical metals in the path towards the decarbonization of the EU energy sector. In *Assessing Rare Metals as Supply-Chain Bottlenecks in Low-Carbon Energy Technologies*; JRC Report EUR 25994 EN; Publications Office of the European Union: Brussels, Belgium, 2013; p. 242.
6. Goodenough, K.; Schilling, J.; Jonsson, E.; Kalvig, P.; Charles, N.; Tuduri, J.; Deady, E.A.; Sadeghi, M.; Schiellerup, H.; Muller, A.; et al. Europe's rare earth element resource potential: An overview of REE metallogenetic provinces and their geodynamic setting. *Ore Geol. Rev.* **2016**, *72 Pt 1*, 838–856. [CrossRef]
7. Guyonnet, D.; Planchon, M.; Rollat, A.; Escalon, V.; Vaxelaire, S.; Tuduri, J. Primary and secondary sources of rare earths in the EU-28: Results of the ASTER project. In Proceedings of the ERES 2014—1st Conference on European Rare Earth Resources, Milos, Greece, 4–7 September 2014; pp. 66–72.
8. Deady, E.; Lacinska, A.; Goodenough, K.M.; Shaw, R.A.; Roberts, N.M.W. Volcanic-Derived Placers as a Potential Resource of Rare Earth Elements: The Aksu Diamas Case Study, Turkey. *Minerals* **2019**, *9*, 208. [CrossRef]
9. Deady, É.; Mouchos, E.; Goodenough, K.; Williamson, B.; Wall, F. A review of the potential for rare-earth element resources from European red muds: Examples from Seydişehir, Turkey and Parnassus-Giona, Greece. *Mineral. Mag.* **2016**, *80*, 43–61. [CrossRef]
10. Pergamalis, F.; Karageorgiou, D.E.; Koukoulis, A. The location of Tl, REE, Th, U, Au deposits in the seafront zones of Nea Peramos-Loutra Eleftheron area, Kavala (N. Greece) using γ radiation. *Bull. Geol. Soc. Greece* **2001**, *34*, 1023–1029.
11. Pergamalis, F.; Karageorgiou, D.E.; Koukoulis, A.; Katsikis, I. Mineralogical and chemical composition of sand ore deposits in the seashore zone N. Peramos-L. Eleftheron (N. Greece). *Bull. Geol. Soc. Greece* **2001**, *34*, 845–850. [CrossRef]
12. Institute of Geological and Mineral Exploration (IGME). *Geological Map of Greece, Nikisiani-Loutra Eleftheron Sheet, 1:50000*; Kronberg, P., Schenk, P.F., Eds.; IGME: Madrid, Spain, 1974.
13. Institute of Geological and Mineral Exploration (IGME). *Geological Map of Greece, Kavala Sheet, 1:50000*; Kronberg, P., Ed.; IGME: Madrid, Spain, 1974.
14. Eliopoulos, D.; Economou, G.; Tzifas, I.; Papatrechas, C. The potential of Rare Earth elements in Greece. In Proceedings of the ERES2014: First European Rare Earth Resources Conference, Milos, Greece, 4–7 September 2014; pp. 308–316.

15. Papadopoulos, A.; Koroneos, A.; Christofides, G.; Stoulos, S. Natural radioactivity distribution and gamma radiation exposure of beach sands close to Kavala pluton, Greece. *Open Geosci.* **2015**, *7*, 64. [CrossRef]
16. Papadopoulos, A.; Koroneos, A.; Christofides, G.; Papadopoulou, L. Geochemistry of beach sands from Kavala, Northern Greece. *Ital. J. Geosci.* **2016**, *135*, 526–539. [CrossRef]
17. Angelatou, V.; Papamanoli, S.; Stouraiti, C.; Papavasiliou, K. REE distribution in the Black Sands in the Area of Loutra Eleftheron, Kavala, Northern Greece: Mineralogical and Geochemical Characterization of Fractions from Grain Size and Magnetic Separation Analysis (doi:10.3390/IECMS2018-05455). Available online: https://sciforum.net/paper/view/conference/5455 (accessed on 25 April 2020).
18. Tzifas, I.; Papadopoulos, A.; Misaelides, P.; Godelitsas, A.; Göttlicher, J.; Tsikos, H.; Gamaletsos, P.N.; Luvizotto, G.; Karydas, A.G.; Petrelli, M.; et al. New insights into mineralogy and geochemistry of allanite-bearing Mediterranean coastal sands from Northern Greece. *Geochemistry* **2019**, *79*, 247–267. [CrossRef]
19. Papadopoulos, A.; Christofides, G.; Koroneos, A.; Stoulos, S. Natural radioactivity distribution and gamma radiation exposure of beach sands from Sithonia Peninsula. *Cent. Eur. J. Geosci.* **2014**, *6*, 229–242. [CrossRef]
20. Papadopoulos, A.; Christofides, G.; Koroneos, A.; Hauzenberger, C. U Th and REE content of heavy minerals from beach sand samples of Sithonia Peninsula (northern Greece). *J. Mineral. Geochem.* **2015**, *192*, 107–116. [CrossRef]
21. Filippidis, A.; Misaelides, P.; Clouvas, A.; Godelitsas, A.; Barbayiannis, N.; Anousis, I. Mineral, chemical and radiological investigation of a black sand at Touzla Cape, near Thessaloniki, Greece. *Environ. Geochem. Health* **1997**, *19*, 83–88. [CrossRef]
22. Papadopoulos, A.; Koroneos, A.; Christofides, G.; Stoulos, S. Natural Radioactivity Distribution and Gamma Radiation exposure of Beach sands close to Maronia and Samothraki Plutons, NE Greece. *Geol. Balc.* **2015**, *43*, 1–3.
23. Papadopoulos, A.; Tzifas, I.; Tsikos, H. The Potential for REE and Associated Critical Metals in Coastal Sand (Placer) Deposits of Greece: A Review. *Minerals* **2019**, *9*, 469. [CrossRef]
24. Eliopoulos, D.; Aggelatou, V.; Oikonomou, G.; Tzifas, I. REE in black sands: The case of Nea Peramos and Strymonikos gulf. In Proceedings of the ERES, Santorini, Greece, 28–31 May 2017; pp. 49–50.
25. Sengupta, D.; Van Gosen, B.S. Placer-type rare earth element deposits. *Rev. Econ. Geol.* **2016**, *18*, 81–100.
26. Jordens, A.; Cheng, Y.P.; Waters, K.E. A review of the beneficiation of rare earth element bearing minerals. *Miner. Eng.* **2013**, *41*, 97–114. [CrossRef]
27. Jordens, A.; Sheridan, R.S.; Rowson, N.A.; Waters, K.E. Processing a rare earth mineral deposit using gravity and magnetic separation. *Miner. Eng.* **2014**, *62*, 9–18. [CrossRef]
28. Jordens, A.; Marion, C.; Langlois, R.; Grammatikopoulos, T.; Sheridan, R.; Teng, C.; Demers, H.; Gauvin, R.; Rowson, N.; Waters, N. Beneficiation of the Nechalacho rare-earth deposit. Part 2: Characterization of products from gravity and magnetic separation. *Miner. Eng.* **2016**. [CrossRef]
29. Yang, X.; Makkonen, H.T.; Pakkanen, L. Rare Earth Occurrences in Streams of Processing a Phosphate Ore. *Minerals* **2019**, *9*, 262. [CrossRef]
30. Caster, S.B.; Hendrick, J.B. Rare Earth Elements. In *Industrial Minerals and Rocks: Commodities, Markets, and Uses*, 7th ed.; Kogel, J.E., Trivedi, N.C., Barker, J.M., Krudowski, S.T., Eds.; SME: Dearborn, MI, USA, 2006; p. 1568.
31. Burg, J.-P.; Ricou, L.-E.; Ivanov, Z.; Godfriaux, I.; Dimov, D.; Klain, L. Syn-metamorphic nappe complex in the Rhodope Massif. Structure and kinematics. *Terra Nova* **1996**, *8*, 6–15. [CrossRef]
32. Ricou, L.-E.; Burg, J.-P.; Godfriaux, I.; Ivanov, Z. Rhodope and Vardar: The metamorphic and the olistostromic paired belts related to the Cretaceous subduction under Europe. *Geodin. Acta* **1998**, *11*, 285–309. [CrossRef]
33. Mposkos, E.; Kostopoulos, D. Diamond, former coesite and supersilicic garnet in metasedimentary rocks from the Greek Rhodope: A new ultrahigh-pressure metamorphic province established. *Earth Planet. Sci. Lett.* **2001**, *192*, 497–506. [CrossRef]
34. Perraki, M.; Proyer, A.; Mposkos, E.; Kaindl, R.; Hoinkes, G. Raman micro-spectroscopy on diamond, graphite and other carbon polymorphs from the ultrahigh-pressure metamorphic Kimi Complex of the Rhodope Metamorphic Province, NE Greece. *Earth Planet. Sci. Lett.* **2006**, *241*, 672–685. [CrossRef]
35. Brun, J.P.; Sokoutis, D. Kinematics of the Southern Rhodope Core Complex (North Greece). *Int. J. Earth Sci.* **2007**, *96*, 1079–1099. [CrossRef]

36. Liati, A.; Gebauer, D.; Fanning, C.M. Geochronology of the Alpine UHP Rhodope Zone: A review of isotopic ages and constraints on the geodynamic evolution. In *Ultrahigh–Pressure Metamorphism: 25 Years after the Discovery of Coesite and Diamond*; Dobrzhinetskaya, L., Faryad, S.W., Wallis, S., Cuthbert, S., Eds.; Elsevier: Amsterdam, The Netherlands, 2011; pp. 295–324.

37. Burg, J.P. Rhodope: From Mesozoic convergence to Cenozoic extension. Review of petro-structural data in the geochronological frame. *J. Virtual Explor.* **2012**, *42*, 1–44. [CrossRef]

38. Tranos, M.D. Slip preference analysis of faulting driven by strike-slip Andersonian stress regimes: An alternative explanation of the Rhodope metamorphic core complex (northern Greece). *J. Geol. Soc.* **2017**, *174*, 129–141. [CrossRef]

39. Dinter, D.A.; Royden, L. Late Cenozoic extension in northeastern Greece: Strymon Valley detachment system and Rhodope metamorphic core complex. *Geology* **1993**, *21*, 45–48. [CrossRef]

40. Papanikolaou, D.; Panagopoulos, A. On the structural style of southern Rhodope, Greece. *Geol. Balc.* **1981**, *11*, 13–22.

41. Dinter, D.A.; MacFarlane, A.; Hames, W.; Isachsen, C.; Bowring, S.; Royden, L. U-Pb and 40Ar/39Ar geochronology of the Symvolon granodiorite: Implications for the thermal and structural evolution of the Rhodope metamorphic core complex, northeastern Greece. *Tectonics* **1995**, *14*, 886–908. [CrossRef]

42. Pe-Piper, G.; Piper, D.J.; Lentz, D.R. *The Igneous Rocks of Greece: The Anatomy of an Orogeny*; Gebruder Borntraeger: Berlin, Germany, 2002; 573p.

43. Hubbard, C.; Snyder, R. RIR—Measurement and Use in Quantitative XRD. *Powder Diffr.* **1988**, *3*, 74–77. [CrossRef]

44. Anthony, J.W.; Bideaux, R.A.; Bladh, K.W.; Nichols, M.C. Handbook of mineralogy. In *Mineralogical Society of America*; Mineral Data Publishing: Chantilly, VA, USA, 2001.

45. Rosenblum, S.; Brownfield, I.K. *Magnetic Susceptibilities of Minerals—Report for US Geological Survey*; U.S. Geological Survey: Reston, VA, USA, 1999; pp. 1–33.

46. Gupta, C.K.; Krishnamurthy, N. *Extractive Metallurgy of Rare Earths*; CRC Press: Boca Raton, FL, USA, 2005; p. 484.

47. Rudnick, R.L.; Gao, S. Composition of the continental crust. *Crust* **2003**, *3*, 1–64.

48. McDonough, W.F.; Sun, S.S. The Composition of the Earth. *Chem. Geol.* **1995**, *120*, 223–253. [CrossRef]

49. Reisman, D.; Weber, R.; McKernan, J.; Northeim, C. *Rare Earth Elements: A Review of Production, Processing, Recycling, and Associated Environmental Issues*; EPA Report EPA/600/R-12/572; U.S. Environmental Protection Agency (EPA): Washington, DC, USA, 2013. Available online: https://nepis.epa.gov/Adobe/PDF/P100EUBC.pdf (accessed on 30 August 2019).

50. Weng, Z.H.; Jowitt, S.M.; Mudd, G.M.; Haque, N. Assessing rare earth element mineral deposit types and links to environmental impacts. *Appl. Earth Sci.* **2013**, *122*, 83–96. [CrossRef]

51. Dutta, T.; Kim, K.-H.; Uchimiya, M.; Kwonc, E.E.; Jeon, B.-H.; Deep, A.; Yun, S.-T. Global demand for rare-earth resources and strategies for green mining. *Environ. Res.* **2016**, *150*, 182–190. [CrossRef]

52. Wall, F.; Rollat, A.; Pell, R.S. Responsible sourcing of critical metals. *Elements* **2017**, *13*, 313–318. [CrossRef]

53. Grammatikopoulos, T.; Mercer, W.; Gunning, C. Mineralogical characterisation using QEMSCAN of the Nechalacho heavy rare earth metal deposit, Northwest Territories, Canada. *Can. Metall. Q.* **2013**, *52*, 265–277. [CrossRef]

54. British Geological Survey. Rare Earth elements profile. In *Mineral Profile Series*; BGS NERC: Keyworth, UK, 2011; p. 53.

55. Yang, X.; Satur, J.V.; Sanematsu, K.; Laukkanen, J.; Saastamoinen, T. Beneficiation studies of a complex REE ore. *Miner. Eng.* **2015**, *71*, 55–64. [CrossRef]

56. Jordens, A.; Marion, C.; Kuzmina, O.; Waters, K.E. Physicochemical aspects of allanite flotation. *J. Rare Earths* **2014**, *32*, 476–486. [CrossRef]

57. Angelatou, V.; Drossos, E. Beneficiation of green black sands for REE recovery. In Proceedings of the ERES, Santorini, Greece, 28 June–1 July 2017; pp. 49–50.

Article

Factors Controlling the Chromium Isotope Compositions in Podiform Chromitites

Maria Economou-Eliopoulos [1,*], Robert Frei [2] and Ioannis Mitsis [1]

[1] Department of Geology and Geoenvironment, University of Athens, 15784 Athens, Greece; mitsis@geol.uoa.gr
[2] Department of Geosciences and Natural Resource Management, University of Copenhagen, Øster Voldgade 10, 1350 Copenhagen K, Denmark; robertf@ign.ku.dk
* Correspondence: econom@geol.uoa.gr

Received: 12 November 2019; Accepted: 19 December 2019; Published: 21 December 2019

Abstract: The application of Cr isotope compositions to the investigation of magmatic and post-magmatic effects on chromitites is unexplored. This study presents and compiles the first Cr stable isotope data (δ^{53}Cr values) with major and trace element, contents from the Balkan Peninsula, aiming to provide an overview of the compositional variations of δ^{53}Cr values in ophiolite-hosted chromitites and to delineate geochemical constraints controlling the composition of chromitites. The studied chromitites exhibit δ^{53}Cr values ranging from −0.184‰ to +0.159‰, falling in the range of so-called "igneous Earth" or "Earth's mantle inventory" with values −0.12 ± 0.11‰ to 0.079 ± 0.129‰ (2sd). A characteristic feature is the slightly positively fractionated δ^{53}Cr values of all chromitite samples from Othrys (+0.043 ± 0.03‰), and the occurrence of a wide range of δ^{53}Cr values spanning from positively, slightly negatively to the most negatively fractionated signatures (Pindos, δ^{53}Cr = −0.147 to +0.009‰; Skyros, δ^{53}Cr = −0.078 to +0.159‰). The observed negative trend between δ^{53}Cr values and Cr/(Cr + Al) ratios may reflect a decrease in the δ^{53}Cr values of chromitites with increasing partial melting degree. Alternatively, it may point to processes related to magmatic differentiation, as can be seen in our data from Mikrokleisoura (Vourinos).

Keywords: chromium isotopes; chromitites; ophiolites; Balkan Pensula

1. Introduction

Chromitites in ophiolite complexes are studied extensively, in particular with respect to mantle-lithospheric slab interactions, post magmatic processes and subduction recycling scenarios [1–8]. Chromitites differ in their major and trace element compositions, including their platinum-group element (PGE) content, and also have been shown to exhibit variations in their chromium isotope signatures. Although the PGE content in chromitites is commonly a few 100 s of ppb, a significant PGE-enrichment, reaching tens of ppm has been reported in high-Cr or high-Al type chromitites from small occurrences, either in Pt and Pd (PPGE; incompatible, with $Di > 1$) or in Os, Ir, Ru (IPGE; compatible, $Di < 1$). Primary magmatic compositional trends recorded in chromite may be obliterated later on during secondary processes, in that Platinum-Group Minenerals (PGM) and silicates may have been substantially modified during subsequent subsolidus re-equilibration and/or by reaction with metasomatic fluids, during extended periods of deformation (including ductile asthenosphere mantle flow), and by shallow crustal, brittle deformation. The factors controlling the trace element distribution in chromitites remain unclear [8–20].

Likewise, limited data on the chromium isotope compositions of chromitites have shown isotopically heavier signatures compared to mantle peridotites [2,21–23]. However, the role of magmatic processes on the Cr isotope system during partial melting and magma differentiation is still unexplored, and there is still extensive debate on the reasons for the diversity of PGE contents in chromitites [22–24].

Although δ^{53}Cr values of chromitites from the Balkan peninsula are still very limited [2], Cr isotope studies have been performed by several authors in the view to evaluate hexavalent chromium (Cr(VI)) contamination in groundwater and rock leachates [25–30]. For example, Cr isotope signatures indicated that Cr(VI) contamination in groundwater from the Friuli Venezia Giulia Region of Italy is related to the oxidation of trivalent (Cr(III)) deposited as a consequence of past industrial activities [25]. The δ^{53}Cr values and Cr(VI) concentrations in contaminated water from the Evia and Assopos Basins (Greece) are compatible with global data ranges that characterize waters contaminated by both natural processes and anthropogenic activities, and probably delineate potential contamination sources [28]. Also, Cr isotope-based studies have shown oxidative mobilization of Cr(VI) from ultramafic host rocks, and successive back-reduction of the thus mobilized Cr(VI) fractions [29]. Furthermore, the record of a 12-month long time-series of δ^{53}Cr values in run-off from a small serpentinite-dominated catchment in Central Europe, revealed that the Cr(VI) export flux during winter was significantly higher than during the summer [29]. These recent studies and respectively supporting data, in the light of the widespread distribution of ophiolite complexes in Europe and in orogenic zones elsewhere around the globe, emphasize the need for further more detailed studies addressing the variation of δ^{53}Cr values in Cr-bearing rocks and ores.

The ophiolite complexes form an important component of the Tethyan metallogenetic belt, extending from the Mirdita zone in Albania to the Vourinos, Pindos and Othrys complexes of the south zone. These ophiolite complexes host large chromite deposits and smaller chromite occurrences, characterized by high-Cr and high-Al, low- and high-IPGE and PPGE compositions, and by PGE-patterns with both negative and positive slopes [10,14,15,31]. Such a compositional diversity in chromitites may provide an opportunity for in detail studies aimed at a better understanding of the role of magmatic processes on the Cr isotope signatures of the respective chromites [21–24]. In our study we present the first systematic chromium isotope data (expressed as δ^{53}Cr values) of massive chromitites hosted in ophiolite complexes of the Balkan Peninsula, along with trace element contents, including PGE, and combined with scanning electron microscopy/energy-dispersive X-ray spectroscopy (SEM/EDS) identifications. We aim at delineating geochemical constraints that potentially control the elemental composition of these chromitites and their Cr isotope compositions, ultimately with the overall intention to contribute to the origin and genesis of chromite and PGE mineralization and deposits formation.

2. Materials and Methods

2.1. Preparation and Analysis of the Chromitites

For the purpose of the present study, 2.5 to 3.0 kg amounts of chromitite sample mostly from locations of the Balkan Peninsula (Table 1) were crushed into pieces of approximately 0.3 × 0.3 cm with the help of a jaw crusher. The crushed piles were mixed well and divided into four equal portions. Subsequently, two fractions were removed and the remains were remixed again. The procedure was repeated until the final split weighed about 500 g. This split, serving as experimental material, was pulverized to about 100 mesh using an agate mortar. A conventional oxidative alkaline fusion (OAF) was carried out in corundum crucibles; about 1 g of sodium peroxide (Na_2O_2) and 0.3 g of Na_2CO_3 were added to about 0.1 to 0.3 g of the powdered sample (amount depended on the concentration of Cr in the sample). The sample and flux materials were mixed and molten at temperatures of 700 to 800 °C in a muffle furnace during 10 min. After cooling, the fused cake was extracted from the crucible and transferred into 100 mL volumetric flasks using deionized water [2]. The solutions were finally filtered through 0.45 μm polyamide membrane filters and aliquots of these solutions were then processed for chromium isotope analysis.

2.2. Chromium Isotope Analysis

Solutions of samples in the amount which would yield about 1 μg of total chromium were pipetted into 13 mL Savillex Teflon beakers. These aliquots were spiked up with adequate amount of a $^{50}Cr-^{54}Cr$ double spike so that a sample to spike ratio of ~3:1 (total chromium concentrations) was achieved. The addition of a $^{50}Cr-^{54}Cr$ double spike of a known isotope composition to a sample before chemical purification allows accurate correction of both the chemical and the instrumental shifts in Cr isotope abundances [1,32]. The mixture was totally evaporated and 1 mL of concentrated aqua regia was subsequently added. After 3 h during which the sample was exposed to aqua regia on a hotplate at 100 °C, the sample was again dried down. The sample was then purified by passing the sample in 0.5 N HCl over an extraction column (BioRad PP columns) charged with 1 mL of 200–400 mesh BioRad AG-50W-X12 cation resin, employing a slightly modified extraction recipe published [33,34]. The Cr yield of this column extraction and purification step is usually ~70%. Samples were loaded onto Re filaments with a mixture of 3 μL silica gel, 0.3 μL 0.5 mol L^{-1} of H_3BO_3 and 0.5 μL 0.5 mol L^{-1} of H_3PO4. The samples were statically measured on a IsotopX "Phoenix" multicollector thermal ionization mass spectrometer (TIMS) at the Department of Geoscience and Natural Resource Management, University of Copenhagen, at temperatures between 1040 and 1150 °C, aiming for beam intensity at atomic mass unit (AMU) 53.9407 of 1 to 2 Volts. Every load was analyzed five times. Titanium, vanadium and iron interferences with Cr isotopes were corrected by comparing with $^{49}Ti/^{50}Ti$, $^{50}V/^{51}V$ and $^{54}Fe/^{56}Fe$ ratios. The final isotope composition of a sample was determined as the average of the repeated analyses and reported relative to the certified (standard reference material) SRM 979 standard as

$$\delta^{53}Cr(‰) = [(^{53}Cr/^{52}Cr_{sample}/^{53}Cr/^{52}Cr_{SRM979}) - 1] \times 1000$$

Repeated analysis of 1 μg loads of unprocessed double spiked SRM 979 standard during the duration of the analysis period yielded an average $\delta^{53}Cr$ value of 0.05 ± 0.06‰ (n = 12; 2σ; ^{52}Cr signal intensity at 2 V) on the "Phoenix" TIMS which we consider as a minimum external reproducibility for a sample run in this study, including separation procedure, double spike correction error, and respective internal analytical errors.

2.3. Mineral Chemistry and Whole Rock Analyses

Polished sections from chromite deposits of Greece (Table 1) were carbon or gold coated and examined by reflected light microscopy and with a SEM using EDS. SEM images and EDS analyses were carried out at the University of Athens (Department of Geology and Geoenvironment) using a JEOL JSM 5600 instrument, equipped with automated energy-dispersive analysis system ISIS 300 OXFORD, with the following operating conditions: accelerating voltage 20 kV, beam current 0.5 nA, time of measurement (dead time) 50 s and beam diameter 1–2 μm. The following X-ray lines were used: OsMα, PtMα, IrMα, AuMα, AgLα, AsLα, FeKα, NiKα, CoKα, CuKα, CrKα, AlKα, TiKα, CaKα, SiKα, MnKα, MgKα, ClKα. Standards used were pure metals for the elements Cr, Fe, Mn, Ni, Co, Ti and Si, MgO for Mg and Al_2O_3 for Al. Contents of Fe_2O_3 and FeO were calculated on the basis of the spinel stoichiometry.

Major and trace elements in massive chromitite samples were determined by inductively coupled plasma mass spectrometry (ICP-MS) analysis, at the ACME Analytical Laboratories Ltd., Vancouver, BC, Canada. The samples were dissolved using a strong multi-acid (HNO_3–$HClO_4$–HF) digestion and the residues dissolved in concentrated HCl. Platinum-group element (PGE) analyses were carried out by Ni-sulphide fire-assay pre-concentration technique, using the nickel fire assay technique from large (30 g) samples. This method allows for complete dissolution of samples. Detection limits are 5 ppb for Ru 2 ppb for Os, Ir, Pt, Pd and 1 ppb for Rh and Au. The CDN-PGMS-23 was used as standard.

3. A Brief Outline of Characteristics of Ophiolites and Hosted Chromitites

3.1. Ophiolites

The geology, petrography, mineral chemistry and geochemistry of ophiolites in the Balkan Peninsula has been a topic of extensive investigation in previous publications [10,35–48]. They are an important component of the Upper Jurassic to Lower Cretaceous Tethyan ophiolite belt, which extends through the Serbian zone of the Dinarides in the north (Mirdita zone) to the Subpelagonian zone (Pindos, Vourinos and Othrys complexes) in the south (Figure 1).

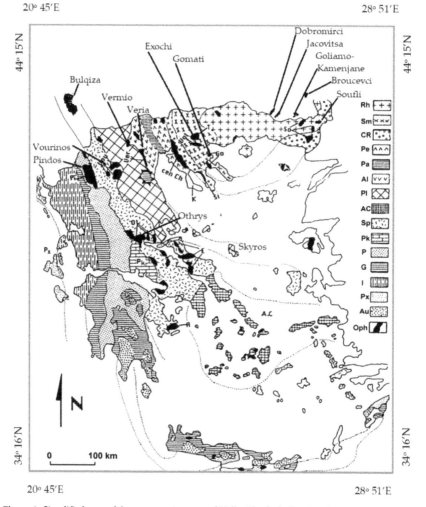

Figure 1. Simplified map of the geotectonic zones of Hellenides [35], showing the ophiolite complexes from which chromitite samples were studied. Symbols: Oph = ophiolites; Rh = Rhodope zone; Sm = Serbo–Macedonian Massif; CR = Circun–Rhodope zone; (Pe, Pa, Al) = Axios zone; Pl = Pelagonian zone; AC = Attico–Cucladic zone; Sp = Pindos zone; P = Parnassos–Giona zone; G = Gavrovou–Tripolis zone; I = Ionian zone; Px = Pakson zone.

These ophiolites are characterized by petrological and geochemical features typical of both Mid-Ocean Ridge (MOR) and Supra-Subduction Zone (SSZ)-type associations (fore-arc and back-arc ridges). They are accompanied by minor dunite bodies, and overlain by gabbroic cumulates, mafic dykes, and an extrusive sequence with a compositional basalt range between MOR and island-arc-tholeiitic to boninitic types [35–48]. Recently, ophiolite studies have focused on the importance of intra-oceanic subduction-initiation processes in ophiolite genesis [49]. Relatively small isolated ophiolite masses, of mostly serpentinized dunite and harzburgite, are located at the western margin of the Axios zone (Vermio–Veria), and the Eohellenic Pre-Cretaceous nape, including Skyros Island (Figure 1). Structural and paleomagnetic studies on those ophiolites have revealed widespread heterogeneous deformation and rotation during their original displacement and subsequent tectonic incorporation into continental margins [35–38]. The origin of major Jurassic ophiolite complexes in Greece in a hydrous SSZ environment has been evidenced by the presence of hydrous silicate inclusions, such as amphibole and phlogopite, in chromitites and by the trace and Rare Earth Elements (REE) data on separated orthopyroxenes and clinopyroxenes from harzburgites of these ophiolites [48]. Pyroxenites with variable modal contents of olivine, garnet and spinel, ranging in composition from orthopyroxenite through websterite to clinopyroxenite have been described in the Veria ophiolites [50,51].

3.2. Characteristics of Chromitites

Chromitites in the Balkan Peninsula are classified on the basis of the variation in the chromitite tonnage, the composition of chromite, the degree of transformation of ores and the associated ophiolites, and on the association of chromitites with sulphides.

The Othrys complex includes two tectonically separated chromite deposits, namely Eretria (Tsagli) and Domokos, in addition to several other occurrences, including Agios Stefanos, the combined tonnage being approximately 3 Mt of high-Al massive chromite ores occurring in moderately depleted harzburgite [10,35,39,44,45]. A salient feature of the Eretria (Tsagli) chromite deposits is the occurrence of Fe–Ni–Cu-sulphide mineralization with dominant minerals pyrrhotite, chalcopyrite and minor cobalt-bearing pentlandite, hosted in serpentinized harzburgites [52]. Massive sulphide mineralization occurs at the peripheral parts of podiform chromite bodies in association with chromitite and magnetite (Figure 2). Chromite is often characterized by brittle deformation and by sulphide-silicate veins (Figure 2d,e). Texture and geochemical characteristics, including PGE content, flat chondrite-normalized PGE-patterns, and very low partition coefficients for Ni and Fe between olivine and sulphides are inconsistent with sulphides having been in equilibrium with Ni-rich host rocks, at magmatic temperature [52,53]. In addition, it has been documented [54] that in the Othrys and Vourinos ophiolite complexes massive and disseminated chromitites host considerable amounts of methane in micro-fractures and in porous serpentine- or chlorite-filled veins.

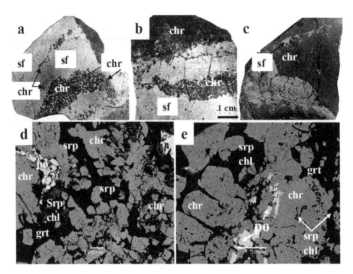

Figure 2. Photographs showing the texture of chromitite associated with sulphides and magnetite from the peripheral parts of podiform chromitite bodies at Eretria. Chromite occurs interstitially in sulphide rich portions (**a,b**) and sulphides occur as veins and irregular masses within zones dominated by chromite (**b,c**). Back-scattered electron images from massive chromitite samples from central parts of podiform bodies, showing fragmented chromite and evidence for the existence of sulphides (**d,e**) even distant from the peripheral parts of the chromitite bodies. Abbreviations: chr = chromite; sf = sulphides; po = pyhhrotite; srp = serpentine; chl = chlorite; grt = garnet. Scale bar: 1 cm is the same for all hand samples.

Chromite ores in the Vourinos complex occur in tectonite and cumulate sequences. Only those in the tectonites are under exploitation. All chromite ores, with a tonnage estimated to be approximately 10 Mt of high-Cr ore are found within dunite bodies or dunite envelopes in harzburgite, but there is no systematic relation between size of the dunite body and that of the ore body. The sizes of the ore bodies vary widely and contain all textural types (massive, schlieren, banded, disseminated and nodular), but usually a single type dominates. High-temperature deformation, superimposed on primary magmatic textures is very common [42,47]. The mantle sequence of the Pindos ophiolite complex resembles that of Vourinos in the presence of extensive and highly depleted harzburgite but, in contrast to Vourinos, there is only a limited number of small chromite occurrences (low potential for exploitation).

The chromitite occurrences in the Pindos ophiolite complex are small (a few tens of meters × a few tens of centimeters) and cover all textural types (massive, schlieren, banded, disseminated and nodular). Chromitites occur within completely serpentinized and weathered, intensively deformed dunite–harzburgite blocks, as a result of strong plastic and brittle deformation that was superimposed on primary magmatic textures. The chromitites are mostly fine- to medium-grained, and consist of aggregates of fractured chromite accompanied by chlorite and traces of tremolite. Primary olivine is preserved only in the form of inclusions within chromian spinel, while abundant remnants of base metal sulphides (BMS), now preserved as alloys, occur both as inclusions and interstitial phases in the silicate matrix [10,18,19]. Chromitites throughout the Pindos complex are of high-Cr and high-Al compositions, often in a spatial association with each other [45,55–57]. The most salient feature of the Pindos chromitites is the PPGE-enrichment in those bodies occurring in the area of Korydallos, with PGE_{total} concentrations reaching 7 ppm [43–45] and reportedly up to 29 ppm [18].

Relatively small massive chromitite bodies are located at the Achladones area on the island of Skyros. They are of high-Al type (average Cr/(Cr + Al) ratio of unaltered chromite is 0.56 while the Mg/(Mg + Fe^{2+}) ratio is 0.64) and contain elevated PGE contents, up to 3 ppm PGE_{total}. However, both high-Cr and high-Al types have been reported from Skyros [46,47,58]. Furthermore, SEM/EDS investigations of Au-coated polished sections of chromitite from Skyros revealed the presence of graphite (Figure 3).

Ophiolites associated with the Serbomacedonian massif (Gomati and Exochi) and Rhodope massif including the ophiolites Soufli–Tsoutoura (Greece), Dobromirci, Jacovitsa, Broucevci and Goliamo–Kamenjane (Bulgaria) (Figure 1) are completely serpentinized, locally sheared and metamorphosed to antigorite–tremolite and/or talc schists, and they are all small (a few thousand tons of ore). A characteristic feature of those chromitite occurrences is the common spatial association of high-Cr and high-Al ores, and the negative slope of the PGE patterns, comparable to those for podiform chromitites in ophiolites elsewhere [59–61]. The area of Ceruja is part of the western Bulqiza ophiolite complex as part of the Mirdita zone (Albanides), and includes harzburgitic to lherzolitic types [62]. Although more than 40 Mt of chromitites have been located in the uppermost part of the mantle harzburgites throughout the Bulqiza complex, in the area of Ceruja there are only small (a few thousand) chromite occurrences.

There is a wide compositional variation in chromite throughout the Bulqiza complex, with Cr/(Cr + Al) ranging from 0.46 to 0.86 and Mg/(Mg + Fe^{2+}) ratios ranging from 0.46 to 0.69. The Cr/Cr + Al) ratio in studied samples from the area of Ceruja is restricted to 0.43–0.56, their PGE content is low and the PGE-patterns show a negative slope, similar to other chromitites hosted in ophiolite complexes [62]. However, in the western Bulqiza massif, mineralization occurring as disseminations in dunite is accompanied by chromite in the transition zone and lower cumulates sequence of the complex and elevated PGE contents, up to 9000 ppb ΣPGE [62–64].

Figure 3. Back-scattered electron images (Au-coated polished sections) of strongly fragmented chromitite from Skyros Island, showing microstructural inclusions of graphite. Partial replacement of serpentine flakes by graphite (**a,b,d**); Overgrowth of fine graphite nodules (white arrows) are revealed on unpolished parts of polished sections (**a,c**); (**f**) Dissolution pits in silicate inclusions of chromite are filled with graphite (**b,d**). Abbreviations: grp = graphite-like chr = chromite; srp = serpentine; chl = chlorite.

4. Results

4.1. Distribution of Selected Trace Elements

Selected trace element contens of Ni, Co, Mn, Zn, V, Ga Ti, and platinum-group elements, which are from previous studies, are contained in Table 1.

Table 1. Chromium isotope (δ^{53}Cr) and selected trace element data for chromitites from the Balkan Peninsula.

No	Location	δ^{53}Cr (‰)	SEM/EDS		PGE Content (ppb)									Trace Element Content (ppm)							wt %	wt %	wt %
			Cr#	Mg#	Os	Ir	Ru	Rh	Pt	Pd	ΣPGE	Pd/Ir	Pt/Pt*	Ni	Co	V	Zn	Mn	Ga	Sc	Fe	Ca	Ti
	Othrys (D), GR																						
1	Eretria	0.023	0.57	0.63	22	13	67	2	10	1	115	0.08	2.26	1600	220	900	340	1250	34	6	9.6	<0.1	0.03
2	Eretria	0.036	0.56	0.66	25	14	25	3	4	1	72	0.07	0.74	1400	210	1100	430	990	32	6	10	0.2	0.03
3	Domokos	0.025	0.53	0.74	15	11	30	5	24	2	87	0.18	2.43	1600	200	960	310	1300	32	7	9.8	<0.1	0.04
4	Domokos	0.088	0.5	0.73	20	10	35	5	8	11	89	1.1	0.34	1200	220	900	400	2500	33	6	9.6	<0.1	0.04
	Vourinos (D), GR																						
5	Kondro		0.77	0.65	26	36	55	17	4	2	140	0.06	0.22	2000	240	500	260	1560	14	6	8.1	<0.1	0.04
6	Voidolakos	-0.12	0.81	0.62	14	17	80	12	6	6	135	0.35	0.23	1580	210	560	550	1540	14	6	7.9	<0.1	0.04
7	Kissavos	-0.093	0.81	0.66	14	11	49	7	4	7	92	0.55	0.55	1900	200	400	300	1000	13	6	8.7	<0.1	0.04
8	Kissavos	-0.078	0.56	0.72	4	3	11	3	3	6	30	0.54	0.23	1800	170	780	360	850	27	6	9.2	<0.1	0.03
9	Mikrokleisoura	-0.056	0.78	0.61	30	11	40	16	8	4	109	0.63	0.26	1800	180	620	280	950	14	8	7.8	<0.1	0.05
10	Mikrokleisoura	-0.032	0.61	0.52	50	8	42	18	8	6	132	0.75	0.96	1000	230	1000	490	1760	20	12	15	0.5	0.1
	Pindos, Low PGE																						
11	P Dako's mine	-0.03	0.52	0.63	4	6	30	1	4	6	51	1.0	0.52	1300	140	600	400	760	23	5	7.8	<0.1	0.04
12	Korydallos	0.009	0.48	0.6	40	20	33	5	25	20	143	1.0	0.8	1500	260	710	410	1200	32	5	9.8	<0.1	0.08
13	Kampos Despoti	-0.096	0.81	0.58	30	18	50	3	76	4	181	0.22	7.01	1450	290	560	490	1470	17	5	10	<0.1	0.03
14	Korydallos	-0.149	0.42	0.71	8	3	11	1	75	19	117	6.33	5.5	1500	260	710	520	1300	34	5.1	7.8	<0.1	0.09
	Pindos, High-PPGE																						
15	Korydallos	-0.098	0.68	0.57	14	11	57	13	3460	1660	5215	151	7.53	1550	230	510	450	1200	30	5.7	7.8	<0.1	0.07
16	Korydallos	0.062	0.67	0.57	47	49	55	104	3020	600	3875	12.2	3.86	750	270	760	520	1280	16	6.8	9.7	<0.1	0.07
17	Korydallos	-0.102	0.69	0.49	62	47	80	112	1460	337	2098	7.2	2.4	720	240	760	620	1340	23	6.7	10	<0.1	0.08
18	Korydallos	-0.06	0.4	0.72	88	74	130	142	1720	525	2679	7.1	2.01	630	240	790	510	1300	25	6.8	9.9	<0.1	0.09
19	Korydallos	-0.086	0.4	0.72	45	38	82	78	4580	1300	6123	34.2	4.6	1630	240	580	460	1400	32	5.4	7.3	<0.1	0.07
	Pindos, High-IPGE																						
20	Milia	-0.147	0.82	0.42	150	320	350	82	150	7	1059	0.02	2.00	1600	340	980	1160	2000	15	5	12	0.1	0.01
	Skyros, high IPGE																						
21	Achladones	-0.078	0.61	0.64	140	480	1200	160	280	39	2300	0.08	1.13	1300	250	1200	540	1300	34	5	11	<0.1	0.04
22	Achladones	0.117	0.55	0.69	60	40	300	13	23	28	464	0.7	0.39	1600	200	1000	400	1650	36	5	9.4	<0.1	0.04
23	Achladones	0.159	0.58	0.51	13	31	159	7	20	21	251	0.67	0.53	1500	240	870	450	1600	40	5	9.6	<0.1	0.04
	Skyros, Low PGE																						
24	Ag. Ioannis	-0.111	0.75	0.69	30	15	85	6	4	5	145	0.33	0.23	1250	220	640	420	2150	14	6	11	<0.1	0.08
25	Parapisti	-0.084	0.79	0.69	25	20	90	5	3	2	145	0.1	0.3	1200	200	620	400	1900	11	5	10	<0.1	0.05
26	Ceruja	-0.051	0.43	0.69	3	7	18	3	25	7	63	1	1.74	2640	170	610	300	960					
27	Ceruja	-0.129	0.52	0.58	3	7	15	2	28	9	64	1.3	2.1	1640	160	1030	420	1210					
28	Ceruja	-0.047	0.56	0.65	3	7	18	3	25	7	63	1	1.74	2150	160	990	340	1100					

Table 1. Cont.

No	Location	δ53Cr (‰)	SEM/EDS		PGE Content (ppb)									Trace Element Content (ppm)							wt %	wt %	wt %
			Cr#	Mg#	Os	Ir	Ru	Rh	Pt	Pd	ΣPGE	Pd/Ir	Pt/Pt*	Ni	Co	V	Zn	Mn	Ga	Sc	Fe	Ca	Ti
	Vermio-Veria																						
29	Veria	−0.075	0.75	0.62	70	15	100	9	3	2	199	0.13	0.23	1500	200	650	700	2800					
30	Vermio	−0.076	0.7	0.65	30	20	55	4	4	31	144	0.05	0.12	880	170	700	500	1400					
31	Rhodope massif, GR	−0.145	0.75	0.66	70	15	100	9	3	2	199	0.13	0.23	1770	150	550	300	1800					
	Rhodope massif, Bu																						
32	Soufli	−0.037	0.7	0.62	20	25	85	7	7	5	150	0.2	0.38	1260	210	610	360	1850					
33	Tsoutoura	−0.033	0.63	0.65	25	6	40	2	3	6	82	1.00	0.28	1430	220	670	360	1270					
34	Dobromirci	−0.184	0.65	0.68	16	32	120	6	4	14	190	0.44	0.14	520	160	590	230	2300					
35	Goliamo Kamenjane	−0.099	0.6	0.62	3	6	14	3	3	1	30			2000	780	500	3700	7500					
36	Broucevci	−0.147	0.47	0.72	5	13	23	4	2	12	60	0.92	0.09	800	180	600	370	2500					
37	Jacovitsa	−0.097	0.75	0.61	22	42	53	7	3	2	129	0.05	0.26	790	240	650	400	3500					
	SMM, GR																						
38	Exochi	0.02	0.76	0.62	23	24	73	6	5	7	140	0.29	0.25	1240	240	670	400	1850					
39	Exochi	−0.057	0.75	0.6	25	10	70	4	4	6	119	0.6	0.26	1030	150	1300	300	1270					
40	Gomati	0.084	0.49	0.67	9	20	60	6	5	4	104	0.25	0.33	1050	330	660	350	2700					

* Present data; PGE data from previous publications [24,34,43,44,50]. Symbols: Cr# = Cr/(Cr + Al); Mg# = (Mg + Fe^{2+}).

The rare earth elements La, Ce, Nd, Sm, Yb, Pr, Eu, Gd, Tb, Dy, Ho, Er, Tm and Lu as well as U, Th, Cs, Sb, Y, Zr, Ge, W, Mo, Sc, Pb, Li, Sr and K were lower than the detection limits in detail specified in the section of the analytical methods, except for La and Ce contents in the chromitite samples from Othrys and Skyros island, which are slightly higher than the respective detection limit. The best defined inverse correlations are those between the Cr/(Cr + Al) ratio and V (r = −0.80 to −0.85) for chromitites from Vourinos, Pindos (low PGE) and Skyros, and between the Cr/(Cr + Al) ratio (Figure 4a), and Ga (r = −0.90 to −0.96) for chromitites from Vourinos, Pindos (low PGE) and Skyros and for high-PGE chromitites (r = −0.65; Figure 4b).

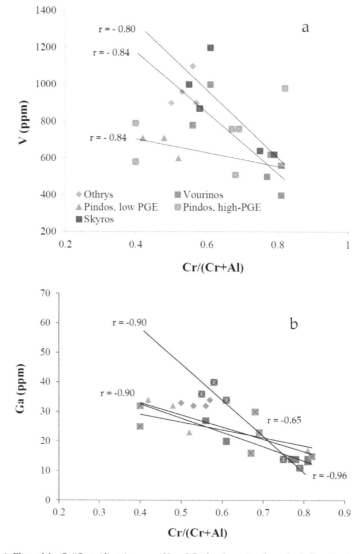

Figure 4. Plots of the Cr/(Cr + Al) ratio versus V and Ga for chromites from the Balkan Peninsula. The best pronounced negative correlations are those between Cr/(Cr + Al) ratio and V for chromitites from Vourinos, Pindos (low PGE) and Skyros (**a**), and Ga for chromitites from Vourinos, Pindos (low PGE) and Skyros (**b**). Data from Table 1.

4.2. Chromium Isotope Results

The analysed chromitites, representative of a wide range in their major element (high-Cr and high-Al) and PGE composition, show a relatively wide range of chromium isotope compositions, expressed as δ^{53}Cr values, even in spatially related occurrences, throughout the entire metallogenic belt. The δ^{53}Cr values range from −0.184‰ (in the Bulgarian Rhodope massif) to +0.159‰ on Skyros Island (Figure 5; Table 1). The lower values (−0.246‰) are comparable to those having been reported by [2]. δ^{53}Cr as low as −0.21‰ are compatible with the δ^{53}Cr isotope range for mantle-derived rocks [1], while the more positively fractionated values of up to +0.159‰ exceed those values around +0.068‰ reported for global chromitites from layered intrusions and podiform-type occurrences [2].

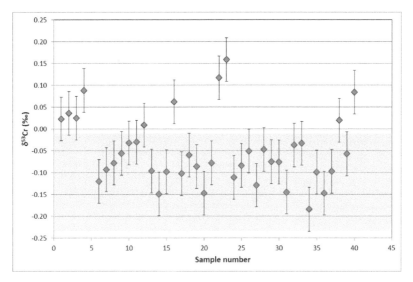

Figure 5. δ^{53}Cr values of the chromitites (n = 40) from the Balkan Peninsula (Table 1) studied herein. The blue coloured band displays the range of high-temperature magmatic signatures as has been defined by [1]. Errors are 2sd. The reproducibility of the (standard reference material) SRM 979 standard under similar measuring conditions is at +/−0.06‰. The horizontal axis corresponds to the number of analyzed samples 1–40 (Table 1).

Along with the δ^{53}Cr values of the chromitites, selected major and trace element compositions reported in previous publications are given (Table 1) in order to delineate potential geochemical constraints that control their composition and the Cr isotope compositions. It is remarkable that all chromite samples from Othrys and certain ores from Skyros exhibit positive δ^{53}Cr values, while chromitites from the Serbo-macedonian-Rhodope massifs and Pindos show a wide range from positive, slightly negative to the most negative δ^{53}Cr values (Figures 5 and 6; Table 1). Also, the results do not show any correlation trend either between total PGE content or Mg/(Mg + Fe^{2+}) ratio and δ^{53}Cr values. There is a slightly negative trend between δ^{53}Cr values and Cr/(Cr + Al) ratio, which is better pronounced for the chromitite samples from Othrys and Vourinos (Figure 6a) and a positive trend between δ^{53}Cr values and Ga content (Figure 6b).

The Pd/Ir values for chromitites from the Balkan Peninsula are plotted *versus* normalized values (Pt/Pt* = (Pt/8.3)/(Rh/1.6)(Pd/4.4) (Figure 7, [65])). The negative Pt/Pt* values (Pt/Pt* < 1) and low Pd/Ir ratios are a characteristic feature of IPGE-elevated chromitites from Skyros and of low-PGE chromitites, in contrast to PPGE-elevated ones from the Pindos complex (Figure 7; Table 1).

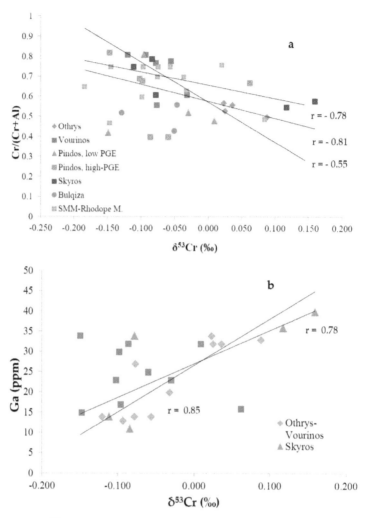

Figure 6. Plots of δ^{53}Cr values *versus* Cr/(Cr + Al) ratios and Ga for certain chromitites from the Balkan Peninsula. There is a good negative trend between δ^{53}Cr values and Cr/(Cr + Al) ratios for chromitites from Othrys and Skyros (**a**), and a positive trend between δ^{53}Cr values and Ga contents for chromitites from Skyros and Vourinos–Othrys (**b**). Data from Table 1.

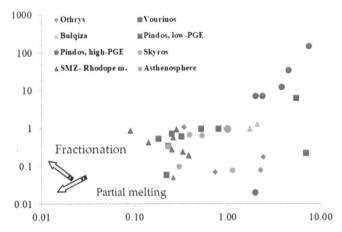

Figure 7. Plots of Pd/Ir *versus* Pt/Pt* normalized values [(Pt/Pt* = (Pt/8.3)/(Rh/1.6) × (Pd/4.4)] after [65] of chromitities studied herein, grouped by locality. Green circle = Asthenosphere mantle, calculated by [66]. Data from Table 1.

5. Discussion

Geochemical compositions and mineral chemistry characteristics of peridotites and chromitites from Greece have shown that the partial melting degree of primitive mantle and the hydrous nature of parent magma are major factors that control the compositional variations in them [10,18,19,42,45,56,57]. In addition, the known widespread geochemical heterogeneities in the bulk-rock compositions of chromitites from the Balkan Peninsula (Table 1) would be inconsistent with their evolution through simple partial melting processes, and suggest that their original geochemical compositions may have been modified by subsequent, post-magmatic processes taking place at shallow mantle depths [8,66–69].

5.1. Potential Pathways Contributing to Chromium Isotope Fractionation

Chromium occurs in different oxidation states in nature, with relatively immobile Cr(III) and very soluble oxidized Cr(VI) being the most abundant species [70]. Redox processes are accompanied by significant isotope fractionation, also in systems where Cr(VI) is transformed to Cr(III) [71]. During reduction, the lighter isotopes are preferentially reduced, resulting in an enrichment of ^{53}Cr relative to ^{52}Cr values in the remaining Cr(VI) pool. This enrichment is measured as the change in the ratio of ^{53}Cr/^{52}Cr, and is expressed as δ^{53}Cr values in units permil (‰) relative to a standard [32]. In the following we consider potential pathways for changes in redox conditions that might affect the Cr isotope signatures and their changes in magmatic to post-magmatic processes: (1) measured δ^{53}Cr signatures in mantle chromitites reflect primary magmatic compositional variations, (2) oxidative fluids may affect solubility of Cr-bearing minerals and release of Cr(VI) during the formation of ferrian-chromite, (3) reductive fluids may cause reduction of Cr(VI) back to Cr(III), deposition of graphite and Fe–Cr alloys and (4) weathering of Cr-bearing minerals [19–21,25,60–64], and related dissolution/re-precipitation processes potentially having a control on the bulk δ^{53}Cr signatures of chromitites. Based on the observation that bulk silicate earth (BSE) carries a slightly lighter δ^{53}Cr signature compared to that of mantle-derived chromitites, it has been suggested that small but discernible high-temperature Cr isotope fractionations might occur within the Earth's mantle, potentially imparted during fractional crystallization and/or partial melting of mantle sources [1].

5.1.1. Magmatic Processes

It has been established that high-Cr and high-Al podiform chromitites are derived from melts formed by high and low degrees of partial melting in the upper mantle, respectively [3,13,14,72].

Considering the data distribution in the scatter-diagram of Pd/Ir versus the Pt/Pt* ratios for chromitites from the Balkan Peninsula, it appears that chromitites from the large deposits (Vourinos and Othrys) exhibit Pd/Ir < 1 and Pt/Pt* < 1, suggesting partial melting of already depleted mantle [65], and an increasing partial melting trend from high-Al toward high-Cr deposits (Table 1; Figure 7). In addition, the relatively high Pd/Ir ratio, that is considered to indicate the fractionation degree of the PGE [66], provides evidence for fractionation to some extent of parent magmas for the PPGE-elevated chromitites from Pindos compared to the IPGE-elevated chromitites from Skyros and the Milia area of Pindos (Table 1; Figure 7).

The closer inspection of the Cr isotope data from chromitites studied herein, ranging from negative (−0.184‰) to positive (+0.159‰) δ^{53}Cr values (Table 1; Figures 5 and 6), may provide an opportunity for a better understanding of the role of magmatic or/and post-magmatic processes on the Cr isotope signatures of chromitites [2,21–23]. A comparison of the Cr isotope data for chromitites from our study (Table 1) with published values from other chromitites reveals a significant overlap in the δ^{53}Cr data. Our lowest δ^{53}Cr signature (−0.184‰) is slightly higher than the most negatively fractionated δ^{53}Cr value of −0.246‰ reported by [19], and the most negatively fractionated value of −0.21 ‰ in the data set published [1,2], while the isotopically most positively fractionated value in our data set (+0.159‰) is somewhat higher than the hitherto highest reported δ^{53}Cr value (+0.068‰) for global chromitites from ophiolite complexes and layered intrusions [1,2].

The diagrams of δ^{53}Cr values versus Cr/(Cr + Al) ratios depict better defined negative trends for chromitites from large deposits such as those from Othrys–Vourinos and Skyros (Figure 6a), than for chromitites from smaller occurrences. Given that the former chromitites are characterized by small degrees of fractionation of the respective parental melts, as exemplified by the low Pd/Ir ratios (Table 1), the observed correlation trends may point to the lowering of δ^{53}Cr values in the melts alongside with the partial melting degree of the respective magma sources. In addition, the most evolved samples from the lower cumulate sequence of the Vourinos ophiolite complex (sample 10) has higher Cr isotope compositions, along with higher Fe contents, compared to the mantle chromitites (represented by Voidolakkos, Kondro and Kissavos (Table 1), that may suggest a coupling of increasing trends of δ^{53}Cr values with the degree of fractionation of the parent magma. Also, with respect to trace elements, including Ni, Co, Zn, V, Mn, Ti and Ga, defined correlations are only depicted between Ga and V concentrations, and Cr/(Cr + Al) ratios. Negative trends are revealed by the Othrys–Vourinos and Skyros chromitites, both of which including high-Cr and high-Al ores (Figure 4). Considering the low Pd/Ir ratios shown by those chromitites (Table 1), these negative trends could be attributed to compositional variations in the parent magmas, rather than to changes inherent during fractionation of these melts.

5.1.2. Post-Magmatic Processes

In general, the lack of any clear correlation between major and trace elements, and δ^{53}Cr values may be due to post-magmatic processes (dehydration, serpentinization, metamorphism), which may modify the primary magmatic compositional trends recorded by chromite and silicates (depending on the chromite/silicate ratio), during an extended period of deformation (ductile asthenospheric mantle flow to shallow crustal brittle deformation) [68,72–80]. With respect to the Rhodope massif, on the basis of major-, minor- and trace-element compositions of chromite and Re–Os isotopic compositions of magmatic PGM on the Dobromirtsi chromitites, a long-term history involving reworking of the ancient Archean mantle, during the Phanerozoic, has been suggested [78–82]. Geochemical and mineral chemistry data demonstrated a complex interplay of substitutions, related to the ability of fluids to infiltrate the chromite and the extent of the interaction between pre-existing cores and rims, and that during metamorphism minor and trace elements in chromite can be strongly modified [68]. Also, a sequence of petrological, deformation and redox events has been described for the upper-mantle rocks of the Othrys ophiolite complex. These events are seen to have affected the chromitite bodies during their obduction, emplacement and subsequent alteration [36]. The compositional variations in the Skyros chromitites in particular [83], may have been caused by redox changes during multistage transformations in the Othrys chromites. Cr oxidation and reduction processes accompanying these deformational events may have caused the variations of Cr isotope signatures measured in those chromitites. (Table 1; Figures 4 and 6).

The variations of δ^{53}Cr signatures in the high-temperature magmatic environment in which the chromitites from the Balkan peninsula were formed, seems to be in a good agreement with the variations measured in some metamorphic Cr-rich minerals such as uvarovites, Cr-tremolite, Cr-diopside, Cr-pyrope. Some of the positively fractionated δ^{53}Cr signatures of these phases have been interpreted as a consequence of heavy Cr isotope enrichments caused by metamorphism [2,21]. Moreover, the investigation of Cr isotopic composition of different types of mantle xenoliths from diverse geological settings (fertile to refractory off-craton spinel and garnet peridotites, pyroxenite veins, metasomatised spinel lherzolites and associated basalts) revealed δ^{53}Cr values of peridotites which also span a range from negatively to positively fractioned values of −0.51 to +0.75‰, and a slightly negative correlation between δ^{53}Cr and Al_2O_3 and CaO contents for most mantle peridotites [23].

5.2. The Role of Sulphides on the δ^{53}Cr Signatures

The most positively fractionated δ^{53}Cr values are depicted by all massive chromitite samples from the large Othrys complex, and by a few samples from small chromite occurrences (Table 1). Chromitites in certain ophiolite complexes, such as Othrys, Lemesos, Cyprus, Shetland, Bulqiza, Oregon, Moa-Baracoa and elsewhere are associated with Fe–Ni–Cu-sulphide mineralization with dominant minerals pyrrhotite, chalcopyrite and pentlandite, hosted in mantle serpentinized peridotites and ultramafic rocks of the cumulate sequence of the ophiolite complexes [9,17,52,53,62,63]. Although an originally primary magmatic origin for these sulphides is not precluded, characteristics of the highly transformed ore at the Eretria chromite deposit (Greece) may indicate that the original magmatic features have been overprinted and that metals were released from the host rocks by a low-level hydrothermal circulation process [52]. Therefore, in addition to a potential release of Cr from the chromite ore affected by post-magmatic processes, such as brittle deformation and fluid circulation, the less alkaline environment which would be controlled by the dissolution of sulphides could be potentially important for the control of redox-mediated Cr mobilizations and accompanying changes in δ^{53}Cr signatures.

5.3. The Potential Role of Abiotic Methane (CH_4)

A salient feature of the presented data that potentially could explain the rather large compositional and isotope variations, could be the presence of graphite (Figure 3), and the observation of methane [54] occurring in fluid inclusions of fracture-filling minerals, along with the presence of chlorite and serpentine. These features are especially present in the Othrys chromitites, which also show the most positively fractionated δ^{53}Cr values (Table 1; Figure 5). Recently, the investigation of microdiamonds and graphite in chromite from ophiolite-type chromitites hosted in the Tehuitzingo serpentinite (southern Mexico), have been attributed a secondary origin and process related to the retrograde evolution of the respective chromitite. These processes apparently took place at relatively low temperatures (520–670 °C), at low pressures and in shallow depths [69]. The presence of minerals indicative of super-reducing phases, such as graphite-like, Fe–Ni–Cr alloys, awaruite (Ni_3Fe), and heazlewoodite (Ni_3S_2) [44,45,82,83], in the chromitite ores from the islands of Othrys and Skyros may be suggestive of the introduction of continental crust derived, reduced C–O–H fluids, during post-magmatic alteration processes. These were probably facilitated during brittle deformation stages in the shallow crust and in the subducted oceanic slab [69,84,85]. The association of graphite with sperrylite and sulphides in magnetite ore from the Skyros ophiolite complex has also been attributed to its deposition from a CO_2^- and CH_4^-dominated reducing fluid at low oxygen fugacity at relatively low temperatures (500–300 °C) during serpentinization [58]. In addition, methane is widespread in microfractures and porous serpentine- or chlorite-filled veins [54]. The widespread occurrence of methane and Ni–V–Co phosphides in the Othrys chromitites [54,84,85] and graphite in the Skyros island ophiolite (Figure 3) seems to be compatible with the positively fractionated δ^{53}Cr values (Table 1), features pointing to the circulation of reducing fluids, probably during serpentinization, which facilitated the back-reduction of mobilized Cr(VI) fractions to secondary immobile Cr(III)-bearing mineral phases [32,86].

Therefore, the relatively elevated, slightly heavy Cr isotope signatures recorded in chromitites may be the result of a combination of (a) fractionation during magmatic processes, i.e., degree of partial melting and differentiation of the parent magma, and (b) multi-stage post-magmatic overprinting, which seem to depend on the size of the respective chromitite occurrence (in terms of tonnage of chromite) and on a probable relationship between the size and degree of shielding towards alteration of the chromitite bodies.

5.4. Weathering of Cr-Bearing Minerals and Environmental Significance

Contamination of groundwater and soil by heavy metals is becoming a serious threat to our environments worldwide. The presence of harmful Cr(VI) in soils and in turn in groundwater may be related to natural processes or/and human activities, such as transfer of weathered material from rocks and primary raw materials, wastes and/or application of large amounts of fertilizers and pesticides for a long time in cultivated areas. The European Soil Data Centre has established a link between the effects of metal bio-availability and metal bio-accumulation, and human health and negative impacts on ecosystems [25–30,87–96]. Dose-dependent differences in toxicities of elements, the particle size, and the oxidation state require serious consideration in health-risk assessments [90]. A very common source of salts in irrigated soils is the irrigation water itself, affected by sea water in low-lying areas along the coast. The salinization of shallow aquifers may have a major effect on plant/crop growth and toxicity. The use of multidisciplinary methods in the study of ecosystem processes in response to groundwater and soil system has great potentials for sustainable developments [87–96].

The application of Cr isotope measurements to evaluate Cr(VI) contamination in groundwater and rock leachates from the central Evia and Assopos basin in Greece has revealed positively fractionated δ^{53}Cr values ranging from +0.56 to +0.96‰ in water leachates of ultramafic rocks, which are comparable to those from Central Europe [26–28]. These results imply oxidative mobilization of Cr(VI) from the ultramafic host rocks, and successive back-reduction of so mobilized Cr(VI) fractions [29]. The widespread distribution of ophiolite complexes in Europe and in orogenic zones elsewhere around the globe, emphasizes the need for further more detailed studies addressing the variation of δ^{53}Cr values in Cr-bearing rocks and ores.

In summary, the original geochemical compositions of chromitites may have been modified by subsequent, post-magmatic processes taking place at shallow mantle depths [35–38,56,57,68,69,72–89,97]. The wide range of δ^{53}Cr values, from positive, slightly negative to the most negatively fractionated signatures, even present in individual relatively large chromitite deposits, together with negative trends observed between δ^{53}Cr values and Cr/(Cr + Al) ratios in chromitites, may reflect a control of Cr isotope compositions in chromitites by degrees of partial melting and by magma fractionation. This is best exemplified by the high-Al chromitites from the cumulate sequence of the Vourinos complex. In addition, post-magmatic redox cycling of Cr-isotopes may occur in response to brittle deformation of chromitites and the subsequent interaction of the Cr-bearing rocks with oxidative fluids, also potentially leading to the formation of Fe-chromite through substitution of Cr, Al and Mg by Fe(III) and Fe(II). The formation of Cr-bearing minerals, such as serpentine, chlorite, and some garnets, and the concomitant release of Cr(VI)-bearing fluids, under alkaline conditions, may take place during serpentinization of respective ore bodies. Mobilized Cr(VI) bearing fluids then are prone to back-reduction promoted by the reduced environment in serpentinized ultramafic rocks, causing the formation of graphite, secondary Fe–Cr-bearing silicates, and other secondary alteration phases that potentially can carry a positively fractionated Cr isotope signal.

6. Conclusions

The δ^{53}Cr values presented herein for chromitites from the Balkan Peninsula offer an opportunity to study the effects of Cr isotope fractionations that potentially could be related to primary magmatic processes and to compare these with compositional trends eventually depicted by the respective chromite deposits and chromite occurrences.

A modification of the magmatic control on the geochemical characteristics of chromitites is revealed by positively fractionated δ^{53}Cr values in ophiolite zones affected by brittle deformation/metamorphism and characterized by the presence of secondary epigenetic Cr-bearing minerals, formed by the circulation of reducing fluids carrying abiotic methane.

Chromitites from the Balkan Peninsula depict a wide range in δ^{53}Cr values. Signatures range from positively-, to slightly negatively fractionated δ^{53}Cr values, even in individual relatively large deposits. Positively fractionated δ^{53}Cr values of all chromitite samples from Othrys and of high-Al chromitites from Skyros, and a negatively correlated trend between δ^{53}Cr and Cr/(Cr + Al), may reflect a control of δ^{53}Cr by degree of partial melting and by magma fractionation. This is best exemplified by high-Al chromitites from the cumulate sequence of the Vourinos complex.

The application of Cr isotopes to the evaluation of Cr(VI) contamination in groundwater and rock leachates relies on oxidative mobilization of Cr(VI) and successive back-reduction to C(III). The widespread distribution of significant natural Cr-sources in the Balkan peninsula, enhanced by industrial activities involving Cr in their processes, present a latently dangerous framework which is prone to have negative effects on the environment, particularly with respect to hexavalent Cr toxicity. Detailed studies addressing the variation of δ^{53}Cr values in Cr-bearing rocks, ores, soils, surface run-off and groundwater are required to understand the complicated mutual interplay between original natural Cr sources and the weathering-induced release of Cr into the environment.

Author Contributions: Conceptualization, R.F. and M.E.-E.; Methodology, M.E.-E., R.F. and I.M.; R.F. provided the chromium isotope analyses and their validation. M.E.-E. and I.M. provided the field information and performed the other analyses. All authors (M.E.-E., R.F. and I.M.) contributed to the elaboration, interpretation of the data and to the writing of the manuscript. All authors have read and agreed to the published version of the manuscript.

Funding: This research was funded by the National and Kapodistrian University of Athens (NKUA) (Grant No. KE_11730).

Acknowledgments: Many thanks are expressed to the two anonymous reviewers for their constructive criticism and suggestions on an earlier draft of the manuscript. Vassilios Skounakis is thanked for his assistance with SEM/EDS analysis of the samples. RF thanks Toby Leeper for always keeping the mass spectrometers at IGN in perfect running condition, and Toni Larsen for help in the ion chromatographic separation of the samples. This paper is dedicated to the memory of Maria Zhelyaskova-Panayiotova, University of Sofia, who provided the chromitite samples from Bulgaria, and passed away in May 2018.

Conflicts of Interest: The authors declare no conflict of interest.

References

1. Schoenberg, R.; Zink, S.; Staubwasser, M.; Von Blanckenburg, F. The stable Cr isotope inventory of solid Earth reservoirs determined by double spike MC-ICP-MS. *Chem. Geol.* **2008**, *249*, 294–306. [CrossRef]

2. Farkaš, J.; Chrastny, V.; Novak, M.; Cadkova, E.; Pasava, J.; Chakrabarti, R.; Jacobsen, S.B.; Ackerman, L.; Bullen, T.D. Chromium isotope variations ($\delta^{53/52}$Cr) in mantle derived sources and their weathering products: Implications for environmental studies and the evolution of $\delta^{53/52}$Cr in the Earth's mantle over geologic time. *Geochim. Cosmochim. Acta* **2013**, *123*, 74–92. [CrossRef]

3. Melcher, F.; Grum, W.; Simon, G.; Thalhammer, T.V.; Stumpfl, E. Petrogenesis of the ophiolitic giant chromite deposit of Kempirsai, Kazakhstan: A study of solid and fluid inclusions in chromite. *J. Petrol.* **1997**, *10*, 1419–1458. [CrossRef]

4. Arai, S. Conversion of low-pressure chromitites to ultrahigh-pressure chromitites by deep recycling: A good inference. *Earth Planet. Sci. Lett.* **2013**, *379*, 81–87. [CrossRef]

5.   Robinson, P.T.; Trumbull, R.B.; Schmitt, A.; Yang, J.S.; Li, J.W.; Zhou, M.F.; Erzinger, J.; Dare, S.; Xiong, F. The origin and significance of crustal minerals in ophiolitic chromitites and peridotites. *Gondwana Res.* **2015**, *27*, 486–506. [CrossRef]

6.   Griffin, W.L.; Afonso, J.C.; Belousova, E.A.; Gain, S.E.; Gong, X.H.; González-Jiménez, J.M.; Howell, D.; Huang, J.X.; McGowan, N.; Pearson, N.J.; et al. Mantle recycling: Transition zone metamorphism of Tibetan ophiolitic peridotites and its tectonit implications. *J. Petrol.* **2016**, *57*, 655–684. [CrossRef]

7.   Ballhaus, C.; Wirth, R.; Fonseca, R.O.C.; Blanchard, H.; Pröll, W.; Bragagni, A.; Nagel, T.; Schreiber, A.; Dittrich, S.; Thome, V.; et al. Ultra-high pressure and ultra-reduced minerals in ophiolites may form by lightning strikes. *Geochem. Perspect. Lett.* **2017**, *5*, 42–46. [CrossRef]

8.   González-Jiménez, G.M.; Camprubì, A.; Colás, V.; Griffin, W.L.; Proenza, J.A.; O'Reilly, S.Y.; Centeno-García, E.; Garcia-Casco, A.; Belousova, E.; Talavera, C.; et al. The recycling of chromitites in ophiolites from southwestern North America. *Lithos* **2017**, *294–295*, 53–72. [CrossRef]

9.   Prichard, H.M.; Neary, C.R.; Potts, P.J. Platinum-group minerals in the Shetland ophiolite complex. In *Metallogeny of Basic and Ultrabasic Rocks*; Gallagher, M.J., Ixer, R.A., Neary, C.R., Prichard, H.M., Eds.; Institution of Mining and Metallurgy: Edinburgh, UK, 1986; pp. 395–414.

10.   Economou-Eliopoulos, M. Platinum-group element distribution in chromite ores from ophiolite complexes: Implications for their exploration. *Ore Geol. Rev.* **1996**, *11*, 363–381. [CrossRef]

11.   Tarkian, M.; Economou-Eliopoulos, M.; Sambanis, G. Platinum-group minerals in chromitites from the Pindos ophiolite complex, Greece. *N. J. Mineral. Mon.* **1996**, *4*, 145–160.

12.   Ohnenstetter, M.; Johan, Z.; Cocherie, A.; Fouillac, A.M.; Guerrot, C.; Ohnenstetter, D.; Chaussidon, M.; Rouer, O.; Makovicky, E.; Makovicky, M.; et al. New exploration methods for platinum and rhodium deposits poor in base-metal sulphides-NEXTRIM. *Tran. Inst. Min. Metall. Sect. B Appl. Earth Sci.* **1999**, *108*, B119–B150.

13.   Garuti, G.; Zaccarini, F. In situ alteration of platinum-group minerals at low temperature evidence from serpentinized and weathered chromitites of the Vourinos complex (Greece). *Can. Mineral.* **1997**, *35*, 611–626.

14.   Proenza, J.A.; Gervilla, F.; Melgarejo, J.C.; Bodinier, J.L. Al- and Cr-rich chromitites from the Mayarí-Baracoa ophiolitic belt (eastern Cuba); consequence of interaction between volatile-rich melts and peridotites in suprasubduction mantle. *Econ. Geol.* **1999**, *94*, 547–566. [CrossRef]

15.   Proenza, J.A.; Zaccarini, F.; Escayola, M.; Cábana, C.; Shalamuk, A.; Garuti, G. Composition and textures of chromite and platinum-group minerals in chromitites of the western ophiolitic belt from Córdoba Pampeans Ranges, Argentine. *Ore Geol. Rev.* **2008**, *33*, 32–48. [CrossRef]

16.   Uysal, I.; Kapsiotis, A.; Akmaz, R.M.; Saka, S.; Seitz, H.M. The Guleman ophiolitic chromitites (SE Turkey) and their link to a compositionally evolving mantle source during subduction initiation. *Ore Geol. Rev.* **2018**, *93*, 98–113. [CrossRef]

17.   Prichard, H.M.; Brough, C. Potential of ophiolite complexes to host PGE deposits. In *New Developments in Magmatic Ni-Cu and PGE Deposits*; Li, C., Ripley, E.M., Eds.; Geological Publishing House: Beijing, China, 2009; pp. 277–290.

18.   Kapsiotis, A.; Grammatikopoulos, T.A.; Tsikouras, B.; Hatzipanagiotou, K. Platinum-group mineral characterization in concentrates from high-grade PGE Al-rich chromitites of Korydallos area in the Pindos ophiolite complex (NW Greece). *Resour. Geol.* **2010**, *60*, 178–191. [CrossRef]

19.   Kapsiotis, A.; Grammatikopoulos, T.; Tsikouras, B.; Hatzipanagiotou, K.; Zaccarini, F.; Garuti, G. Chromian spinel composition and Platinum-group element mineralogy of chromitites from the Milia area, Pindos ophiolite complex, Greece. *Can. Mineral.* **2009**, *47*, 1037–1056. [CrossRef]

20.   O'Driscoll, B.; González-Jiménez, J.M. Petrogenesis of the Platinum-group minerals. *Rev. Mineral. Geochem.* **2016**, *81*, 489–578. [CrossRef]

21.   Shen, J.; Liu, J.; Qin, L.; Wang, S.J.; Li, S.; Xia, J.; Zhang, Q.; Yang, J. Chromium isotope signature during continental crust subduction recorded in metamorphic rocks. *Geochem. Geophys. Geosyst.* **2015**, *16*, 3840–3854. [CrossRef]

22.   Xia, C.; Guodong, L.; Yue, H.; Yuchuan, M. Precipitation Stable Isotope Variability in Tropical Monsoon Climatic Zone of Asia. *IOP Conf. Ser. Mater. Sci. Eng.* **2018**, *392*, 042028. [CrossRef]

23.   Xia, J.; Qina, L.; Shena, J.; Carlsonb, R.W.; Ionovd, D.A.; Mockbet, T.D. Chromium isotope heterogeneity in the mantle. *Earth Planet Sci. Lett.* **2017**, *464*, 103–115. [CrossRef]

24. Bonnand, P.; Williams, H.M.; Parkinson, I.J.; Wood, B.J.; Halliday, A.N. Stable chromium isotopic composition of meteorites and metal-silicate experiments: Implications for fractionation during core formation. *Earth Planet. Sci. Lett.* **2016**, *3*, 4–7. [CrossRef]

25. Slejko, F.F.; Petrini, R.; Lutman, A.; Forte, C.; Ghezzi, L. Chromium isotopes tracking the resurgence of hexavalent chromium contamination in a past-contaminated area in the Friuli Venezia Giulia Region, northern Italy. *Isot. Environ. Health Stud.* **2019**, *55*, 56–69. [CrossRef] [PubMed]

26. Novak, M.; Chrastny, V.; Cadkova, E.; Naldras, J.; Bullen, T.D.; Tylcer, J.; Szurmanova, Z.; Cron, M.; Prechova, E.; Curik, J.; et al. Common occurrence of a positive δ^{53}Cr shift in Central European waters contaminated by geogenic/industrial chromium relative to source values. *Environ. Sci. Technol.* **2014**, *48*, 6089–6096. [CrossRef] [PubMed]

27. Novak, M.; Kram, P.; Sebek, O.; Kurik, J.; Andronikov, A.; Veselovsky, F.; Chrastny, V.; Martinkova, E.; Stevanova, M.; Prechova, E.; et al. Temporal changes in Cr fluxes and δ^{53}Cr values in runoff from a small serpentinite catchment (Slavkov Forest, Czech Republic). *Chem. Geol.* **2017**, *472*, 22–30. [CrossRef]

28. Economou-Eliopoulos, M.; Frei, R.; Atsarou, C. Application of chromium stable isotopes to the evaluation of Cr (VI) contamination in groundwater and rock leachates from central Euboea and the Assopos basin (Greece). *Catena* **2014**, *122*, 216–228. [CrossRef]

29. Novak, M.; Martinkova, E.; Chrastny, V.; Stepanova, M.; Sebek, O.; Andronikov, A.; Curik, J.; Veselovsky, F.; Prechova, E.; Houskova, M. The fate of Cr(VI) in contaminated aquifers 65 years after the first spillage of plating solutions: A δ^{53}Cr study at four Central European sites. *Catena* **2017**, *158*, 371–380. [CrossRef]

30. Qin, L.; Wang, X. Chromium isotope geochemistry. *Rev. Mineral. Geochem.* **2017**, *82*, 379–414. [CrossRef]

31. Economou-Eliopoulos, M.; Eliopoulos, D.G.; Tsoupas, G. On the diversity of the PGE content in chromitites hosted in ophiolites and in porphyry-Cu systems: Controlling factors. *Ore Geol. Rev.* **2017**, *88*, 156–173. [CrossRef]

32. Ellis, A.S.; Johnson, T.M.; Bullen, T.D. Chromium isotopes and the fate of hexavalent chromium in the environment. *Science* **2002**, *295*, 2060–2062. [CrossRef]

33. Trinquier, A.; Elliott, T.; Ulfbeck, D.; Coath, C.; Krot, A.N.; Bizzarro, M. Origin of nucleosynthetic isotope heterogeneity in the solar protoplanetary disk. *Science* **2009**, *324*, 295–424. [CrossRef]

34. Bonnand, P.; Parkinson, I.J.; James, R.H.; Karjalainen, A.M.; Fehr, M.A. Accurate and precise determination of stable Cr isotope compositions in carbonates by double spike MC-ICP-MS. *J. Anal. At. Spectrom.* **2011**, *26*, 528–535. [CrossRef]

35. Moudrakis, D. Introduction to the geology and Macedonia and Trace: Aspects of the geotectonic evolution of the Hellenides. *Bull. Geol. Soc. Greece* **1992**, *30*, 31–46.

36. Jones, G.; Robertson, A. Tectono-stratigraphy and evolution of the Pindos ophiolite and associated units. *J. Geol. Soc.* **1991**, *148*, 267–288. [CrossRef]

37. Rassios, A.; Dilek, Y. Rotational deformation in the Jurassic Mesohellenic Ophiolites, Greece, and its tectonic significance. *Lithos* **2009**, *108*, 207–223. [CrossRef]

38. Rassios, A.H.E.; Moores, E.M. Heterogenous mantle complex, crustal processes, and obduction kinematics in a unified Pindos–Vourinos ophiolitic slab (northern Greece). In *Tectonic Development of the Eastern Mediterranean Region*; Robertson, A.H.F., Mountrakis, D., Eds.; Geological Society of London, Special Publications: London, UK, 2006; Volume 260, pp. 237–266.

39. Rassios, A.; Konstantopoulou, G. Emplacement tectonism and the position of chrome ores in the Mega Isoma peridotites, SW Othris, Greece. *Bull. Geol. Soc. Greece* **1993**, *28*, 463–474.

40. Saccani, E.; Photiades, A.; Santato, A.; Zeda, O. New evidence for supra-subduction zone ophiolites in the Vardar Zone from the Vermion Massif (northern Greece): Implication for the tectono-magmatic evolution of the Vardar oceanic basin. *Ofioliti* **2008**, *33*, 17–37.

41. Pearce, J.A.; Lippard, S.J.; Roberts, S. Characteristics and Tectonic Significance of Supra-Subduction Zone Ophiolites. *Geol. Soc. Lond.* **1984**, *16*, 77–94. [CrossRef]

42. Konstantopoulou, G.; Economou-Eliopoulos, M. Distribution of Platinum-group Elements and Gold in the Vourinos Chromitite Ores, Greece. *Econ. Geol.* **1991**, *86*, 1672–1682. [CrossRef]

43. Beccaluva, L.; Coltorti, M.; Saccani, E.; Siena, F. Magma generation and crustal accretion as evidenced by supra-subduction ophiolites of the Albanide-Hellenide Sub Pelagonian zone. *Isl. Arc* **2005**, *14*, 551–563. [CrossRef]

44. Kapsiotis, A.; Rassios, A.; Antonelou, A.; Tzamos, E. Genesis and Multi-episodic Alteration of Zircon-bearing Chromitites from Ayios Stefanos, Othris Massif, Greece: Assessment of an Unconventional Hypothesis on the Origin of Zircon in Ophiolitic Chromitites. *Minerals* **2016**, *6*, 124. [CrossRef]

45. Kapsiotis, A.; Economou-Eliopoulos, M.; Zhenga, H.; Sud, B.X.; Lenaz, D.; Jing, J.J.; Antoneloug, A.; Velicogna, M.; Xia, B. Refractory chromitites recovered from the Eretria mine, East Othris massif (Greece): Implications for metallogeny and deformation of chromitites within the lithospheric mantle portion of a forearc-type ophiolite. *Chemie der Erde* **2019**, *79*, 139–152. [CrossRef]

46. Barth, M.G.; Mason, P.R.D.; Davies, G.R.; Drury, M.R. The Othris Ophiolite, Greece: A snapshot of subduction initiation at a mid-ocean ridge. *Lithos* **2008**, *100*, 234–254. [CrossRef]

47. Economou, M.; Dimou, E.; Economou, G.; Migiros, G.; Vacondios, I.; Grivas, E.; Rassios, A.; Dabitzias, S. *Chromite Deposits of Greece: Athens*; Theophrastus Publications: Athens, Greece, 1986; pp. 129–159.

48. Bizimis, M.; Salters, V.J.M.; Bonatti, E. Trace and REE content of clinopyroxenes from supra-subduction zone peridotites. Implications for melting and enrichment processes in island arcs. *Chem. Geol.* **2000**, *165*, 67–85. [CrossRef]

49. Stern, R.J. The anatomy and ontogeny of modern intra-oceanic arc systems. In *the Evolving Continents: Understanding Processes of Continental Growth*; Kusky, T.M., Zhai, M.G., Xiao, W., Eds.; Geological Society of London: London, UK, 2010; Volume 338, pp. 7–34.

50. Tsoupas, G.; Economou-Eliopoulos, M. High PGE contents and extremely abundant PGE-minerals hosted in chromitites from the Veria ophiolite complex, northern Greece. *Ore Geol. Rev.* **2008**, *33*, 3–19. [CrossRef]

51. Rogkala, A.; Petrounias, P.; Tsikouras, B.; Hatzipanagiotou, K. New occurrence of pyroxenites in the Veria-Naousa ophiolite (North Greece): Implications on their origin and petrogenetic evolution. *Geosciences* **2017**, *7*, 92. [CrossRef]

52. Economou, M.; Naldrett, A.J. Sulfides associated with podiform bodies of chromite at Tsangli, Eretria, Greece. *Mineral. Depos.* **1984**, *19*, 289–297. [CrossRef]

53. Foose, M.P.; Economou, M.; Panayotou, A. Compositional and mineralogic constraints in the Limassol Forest portion of the Troodos ophiolite complex, Cyprus. *Mineral. Depos.* **1985**, *20*, 234–240. [CrossRef]

54. Etiope, G.; Ifandi, E.; Nazzari, M.; Procesi, M.; Tsikouras, B.; Ventura, G.; Steele, A.; Tardini, R.; Szatmari, P. Widespread abiotic methane in chromitites. *Scient. Rep.* **2018**, *8*, 8728. [CrossRef]

55. Economou-Eliopoulos, M.; Vacondios, I. Geochemistry of chromitites and host rocks from the Pindos ophiolite complex, northwestern Greece. *Chem. Geol.* **1995**, *122*, 99–108. [CrossRef]

56. Economou-Eliopoulos, M.; Sambanis, G.; Karkanas, P. Trace element distribution in chromitites from the Pindos ophiolite complex, Greece: Implications for the chromite exploration. In *Mineral Deposits*; Stanley, C.J., Ed.; Balkema: Rotterdam, The Netherlands, 1999; pp. 713–716.

57. Prichard, H.; Economou-Eliopoulos, M.; Fisher, P.C. Contrasting Platinum-group mineral assemblages from two different podiform chromitite localities in the Pindos ophiolite complex, Greece. *Can. Mineral.* **2008**, *46*, 329–341. [CrossRef]

58. Tarkian, M.; Economou-Eliopoulos, M.; Eliopoulos, D. Platinum-group minerals and tetraauricuprite in ophiolitic rocks of the Skyros Island, Greece. *Mineral. Petrol.* **1992**, *47*, 55–66. [CrossRef]

59. Scarpelis, N.; Economou, M. Genesis and metasomatism of chromite ore from the Gomati area, Chalkidiki, Greece. *Ann. Geol. Pays Hell.* **1978**, *29*, 716–728.

60. Magganas, A.; Economou, M. On the chemical composition of chromite ores from the ophiolitic complex of Soufli, NE Greece. *Ofioliti* **1988**, *13*, 15–27.

61. Zhelyaskova-Panayiotova, M.; Economou-Eliopoulos, M. Platinum-group element and gold concentrations in oxide and sulfide mineralizations from ultramafic rocks of Bulgaria. *Ann. Univ. Sofia Geol. Geogr.* **1994**, *86*, 196–218.

62. Cina, A. Pentlandite Mineralization Related to Albanian Ophiolites. In Proceedings of the XIX CBGA Congress Proceedings, Thessaloniki, Greece, 23–26 September 2010; Volume 100, pp. 317–323.

63. Tashko, A.; Economou-Eliopoulos, M. An overview of the PGE distribution in the Bulqiza ophiolite complex, Albania. *Bull. Geol. Soc. Greece* **1994**, *32*, 193–201.

64. Karaj, N. Repartition des platinoides chromites et sulphures dans le massif de Bulqiza, Albania. In *Incidence sur le Processus Metallogeniques Dans les Ophiolites (These)*; Universite de Orleans: Orleans, France, 1992; p. 379.

65. Garuti, G.; Fershtater, G.; Bea, F.; Montero, P.; Pushkarev, E.V.; Zaccarini, F. Platinum-group elements as petrological indicators in mafic-ultramafic complexes of central and southern Urals: Preliminary results. *Tectonophysics* **1997**, *276*, 181–194. [CrossRef]

66. Barnes, S.-L.; Naldrett, A.J.; Gorton, M.P. The origin of the fractionation of the platinum-group elements in terrestrial magmas. *Chem. Geol.* **1985**, *53*, 303–323. [CrossRef]

67. Garuti, G.; Zaccarini, F.; Economou-Eliopoulos, M. Paragenesis and composition of laurite from the chromitites of Othrys (Greece): Implications for Os-Ru fractionation in ophiolitic upper mantle of the Balkan Peninsula. *Miner. Depos.* **1999**, *34*, 312–319. [CrossRef]

68. Colás, V.; González-Jiménez, J.M.; Griffin, W.L.; Fanlo, I.; Gervilla, F.; O'Reilly, S.Y.; Pearson, N.J.; Kerestedjian, T.; Proenza, J.A. Fingerprints of metamorphism in chromite: New insights from minor and trace elements. *Chem. Geol.* **2014**, *389*, 137–152. [CrossRef]

69. Pujol-Solà, N.; Proenza, I.A.; Garcia-Casco, A.; González-Jiméne, J.M.; Andreazini, A.; Melgarejo, J.C.; Gervilla, F. An Alternative Scenario on the Origin of Ultra-High Pressure (UHP) and Super-Reduced (SuR) Minerals in Ophiolitic Chromitites: A Case Study from the Mercedita Deposit (Eastern Cuba). *Minerals* **2018**, *8*, 433. [CrossRef]

70. Losi, M.E.; Amrhein, C.; Frankenberger, W.T. Environmental biochemistry of chromium. *Rev. Environ. Contam. Toxicol.* **1994**, *136*, 91–121. [PubMed]

71. Schauble, E.; Rossman, G.R.; Taylor, H.P., Jr. Theoretical estimates of equilibrium chromium-isotope fractionations. *Chem. Geol.* **2004**, *205*, 99–114. [CrossRef]

72. Zhou, M.F.; Robinson, P.T.; Su, B.X.; Gao, J.F.; Li, J.W.; Yang, J.S.; Malpas, J. compositions of chromite, associated minerals, and parental magmas of podiform chromite deposits: The role of slab contamination of asthenospheric melts in suprasubduction zone environments. *Gondwana Res.* **2014**, *26*, 262–283. [CrossRef]

73. Tzamos, E.; Filippidis, A.; Rassios, A.; Grieco, G.; Michailidis, K.; Koroneos, A.; Gamaletsos, P.N. Major and minor element geochemistry of chromite from the Xerolivado–Skoumtsa mine, Southern Vourinos: Implications for chrome ore exploration. *J. Geochem. Explor.* **2016**, *165*, 81–93. [CrossRef]

74. Tzamos, E.; Kapsiotis, A.; Filippidis, A.; Koroneos, A.; Grieco, G.; Rassios, A.E.; Godelitsas, A. Metallogeny of the Chrome Ores of the Xerolivado–Skoumtsa Mine, Vourinos Ophiolite, Greece: Implications on the genesis of IPGE-bearing high-Cr chromitites within a heterogeneously depleted mantle section. *Ore Geol. Rev.* **2017**, *90*, 226–242. [CrossRef]

75. Grieco, G.; Bussolesi, M.; Tzamos, E.; Rassios, A.E.; Kapsiotis, A. Processes of primary and re-equilibration mineralization affecting chromitite ore geochemistry within the Vourinos ultramafic sequence, Vourinos ophiolite (West Macedonia, Greece). *Ore Geol. Rev.* **2018**, *95*, 537–551. [CrossRef]

76. Kamenetsky, V.S.; Crawford, A.J.; Meffre, S. Factors controlling chemistry of magmatic spinel: An empirical study of associated olivine, Cr-spinel and melt inclusions from primitive rocks. *J. Petrol.* **2001**, *42*, 655–671. [CrossRef]

77. Zaccarini, F.; Proenza, A.J.; Ortega-Gutierrez, F.; Garuti, G. Platinum group minerals in ophiolitic chromitites from Tehuitzingo (Acatlan complex, southern Mexico): Implications for post-magmatic modification. *Mineral. Petrol.* **2005**, *84*, 147–168. [CrossRef]

78. Gervilla, F.; Padrón-Navarta, J.A.; Kerestedjian, T.; Sergeeva, I.; González-Jiménez, J.M.; Fanlo, I. Formation of ferrian chromite in podiform chromitites from the Golyamo Kamenyane serpentinite, Eastern Rhodopes, SE Bulgaria: A two-stage process. *Contr. Mineral. Petrol.* **2012**, *164*, 643–657. [CrossRef]

79. González-Jiménez, J.M.; Locmelis, M.; Belousova, E.; Griffin, W.L.; Gervilla, F.; Kerestedjian, T.N.; Pearson, N.J.; Sergeeva, I. Genesis and tectonic implications of podiform chromitites in the metamorphosed Ultramafic Massif of Dobromirtsi (Bulgaria). *Gondwana Res.* **2015**, *27*, 555–574. [CrossRef]

80. Grammatikopoulos, T.A.; Kapsiotis, A.; Tsikouras, B.; Hatzipanagiotou, K.; Zaccarini, F.; Garuti, G. Spinel composition, PGE geochemistry and mineralogy of the chromitites from the Vourinos ophiolite complex, northwestern Greece. *Can. Mineral.* **2011**, *49*, 1571–1598. [CrossRef]

81. Mposkos, E.; Krohe, A. Pressure–temperature–deformation paths of closely associated ultra-high-pressure (diamond-bearing) crustal and mantle rocks of the Kimi complex: Implications for the tectonic history of the Rhodope Mountains, northern Greece. *Can. J. Earth Sci.* **2006**, *43*, 1755–1776. [CrossRef]

82. Mposkos, E.; Baziotis, I.; Proyer, A. Pressure–temperature evolution of eclogites from the Kechros complex in the Eastern Rhodope (NE Greece). *Int. J. Earth Sci.* **2012**, *101*, 973–996. [CrossRef]

83. Economou-Eliopoulos, M. On the origin of the PGE-enrichment in chromitites associated with ophiolite complexes: The case of Skyros Island, Greece. In *Digging Deeper Proceedings of the Ninth Biennial SGA Meeting, Dublin 2007*; Andrew, C.J., Ed.; Irish Association of Economic Geology: Dublin, Ireland, 2007; Volume 2, pp. 1611–1614.

84. Ifandi, E.; Zaccarini, Z.; Tsikouras, B.; Grammatikopoulos, T.; Garuti, G.; Karipi, S. First occurrences of Ni-V-Co phosphides in chromitite from the Agios Stefanos Mine, Othrys Ophiolite, Greece. *Ofioliti* **2018**, *43*, 131–145.

85. Tsikouras, B.; Etiope, G.; Ifandi, E.; Kordella, S.; Papatheodorou, G.; Hatzipanagiotou, K. Petrological Implications for the Production of Methane and Hydrogen in Hyperalkaline Springs from the Othrys Ophiolite, Greece. In Proceedings of the 13th International Congress, Chania, Greece, 5–8 September 2013; Bulletin of the Geological Society of Greece: Athens, Greece, 2013; Volume XLVII, pp. 449–457.

86. Economou-Eliopoulos, M.; Tsoupas, G.; Skounakis, V. Occurrence of Graphite-Like Carbon in Podiform Chromitites of Greece and Its Genetic Significance. *Minerals* **2019**, *9*, 152. [CrossRef]

87. Xu, X.-Z.; Cartigny, P.; Yang, J.-S.; Dilek, Y.; Xiong, F.; Guo, G. Fourier trans-form infrared spectroscopy data and carbon isotope characteristics of the ophiolite-hosted diamonds from the Luobusa ophiolite, Tibet, and Ray-Iz ophiolite, Polar Urals. *Lithosphere* **2018**, *10*, 156–169. [CrossRef]

88. Apollaro, C.; Marini, L.; Critelli, T.; Barca, D.; Bloise, A.; De Rosa, R.; Liberi, F.; Miriello, D. Investigation of rock-to-water release and fate of major, minor, and trace elements in the metabasaltserpentinite shallow aquifer of Mt. Reventino (CZ, Italy) by reaction path modeling. *Appl. Geochem.* **2011**, *26*, 1722–1740. [CrossRef]

89. Apollaro, C.; Fuoco, I.; Brozzo, G.; De Rosa, R. Release and fate of Cr(VI) in the ophiolitic aquifers of Italy: The role of Fe(III) as a potential oxidant of Cr(III) supported by reaction path modelling. *Sci. Total Environ.* **2019**, *660*, 1459–1471. [CrossRef]

90. Van der Putten, W.; Ramirez, K.; Poesen, J.; Lenka, L.; Šimek, M.; Anne, W.; Mari, M.; Philippe, L.; Heikki, S.; Andrey, Z.; et al. *Opportunities for Soil Sustainability in Europe*; European Academies' Science Advisory Council: Halle, Germany, 2018; 41p.

91. Vithanage, M.; Kumarathilaka, P.; Oze, C.; Karunatilake, S.; Seneviratne, M.; Hseu, Z.Y.; Gunarathne, V.; Dassanayake, M.; Ok, Y.S.; Rinklebe, J. Occurrence and cycling of trace elements in ultramafic soils and their impacts on human health: A critical review. *Environ. Int.* **2019**, *131*, 104974. [CrossRef]

92. Izbicki, J.A.; Ball, J.W.; Bullen, T.D.; Sutley, S.J. Chromium, chromiumisotopes and selected trace elements, western Mojave Desert, USA. *Appl. Geochem.* **2008**, *23*, 1325–1352. [CrossRef]

93. Kazakis, N.; Kantiranis, N.; Kalaitzidou, K.; Kaprara, M.; Mitrakas, M.; Frei, R.; Vargemezis, G.; Tsourlos, P.; Zouboulis, A.; Filippidis, A. Origin of hexavalent chromium in groundwater: The example of the Sarigkiol Basin, Northern Greece. *Sci. Total Environ.* **2017**, *593–594*, 552–566. [CrossRef] [PubMed]

94. Sun, Z.; Wang, X.; Planavsky, N. Cr isotope systematics in the Connecticut River estuary. *Chem.Geol.* **2019**, *506*, 29–39. [CrossRef]

95. Frei, R.; Gaucher, C.; Døssing, L.N.; Sial, A.N. Chromium isotopes in carbonates—A tracer for climate change and for reconstructing the redox state of ancient seawater. *Earth Planet. Sci. Lett.* **2011**, *312*, 114–125. [CrossRef]

96. Frei, R.; Poire, D.; Frei, K.M. Weathering on land and transport of chromium to the ocean in a subtropical region (Misiones, NW Argentina): A chromium stable isotope perspective. *Chem. Geol.* **2014**, *381*, 110–124. [CrossRef]

97. Filippidis, A. Chemical variation of chromite in the central sector of xerolivado chrome mine of Vourinos, Western Macedonia, Greece. *Neues Jahrb. Mineral. Mon.* **1997**, *8*, 354–370. [CrossRef]

 *minerals*

Article

Factors Controlling the Gallium Preference in High-Al Chromitites

Ioannis-Porfyrios D. Eliopoulos * and George D. Eliopoulos

Department of Chemistry, University of Crete, Heraklion GR-70013, Crete, Greece; giorgoshliop@yahoo.gr
* Correspondence: disaca007@hotmail.com; Tel.: +30-281-054-5136

Received: 6 August 2019; Accepted: 7 October 2019; Published: 10 October 2019

Abstract: Gallium (Ga) belongs to the group of critical metals and is of noticeable research interest. Although Ga^{3+} is highly compatible in high-Al spinels a convincing explanation of the positive Ga^{3+}–Al^{3+} correlation has not yet been proposed. In the present study, spinel-chemistry and geochemical data of high-Al and high-Cr chromitites from Greece, Bulgaria and the Kempirsai Massif (Urals) reveals a strong negative correlation (R ranges from −0.95 to −0.98) between Cr/(Cr + Al) ratio and Ga in large chromite deposits, suggesting that Ga hasn't been affected by re-equilibration processes. In contrast, chromite occurrences of Pindos and Rhodope massifs show depletion in Ga and Al and elevated Mn, Co, Zn and Fe contents, resulting in changes (sub-solidus reactions), during the evolution of ophiolites. Application of literature experimental data shows an abrupt increase of the inversion parameter (x) of spinels at high temperature, in which the highest values correspond to low-Cr^{3+} samples. Therefore, key factors controlling the preference of Ga^{3+} in high-Al chromitites may be the composition of the parent magma, temperature, redox conditions, the disorder degree of spinels and the ability of Al^{3+} to occupy both octahedral and tetrahedral sites. In contrast, the competing Cr^{3+} can occupy only octahedral sites (due to its electronic configuration) and the Ga^{3+} shows a strong preference on tetrahedral sites.

Keywords: chromite; gallium; spinel; structure; composition; correlation; ophiolites; disorder

---

1. Introduction

Gallium (Ga) is a vital metal for the economy, due to its use in high-technology applications, such as electronics industry, electric cars, solar panels. Although bauxite deposits are traditionally mined for their Al content and are important sources of Ga as a byproduct commodity [1], the distribution of Ga in chromite ores may be of particular research interest, due to its relationship with the major element composition of chromite.

Chromite belongs to the subgroup of spinels, which accommodate a wide variety of cations in their structure with the general formula AB_2O_4. Many authors emphasized that despite their simple structure, many spinels exhibit complex disordering phenomena involving the two cation sites, which play an important role both in their thermochemical and their physical properties [2–6]. The movements of cations between tetrahedral and octahedral sites, as a result of cation substitution, have been discussed under the aspect of structural parameters, such as tetrahedral and octahedral bond lengths, cation-cation and cation-anion distances, bond angles and hopping lengths, which were calculated by experimental lattice constants and oxygen parameters [2–4,7–12]. The ability of Ga^{3+} (r = 0.62 Å) to replace Al^{3+} (r = 0.54 Å) in aluminum minerals is related to their geochemistry (Group III of the periodic table), while Ga might be expected to behave in a similar way in chromite and magnetite as they share similar ionic radii (Cr^{3+}, r = 0.62 Å; Fe^{3+} = 0.64 Å) for octahedral and tetrahedral coordination (Al^{3+}, r = 0.39; Ga^{3+} = 0.47; Fe^{3+} = 0.49) [3]. Gallium levels reported in Cr-spinel grains from ophiolites, varying from 10 to 50 ppm [13,14] are consistent with experimental mineral-melt data on the partition coefficient (D_{Ga} = 0.9–11.2) [15]. Although mineral–melt partition coefficients are not constants, depending on a number of factors (pressure, temperature, oxygen fugacity or mineral and melt composition) on the basis of experimental data it has been suggested that Ga is volatile and there is no significant effect of temperature, magma composition and at very low oxygen fugacity conditions [15].

A tectonic discrimination of peridotites, using the oxygen fugacity (fO_2)–Cr#[Cr/(Cr + Al)] diagram, and the Ga–Ti–Fe^{3+}# [Fe^{3+}/(Fe^{3+} + Cr + Al)] systematics in chrome-spinels, has been proposed [16]. A negative correlation between Ga and Cr in chromitites has been established, that may be related to the composition of parental magmas [17–24] or to the outer electronic structure of Ga that is similar to that of Al [25]. Also, the investigation of spinels in lithospheric mantle xenoliths from distinct tectonic settings has demonstrated that trace elements contribute in discriminating between spinels hosted in peridotites and those crystallized from magmas [11]. However, a convincing explanation of the positive correlation between Ga^{3+} and Al^{3+} has not yet been offered.

In the present study we characterize the spinel chemistry, bulk ore composition, including Ga, from chromitite samples of selected ophiolite complexes in Greece (Pindos, Central Vourinos and Skyros), the Rhodope–Serbo–Macedonian zone (SMZ) massifs (all of Mesozoic age) and the Kempirsai Massif (Kazakhstan) in the Urals (Palaeozoic age), all hosting both high-Cr and high-Al chromitites. The investigated samples are representative of large chromite deposits and small occurrences, in order to define potential relationships between major, minor or trace elements and Ga and the effect of re-equilibration processes, during a long evolutionary time of the ophiolites. We apply available platinum-group element (PGE) data to define potential correlations between the Ga content and fractional crystallization, and experimental literature data for the structure of spinels, aiming to investigate the role of intra-crystalline cation exchange, and contribute to still uncertain factors controlling the positive Al-Ga correlation of chromitites.

2. Materials and Methods

2.1. Mineral Analysis

Polished sections of all chromitite samples were examined using a reflected light microscope and a scanning electron microscope (SEM), equipped with energy-dispersive spectroscopy (EDS). The SEM-EDS back-scattered electron images (Figure 1) and analyses of chromite ores (Table 1) were carried out at the Faculty of Geology and Geoenvironment, National and Kapodistrian University of Athens (NKUA), using a JEOL JSM 5600 (Tokyo, Japan), scanning electron microscope, equipped with ISIS 300 OXFORD (Oxford shire, UK), automated energy dispersive analysis system. Analytical conditions were 20 kV accelerating voltage, 0.5 nA beam current, <2 μm beam diameter and 50 s count times. The following X-ray lines were used: FeKα, NiKα, CoKα, CuKα, CrKα, AlKα, TiKα, CaKα, SiKα, MnKα and MgKα. Cr, Fe, Mn, Ni, Co, Ti and Si, MgO for Mg and Al_2O_3 for Al. Contents of Fe_2O_3 and FeO were calculated on the basis of the spinel stoichiometry.

Table 1. Electron scanning electron microscope (SEM)/energy-dispersive spectroscopy (EDS) analyses of chromite from chromitites of Greece, Bulgaria and Kempirsai (Urals).

wt%	Vourinos					Pindos							Skyros		
	Vour. 1	Vour. 2	Vour. 3	Vour. 4	Vour. 5	Pi.1	Pi.2	Pi. 3	Pi. 4	Pi. 5	Pi.6	Pi.7	Sky. 1	Sky. 2	Sky. 3
TiO_2	0.3	0.2	0.2	0.1	0.2	0.2	0.3	0.2	0.2	0.1	0.2	0.1	0.2	0.2	0.2
Al_2O_3	11.4	9.4	9.8	23.9	11.3	26.2	27.2	32.9	34.5	16.5	15.5	8.2	20.11	24.5	22.1
Cr_2O_3	59.7	63.2	61.3	44.8	60.4	41.5	39.5	35.9	34.8	52.8	52.1	61.4	48.8	45.3	47.1
MgO	13.6	12.1	13.6	16.2	12.4	14.5	13.6	16.4	16.7	12.4	10.1	11.7	13.9	15.7	10.2
FeO	13.1	14.1	12.7	11.1	14.6	13.5	15.7	12.3	12.5	15.6	18.7	14.9	13.6	11.8	19.7
Fe_2O_3	1.7	0.1	2.1	4.2	.5	3.4	3.1	2.3	2.6	2.9	2.7	2.8	2.2	2.3	0.3
MnO	0.3	0.2	0.2	0.2	0.2	0.3	n.d.	0.1	n.d.	0.2	0.3	0.3	0.1	0.1	0.2
NiO	0.2	0.1	n.d.	n.d.	0.2	0.3	n.d.	0.2	n.d.	0.2	0.1	0.2	0.1	0.2	0.2
Total	100.3	99.5	99.9	100.5	99.8	99.9	99.4	100.1	101.2	100.7	99.7	99.6	99.01	100.1	100
$Cr/(Cr+Al)$	0.77	0.81	0.81	0.56	0.78	0.52	0.49	0.42	0.4	0.70	0.69	0.81	0.62	0.55	0.58
$Mg/(Mg+Fe^{2+})$	0.65	0.62	0.66	0.72	0.60	0.65	0.61	0.71	0.72	0.59	0.49	0.58	0.64	0.69	0.51
$Fe^{3+}/(Cr+Al+Fe^{3+})$	0.031	0.000	0.025	0.036	0.0055	0.038	0.036	0.025	0.027	0.0413	0.028	0.036	0.027	0.026	0.0033
Numbers of Cations on the Basis of 32 Oxygens															
Ti	0.019	0.039	0.040	0.018	0.039	0.036	0.054	0.035	0.014	0.019	0.039	0.020	0.037	0.036	0.038
Al	3.426	2.889	2.997	6.736	3.451	7.415	7.758	8.966	9.241	4.650	4.728	2.555	5.896	6.922	6.532
Cr	12.037	13.031	12.523	8.471	12.383	7.891	7.558	6.563	6.254	10.645	10.662	12.835	9.602	8.586	9.340
Mg	5.170	4.703	5.250	5.774	4.790	5.210	4.906	5.652	5.658	4.712	3.896	4.611	5.156	5.610	3.813
Fe^{2+}	2.744	3.271	2.744	2.203	3.164	2.706	3.149	2.326	2.377	3.222	4.056	3.300	2.840	2.367	4.142
Fe^{3+}	0.498	0.001	0.400	0.757	0.087	0.620	0.575	0.401	0.436	0.659	0.533	0.570	0.427	0.420	0.053
Mn	0.065	0.044	0.045	0.040	0.044	0.061	0.000	0.020	0.000	0.043	0.065	0.067	0.021	0.020	0.042
Ni	0.157	0.021	0.000	0.000	0.042	0.058	0.000	0.037	0.000	0.041	0.020	0.042	0.020	0.039	0.040

wt%	Skyros		Rhodope Massif					Bulgaria			Urals Kempirsai	
			Greece						Goliamo		Northern Part	Southern Part
	Sky. 4	Sky. 5	Soufli 1	Soufli 2	Gomati	Broucevci	Jacovitsa	Pletena	Kamenyane 1	Kamenyane 2	Batamshinsk	Main Ore Field
TiO_2	0.1	0.2	n.d.	0.2	0.2	0.4	0.2	0.5	0.4	0.3	0.3	0.1
Al_2O_3	11.2	12.6	15.8	19.6	30.8	27.9	5.5	10.4	4.5	1.2	24.4	9.3
Cr_2O_3	59.1	58.1	53.9	47.8	35.5	37.3	58.8	49.8	33.2	24.9	46.6	61.3
MgO	13.8	14.4	13.4	14.1	15.3	16.6	10.2	8.2	11.9	7.2	14.3	15.5
FeO	12.1	11.5	13.4	13.5	13.9	10.4	16.2	20.7	13.6	20.2	14.1	9.9
Fe_2O_3	2.4	2.2	3.1	5.0	5.1	6.0	7.2	9.5	35.1	45.6	0.3	4.5
MnO	0.2	0.2	0.2	0.2	n.d.	0.2	0.6	0.5	n.d.	0.2	n.d.	n.d.

Table 1. Cont.

	Rhodope Massif										Urals Kempirsai			
	Skyros		Greece					Bulgaria			Northern Part		Southern Part	
									Goliamo					
wt%	Sky. 4	Sky. 5	Soufli 1	Soufli 2	Gomati	Broucevci	Jacovitsa	Pletena	Kamenyane 1	Kamenyane 2	Batamshinsk		Main Ore Field	
NiO	0.2	0.2	n.d.	0.2	n.d.	0.2	0.2	0.2	0.3	0.4	0.2	0.2	0.1	0.3
Total	99.1	99.7	100.2	99.8	100.3	99.1	99.1	100	99.2	99.9	100.4	99.9	100.8	100.3
$Cr/(Cr+Al)$	0.75	0.79	0.70	0.63	0.44	0.47	0.88	0.74	0.83	0.94	0.56	0.58	0.82	0.80
$Mg/(Mg+Fe^{2+})$	0.68	0.69	0.62	0.65	0.66	0.72	0.52	0.37	0.52	0.38	0.64	0.65	0.74	0.73
$Fe^{3+}/(Cr+Al+Fe^{3+})$	0.03	0.026	0.036	0.059	0.057	0.068	0.093	0.123	0.455	0.62	0.0015	0.009	0.053	0.052
Numbers of cations on the basis of 32 oxygens														
Ti	0.019	0.038	0.019	0.036	0.035	0.035	0.041	0.102	0.082	0.062	0.054	0.054	0.019	0.038
Al	3.414	3.813	4.636	5.591	8.475	7.828	1.767	3.277	1.452	0.386	6.987	6,675	2.787	2.947
Cr	12.066	11.695	10.745	9.396	6.553	7.021	12.670	10.582	7.168	5.648	8.880	9.070	12.323	12.143
Mg	5.321	5.512	5.036	5.231	5.324	5.890	4.143	3.291	4.858	3.088	5.137	5.151	5.874	5.780
Fe^{2+}	2.613	2.442	2.940	2.805	2.711	2.067	3.715	4.653	3.114	4.839	2.879	2.864	2.125	2.197
Fe^{3+}	0.480	0.414	0.580	0.940	0.901	1.079	1.481	1.936	7.215	9.844	0.024	0.146	0.852	0.833
Mn	0.044	0.044	0.043	0.000	0.000	0.040	0.138	0.115	0.000	0.046	0.000	0.000	0.000	0.000
Ni	0.041	0.041	0.000	0.000	0.000	0.038	0.044	0.044	0.114	0.088	0.039	0.038	0.020	0.062

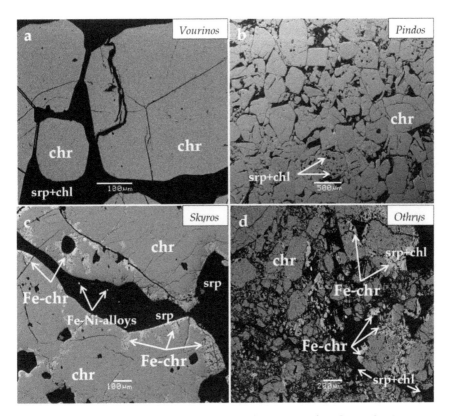

Figure 1. Representative back-scattered electron images of chromite ores from Greece, showing texture relationships between chromite and silicates, the presence of homogeneous chromite (**a**) and also abundant silicate inclusions (chlorite and serpentine) in the host chromite (**b**), porous texture and alteration to Fe-chromite (**c,d**). Abbreviations: chr = chromite; Fe-chr = iron-chromite; srp = serpentine; chl = chlorite.

2.2. Whole Rock Analysis

Major and trace elements in massive chromitite samples (more than 95 vol %) were determined by ICP-MS analysis, at the ACME Analytical Laboratories Ltd., Vancouver, BC, Canada (currently Bureau Veritas Commodities Canada Ltd.). The samples were dissolved using an acid mix (HNO_3–$HClO_4$–HF) digestion and then the residues were dissolved in concentrated HCl. The rare earth elements La, Ce, Pr, Nd, Sm, Eu, Gd, Tb, Dy, Ho, Er, Tm, Yb and Lu as well as Li, K, Ge, Sr, Y, Zr, Mo, Sb, Cs, W, Pb, Th, and U were lower than the detection limits of the analytical methods. The detection limits of the method for the presented elements are 1 ppm for Ga and V, 0.2 ppm for Co and Zn, 0.1 for Ni and 0.01 wt % for Fe. On the basis of the quality control report provided by the Analytical Labs, the results of analyses of the reference material in comparison to expected values, and the results from multistage analysis of certain samples, showed accuracy and precision in good agreement with accepted values for international standards. The analytical error, for Ga, for example, was <5%. Although the PGE data reported in Table 2 have been published previously (Table 2) a portion from the same samples was used for the presented trace element analyses.

Table 2. Trace element contents of high-Cr and high-Al chromitites.

Location	SEM/EDS		Trace Element (ppm)					wt%	ppb	
	Cr/(Cr+Al)	Mg/(Mg+Fe^{2+})	Ni	Co	V	Zn	Ga	Fe	ΣPGE *	Pd/Ir *
Vourinos 1	0.77	0.65	2000	240	500	260	14	8.1	140	0.06
Vourinos 2	0.81	0.62	1580	210	560	550	15	7.9	135	0.35
Vourinos 3	0.81	0.66	1900	200	400	300	13	8.72	92	0.55
Vourinos 4	0.56	0.72	1800	170	780	360	27	9.2	30	0.54
Vourinos 5	0.78	0.61	1800	180	620	280	14	7.76	109	0.63
Pindos 1	0.52	0.63	1300	140	600	400	23	7.84	51	1.0
Pindos 2	0.48	0.6	1500	260	710	410	32	9.77	143	1.0
Pindos 3	0.42	0.71	1500	260	710	520	34	7.8	117	6.33
Pindos 4	0.4	0.72	1630	240	580	460	32	7.3	6123	34.2
Pindos5	0.67	0.57	750	270	760	520	16	9.7	3875	12.2
Pindos 6	0.69	0.49	720	240	760	620	23	10.1	2098	7.2
Pindos 7	0.81	0.58	1450	290	560	490	17	10.2	181	0.22
Skyros 1	0.61	0.64	1300	250	1200	540	34	10.9	2300	0.08
Skyros 2	0.55	0.69	1600	200	1000	400	36	9.45	464	0.7
Skyros 3	0.58	0.51	1500	240	870	450	40	9.6	251	0.67
Skyros 4	0.75	0.69	1250	220	640	420	14	10.7	145	0.33
Skyros 5	0.79	0.69	1200	200	620	400	11	10.1	145	0.1
Othrys (n = 4)	0.54	0.69	1400	210	960	370	33	9.8	91	0.36
Rhodope Massif										
Greece										
Soufli1	0.70	0.62	1700	230	380	580	16	13.2	150	0.2
Soufli2	0.63	0.65	1150	220	460	280	12	10.6	82	1.0
Gomati	0.49	0.67	1030	130	730	280	24	9.6	104	0.25
Bulgaria										
Broucevci	0.47	0.72	1300	230	790	420	45	11	60	0.92
Jacovitsa	0.88	0.52	2250	310	240	760	9	13.9	197	0.46
Pletena	0.74	0.37	890	290	330	1030	13	17.9	563	0.07
Goliamo Kamenyane 1	0.83	0.52	1550	80	1000	450	12	11.6	87	0.14
Goliamo Kamenyane 2	0.94	0.38	2260	970	370	4030	6	64.3	40	1.93
Kemprsai (Urals)										
Northern	0.80	0.73	1600	210	160	160	14	9.1		
Batamshinsk	0.82	0.74	1600	230	200	190	16	9.5		
Southern	0.56	0.58	1500	240	680	480	48	12.6		
XL Let Kazakhstan	0.64	0.65	1700	230	730	340	49	10.1		

Symbol * = Data on PGE from literature [26–32].

3. A Brief Outline of Characteristics for the Studied Chromitites

All chromitite samples selected for the present study come from deposits and occurrences, which have been the subject of detailed geological, mineralogical and geochemical investigation [26–36] and references therein. The main ophiolite complexes of Greece (Vourinos, Othrys and Pindos) belong to the Upper Jurassic to Lower Cretaceous Tethyan ophiolite belt, and are characterized by heterogeneous deformation and rotation, during their original displacement and subsequent tectonic incorporation into continental margins [33]. The studied samples of chromitites are massive (Figure 1) and exhibit variations in the chromitite tonnage, the composition of chromite (Tables 1 and 2), the degree of transformation of ores and the associated ophiolites [27–37].

Chromite ores in the Vourinos complex occur in the mantle and cumulate sequences, with a tonnage estimated to approximately 10 Mt of high-Cr type, but at the central part of the complex there are high-Cr and high-Al ores in a spatial association, with low PGE contents [26]. The Othrys complex has a relatively high tonnage (approximately 3 Mt) of high-Al massive chromite ores and low PGE content [27].

The chromitite occurrences in the Pindos ophiolite complex are small (a few tens of m (x) a few tens of cm) and are hosted within completely serpentinized and weathered, intensively deformed dunite-harzburgite blocks, due to a strong plastic and brittle deformation that was superimposed on primary magmatic textures [17,18,34]. Chromitites throughout the Pindos complex are high-Cr and high-Al, often in a spatial association. The most salient feature of the Pindos chromitites is the enrichment in Pt and Pd at the area of Korydallos, at a level of 7 ppm PGE_{total} [28,29] and up to 29 ppm [35]. In the Achladones area on the Skyros island small massive chromitite bodies are of high-Al type and have elevated PGE contents, up to 3 ppm ΣPGE, although both high-Cr and high-Al types having low PGE content are found on the entire island [30].

Ophiolites associated with the Serbomacedonian massif (Gomati) and Rhodope massif including the ophiolites of Soufli (Greece), Dobromirci, Jacovitsa, Broucevci and Goliamo-Kamenjane (Bulgaria) host small (a few thousand tons) high-Cr and high-Al ores in a spatial are association, which occasionally contain elevated PGE concentrations. They are completely serpentinized, locally sheared and metamorphosed to antigorite-tremolite and/or talc schists. Detailed description of the characteristic mineralogy and texture of those chromitites have been published in previous studies [17,18,31,32,36].

The Kempirsai massif, covering an area of 2000 km^2, is divided by a shear zone into two parts: the southeastern part that is called Main Ore Field (MOF), hosting large high-Cr chromite deposits, and the northwestern area, the so-called Batamshinsk Ore Field (BOF), hosting much smaller high-Al chromite deposits [37]. An excellent description of the petrography and mineral chemistry, including mineral inclusions in the chromite of the giant chromite deposit of Kempirsai has been provided and discussed by Melcher et al [37]. These authors have interpreted their formation by a multistage process: High-Al chromitites may be derived from MORB-type tholeiitic melts, and high-Cr ones from boninitic magmas, during a second stage by interaction of hydrous high-Mg melts and fluids with depleted mantle in a supra-subduction zone setting.

4. Results

4.1. Compositional Variations in Chromite

The chromite samples from the central part of the Vourinos complex, the Pindos, Skyros island, Serbomacedonian, Rhodope and Kempirsai massifs show a wide variation in major elements from high-Cr, with the Cr/(Cr + Al) atomic ratio ranging from 0.81 to 0.69, to high-Al with the Cr/Cr + Al) ratio ranging from 0.63 to 0.4 (Table 1), falling in the range of metallurgical and refractorytype, respectively [21]. In addition, in the Bulgarian Rhodope massif (Jacovitsa, Pletena and Goliamo Kamenyane areas) altered chromite grains are dominant, having relatively high FeO and low Al_2O_3 and MgO contents (Table 1). As a consequence, the Cr/(Cr + Al) atomic ratios of those chromitites are significantly higher than those of high-Cr chromitites from the Vourinos complex (Table 1).

4.2. Distribution of Trace Elements in Chromitites

The geochemical data from whole rock analyses show a wide variation in major and trace element contents (Table 2). Gallium contents are lower in high-Cr chromitites (11 to 23 ppm) compared to high-Al ones (27–49 ppm), that seems to be independent on the degree of fractionation of parent magma, as exemplified by the Pd/Ir ratio [38]. The highest Co, Mn, Zn, Fe and lowest Ga were mainly recorded in strongly altered small chromite occurrences from the Rhodope massif in Bulgaria. They are in a good agreement with other chromitites [21–25] and are independent of the age of the associated ophiolites. Platinum-group elements (PGE) show total contents ranging from 30 to 6120 ppb and Pd/Ir ratios from 0.06 to 34, which are independent of the major element composition of chromitites (Table 2).

The results show a strong negative correlation (R ranges from −0.98 to −0.95) between the Cr/(Cr + Al) atomic ratio and Ga for the relatively large chromite deposits of Vourinos, Kempirsai massif (Urals) and the Skyros island. In addition, there is a less strong negative correlation for small chromite occurrences from the Pindos and Rhodope massifs (R ≥ −0.76 and R ≥ −0.83, respectively). Apart from Ga, the best correlation is found between Cr/(Cr + Al) and V for the Vourinos (R = −0.84), Skyros and Kempirsai (R ≥ −0.93), whereas no significant relationship for chromitites from the Pindos and Rhodope massifs (Figure 2b) or between Cr/(Cr + Al) and other minor and trace elements is observed.

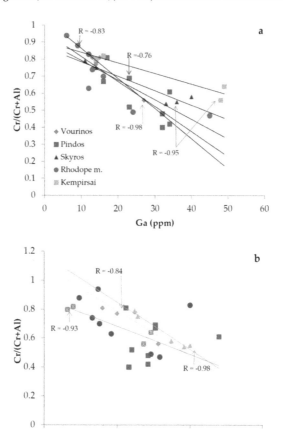

Figure 2. *Cont.*

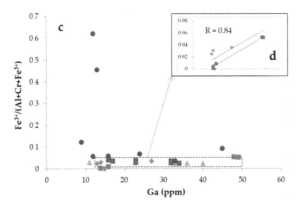

Figure 2. Plots of the Cr/(Cr + Al) atomic ratio versus Ga content (**a**); the Cr/(Cr + Al) ratio versus V content (**b**) and the Fe^{3+}/(Al + Cr + Fe^{3+}) ratio versus Ga content (**c**), including a detailed plot for the Vourinos and Kempirsai samples (**d**). Data from the Table 2.

5. Discussion

5.1. Factors Controlling the Spinel Chemistry

5.1.1. Magmatic Versus Post-Magmatic Processes

The wide variation of the Cr/(Cr + Al) atomic ratio for the chromitite samples from Greece, Bulgaria and Kempirsai massifs (Tables 1 and 2) fall in the range of metallurgical and refractory types. Differences in the trace element content (Table 2) may suggest trace element partitioning, depending on a number of factors, such as temperature, pressure, oxygen fugacity and the chemical composition of parent magmas [17–47].

As the partition coefficient of elements is defined as the ratio of the element content in a mineral and the melt [15], the Pd/Ir ratio can be used as an indicator of the degree of fractionation of parent magma for chromitites [38] and the presence of very low Pd/Ir values (low degree of fractionation) for both high-Cr and high-Al chromitites, suggest their origin from different magma sources. It has been argued that high-Cr chromitites, which have higher Sc, Mn, Co and Ni, and lower Ti, V, Zn and Ga contents may be derived from boninitic magmas, while high-Al ores may be derived from MORB-type tholeiitic magmas [17–24]. Experimental data at high temperature have shown that Ga is compatible in spinel with D values ranging between 0.9 and 11.2, and slightly lower D values in the most reducing experiments, while experimental data at temperatures >1300 °C and low oxygen fugacity have shown that there is no significant effect of temperature, composition and redox conditions [15]. However, the negative correlation between the Cr/(Cr + Al) atomic ratio and Ga content in natural chromitites points to the potential effect of the composition of the parent magma, while a positive trend between the Fe^{3+}/(Al + Cr + Fe^{3+}) atomic ratio and Ga content for large chromite deposits (Figure 2c,d) may suggest the effect of the redox conditions on the Ga distribution in chromitites. Specifically, high-Cr chromitites formed earlier from a primary magma (under relatively reducing conditions), compared to high-Al ones formed later from an evolved magma (and more oxidized conditions) magma [47].

In addition, differences in the negative correlations between the Cr/(Cr + Al) atomic ratio and Ga content and the slope of correlation lines for the different occurrences (Figure 2) may suggest that in addition to the composition of parent magmas, which is a major factor for large deposits (like Vourinos and Kempirsai massif, Urals) other factors such as temperature, pressure or redox conditions may be responsible for the observed deviation from linearity for small metamorphosed occurrences of chromitites, such as those from the Pindos and Rhodope massifs (Figure 2b). The lack of significant relationships between major and trace elements, in small chromitite occurrences from the Pindos and Rhodope massifs (Figure 2) may be related with post-magmatic processes. The elevated Mn, Co, Zn and Fe contents and depletion in Ga (Table 2) is consistent with the spinel chemistry in the Rhodope massif of Bulgaria, showing a trend of depletion in Ga, in the metamorphic Fe-chromite rims surrounding the cores of chromite grains, implying that most tetrahedral sites are still occupied by Fe^{2+} [17]. In addition, it has been suggested that the Mg cations can be replaced by Mn, Zn or Co, whereas Al and Fe^{3+} compete for the octahedral sites, hampering the entry of Ga [17,18,24].

Despite the recorded modification in trace elements by re-distribution during post-magmatic processes, as exemplified by bulk analysis (Tables 1 and 2) and spinel chemistry [17,18,24] limited only to relatively small chromitite occurrences, the well-established relationship (R = −0.95 to −0.98) between Cr/(Cr + Al) ratio and Ga (Table 2; Figure 1a) that is comparable to literature data for chromitites hosted in other ophiolite complexes [21–24,40–47] seems to be a salient feature.

5.1.2. Spinel Structure

The structure of spinel is a cubic close-packed array of 32 oxygen ions, with 64 tetrahedral vacancies and 32 octahedral vacancies in one unit cell each, containing 8 formula units, with the general formula: $A^{2+}B^{3+}_2O_4$, where A = Fe^{2+}, Mn^{2+}, Mg^{2+}, Co^{2+}, Zn^{2+}, Ni^{2+} and B = Fe^{3+}, Cr^{3+}, Al^{3+}, Ga^{3+}, V^{3+} [36]. Spinels are traditionally denoted as either "normal", where the A cation occupies T sites, the B cation occupies M sites, whereas in the "inverse" type cation B occupies the T site and the M site is occupied by both cations A and B [4,8–12]. The degree of inversion x characterizing the cation distribution can show values between $x = 0$ (normal spinel) and $x = 1$ (inverse spinel). The spinel structure is able to accommodate many cations (at least 36) by enlarging and decreasing its tetrahedral and octahedral bond distances, while the oxygen positional parameter (u) should be regarded as a measure of distortion of the spinel structure from cubic close packing or as the angular distortion of the octahedron [10,12]. The movements of cations between tetrahedral T and octahedral M sites, as a result of Mg^{2+} substitution, can be discussed based on structural parameters, such as bond lengths, cation-cation and cation-anion distances, bond angles and hopping lengths, which were calculated using experimental lattice constants and oxygen parameters [4,8–12].

5.1.3. Applications to Natural Spinels

Despite post-magmatic compositional changes in Fe-chromite within the chromitite ores, resulting in the remobilization of cations during metamorphism (700 °C to 450 °C) of chromitites [17,18,21,24,32], the structural incorporation of Ga into the chromite lattice is evidenced by the progressive and linear increase of Al or decreasing Cr/(Cr + Al) atomic ratio (Figure 2a). Experimental data on cation distribution versus temperature may provide valuable information related to the preferences of cations in the spinel lattice. $MgAl_2O_4$ is the most prominent example of a normal spinel because Mg^{2+} is much larger than Al^{3+} [2,6,48]. Gallium (Ga^{3+}) is smaller than Mg^{2+}, but significantly larger compared to Al^{3+}, leading to an ordering which is called mainly inverse, at least for the end-member composition $MgGa_2O_4$ [4].

The intra-crystalline exchange reaction in spinels has been modeled [49,50] and the order-disorder process has been described by the following exchange reaction: $^TAl + {}^MMg = {}^TMg + {}^MAl$ (where T = tetrahedral and M = octahedral site) in which the forward reaction implies an exchange of Mg with Al at the M site (ordering process), and backwards (disordering process). The cation distributions for both disordering and ordering experiments were obtained by measuring the oxygen positional parameter (u), the inversion parameter (x) (Al in T) site, using samples with varying composition. The Mueller kinetic model was satisfactorily applied to the experimental data and allowed the calculation of the kinetic ordering constants K, linearly related to temperature by means of Arrhenius equations [48–51].

Martignago et al. [52] performed crystal structure refinements on three natural spinels, on low-Cr spinel containing small Fe^{3+} quantities and two other samples with high Cr (8.4 wt % Cr_2O_3) and low Cr (3.3 wt % Cr_2O_3) contents [52]. These experiments have shown that both parameters (u) and

(x) remained constant for the three different samples up to 600 °C, independently of Cr^{3+} contents. The distortion of spinels, started at higher temperature, near to 650 °C (Figure 3). The degree of distortion at the highest temperatures is inversely correlated with Cr^{3+} contents. Since Cr^{3+} has a tendency to be completely partitioned on the octahedral site, due to its electronic configuration and size [2,52–54], Al^{3+} cation is unable to substitute Cr^{3+}. The most salient feature derived from the above experimental data [52] is the abrupt increase of the inversion parameter (x) having the highest values for the sample with the lowest Cr content (L-Cr sample), and the lowest values for the sample with the highest Cr content (H-Cr) [52].

Such structural changes should cause modifications of the structural, physical and thermal properties of the spinels. Although the spinel structure is complicated and determination of several parameters may be required, the above results [52], which show that the Cr content in spinels affects the occupancy of Al in the tetrahedral site, may suggest that the co-existence of Ga^{3+} in high-Al chromitites (Table 2) [2,4–12] is related to the degree of disorder that is inversely correlated with Cr contents.

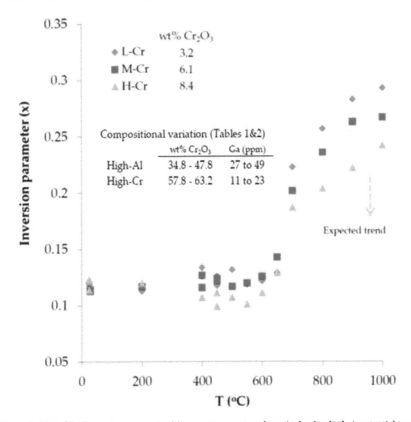

Figure 3. Plot of the Inversion parameter (x) *versus* temperature for spinels after [52]. A potential trend for chromitites with much higher Cr content (Table 1) is presented by the green arrow.

Therefore, potential controlling factors on the Ga preference in high-Al chromitites are (a) the composition of the parent magma (geotectonic setting), temperature and redox conditions, (b) the electronic configuration of Cr^{3+} resulting in occupation of M sites only, the ability of Al^{3+} to occupy T and M sites and the strong preference of Ga^{3+} to T sites, and (c) elevated values of the inversion parameter (x) that is inversely correlated with Cr^{3+} content and favors the co-existence of Ga^{3+} in high-Al chromitites.

6. Conclusions

The presented geochemical and mineral chemistry data on chromitites associated with ophiolite complexes, in conjunction with experimental literature data allowed us to draw the following conclusions:

(1) The lower Ga contents in high-Cr chromitites (11 to 23 ppm) compared to high-Al ones (27–49 ppm) suggest that the composition of the parent magma may be a major factor controlling the preference of Ga in high-Al chromitites.

(2) The positive trend between the $Fe^{3+}/(Al + Cr + Fe^{3+})$ atomic ratio and Ga content for large chromite deposits may suggest the effect of the redox conditions on the Ga distribution in chromitites.

(3) Plot of the $Cr/(Cr + Al)$ atomic ratios versus Ga content exhibits differences in terms of the slope of correlation lines for the different occurrences, suggesting that, in addition to the composition of parent magmas, other factors such as temperature, pressure or redox conditions may affect the observed deviation from linearity for small metamorphosed chromitite bodies.

(4) The depletion of Ga and Al, and elevated Mn, Co, Zn and Fe contents in certain small chromitite occurrences, transformed during post-magmatic metamorphism, suggest potential change of the Ga content in Cr-spinel during sub-solidus reactions.

(5) Assuming that low-Cr spinel is characterized by the highest value of the inversion parameter (x) at higher than 650 °C, then the high Al content in spinels may be a driving force for the degree of inversion in the structure that facilitate the substitution of Al^{3+} for Ga^{3+} at magmatic conditions.

Author Contributions: I.-P.D.E. provided the SEM/EDS analyses. Both authors I.-P.D.E. and G.D.E. contributed to the elaboration and interpretation of the data, and carried out the final revision of the manuscript.

Funding: This research was not funded because it is mostly based on valuable literature data.

Acknowledgments: Many thanks are expressed to the National University of Athens for the donations the chromitite samples and the access to the analytical facilities (electron microprobe analysis). Many thanks are expressed to Heinz-Gunter Stosch, Karlsruhe Institute of Technology (KIT), Institute for Applied Geosciences, Germany, Maria Perraki, Technical University of Athens and the anonymous reviewers for the constructive criticism and suggestions on an earlier draft of the manuscript. The linguistic improvement of this work by H.G. Stosch is greatly appreciated.

Conflicts of Interest: The authors declare no conflict of interest.

References

1. Liu, Z.; Li, H. Metallurgical process for valuable elements recovery from red mud—A review. *Hydrometallurgy* **2015**, *155*, 29–43. [CrossRef]
2. Navrotsky, A.; Kleppa, O.J. The thermodynamics of cation distributions in simple spinels. *J. Inorg. Nucl. Chem.* **1967**, *29*, 2701–2714. [CrossRef]
3. Shannon, R.D. Revised effective ionic radii and systematic studies of interatomic distances in halides and chalcogenides. *Acta Crystallogr.* **1976**, *32*, 751–767. [CrossRef]
4. O'Neill, H.S.C.; Navrotsky, A. Simple spinels: Crystallographic parameters, cation radii, lattice energies, and cation distribution. *Am. Mineral.* **1983**, *68*, 181–194.
5. Sickafus, K.E.; Wills, J.M.; Grimes, N.W. Structure of spinel. *J. Am. Ceram. Soc.* **1999**, *82*, 3279–3292. [CrossRef]
6. Atkins, P.; Overton, T.; Rourke, J.; Weller, M.; Hagerman, M. *Inorganic Chemistry*, 5th ed.; W.H. Freeman and Company: New York, NY, USA, 2013.
7. Princivalle, F.; Della Giusta, A.; Carbonin, S. Comparative crystal chemistry of spinels from some suites of ultramafic rocks. *Mineral. Petrol.* **1989**, *40*, 117–126. [CrossRef]
8. Bosi, F.; Andreozzi, G.B.; Hålenius, U.; Skogby, H. Zn-O tetrahedral bond length variations in normal spinel oxides. *Am. Mineral.* **2011**, *96*, 594–598. [CrossRef]
9. Fregola, R.A.; Bosi, F.; Skogby, S.; Hålenius, U. Cation ordering over short-range and long-range scales in the $MgAl_2O_4$-$CuAl_2O_4$ series. *Am. Mineral.* **2012**, *97*, 1821–1827. [CrossRef]
10. Bosi, F. Chemical and structural variability in cubic spinel Oxides. *Acta Crystallogr.* **2019**, *B75*, 279–285. [CrossRef]

11. Lenaz, D.; Musco, M.E.; Petrelli, M.; Caldeira, R.; De Min, A.; Marzoli, A.; Mata, J.; Perugini, D.; Princivalle, F.; Boumehdi, M.A.; et al. Restitic or not? Insights from trace element content and crystal—Structure of spinels in African mantle xenoliths. *Lithos* **2017**, *278*, 464–476. [CrossRef]

12. Wei, C.; Feng, Z.; Scherer, G.G.; Barber, J.; Shao-Horn, Y.; Xu, Z.J. Cations in Octahedral Sites: A Descriptor for Oxygen Electrocatalysis on Transition-Metal Spinels. *Adv. Mater.* **2017**, *29*. [CrossRef] [PubMed]

13. Burton, J.D.; Culkin, F.; Riley, J.P. The abundances of gallium and germanium in terrestrial materials. *Geochim. Cosmochim. Acta* **1959**, *16*, 151–180. [CrossRef]

14. Paktunc, A.D.; Cabri, L.J. A proton- and electron-microprobe study of gallium, nickel and zinc distribution in chromian spinel. *Lithos* **1995**, *35*, 261–282. [CrossRef]

15. Wijbrans, C.H.; Klemme, S.; Berndt, J.; Vollmer, C. Experimental determination of trace element partition coefficients between spinel and silicate melt: The influence of chemical composition and oxygen fugacity. *Contrib. Mineral. Petrol.* **2015**, *169*, 1–33. [CrossRef]

16. Dare, S.A.S.; Pearce, J.A.; McDonald, I.; Styles, M.T. Tectonic discrimination of peridotites using fO2-Cr# and Ga–Ti–FeIII systematic in chrome-spinel. *Chem. Geol.* **2009**, *261*, 199–216.

17. Gervilla, F.; Padrón-Navarta, J.A.; Kerestedjian, T.; Sergeeva, I.; González-Jiménez, J.M.; Fanlo, I. Formation of ferrian chromite in podiform chromitites from the Golyamo Kamenyane serpentinte, Eastern Rhodopes, SE Bulgaria: A two-stage process. *Contrib. Mineral. Petrol.* **2012**, *164*, 643–657. [CrossRef]

18. Colás, V.; González-Jiménez, J.M.; Griffin, W.L.; Fanlo, I.; Gervilla, F.; O'Reilly, S.Y.; Pearson, N.J.; Kerestedjian, T.; Proenza, J.A. Fingerprints of metamorphism in chromite: New insights from minor and trace elements. *Chem. Geol.* **2014**, *389*, 137–152. [CrossRef]

19. Scowen, P.; Roeder, P.L.; Helz, R. Re-equilibration of chromite within Kilauea Iki lava lake, Hawaii. *Contrib. Mineral. Petrol.* **1991**, *107*, 8–20. [CrossRef]

20. Prasad, S.R.M.; Prasad, B.B.V.S.V.; Rajesh, B.; Rao, K.H.; Ramesh, K.V. Structural and dielectric studies of Mg^{2+} substituted Ni–Zn ferrite. *Mater. Sci. Pol.* **2015**, *33*, 806–815. [CrossRef]

21. Zhou, M.F.; Robinson, P.T.; Su, B.X.; Gao, J.F.; Li, J.W.; Yang, J.S.; Malpas, J. Compositions of chromite, associated minerals, and parental magmas of podiform chromite deposits: The role of slab contamination of asthenospheric melts in suprasubduction zone environments. *Gondwana Res.* **2014**, *26*, 262–283. [CrossRef]

22. Uysal, İ.; Tarkian, M.; Sadiklar, M.B.; Zaccarini, F.; Meisel, T.; Garuti, G.; Heidrich, S. Petrology of Al- and Cr-rich ophiolitic chromitites from the Muğla, SW Turkey: Implications from composition of chromite, solid inclusions of platinum-group mineral, silicate, and base-metal mineral, and Os-isotope geochemistry. *Contrib. Mineral. Petrol.* **2009**, *158*, 659–674. [CrossRef]

23. Proenza, J.; Gervilla, F.; Melgarejo, J.; Bodinier, J.L. Al- and Cr-rich chromitites from the Mayarí-Baracoa ophiolitic belt (eastern Cuba); consequence of interaction between volatile-rich melts and peridotites in suprasubduction mantle. *Econ. Geol.* **1999**, *94*, 547–566. [CrossRef]

24. González-Jiménez, J.M.; Locmelis, M.; Belousova, E.; Griffin, W.L.; Gervilla, F.; Kerestedjian, T.N.; Pearson, N.J.; Sergeeva, I. Genesis and tectonic implications of podiform chromitites in the metamorphosed Ultramafic Massif of Dobromirtsi (Bulgaria). *Gondwana Res.* **2015**, *27*, 555–574. [CrossRef]

25. Brough, C.P.; Prichard, H.M.; Neary, C.R.; Fisher, P.C.; McDonald, I. Geochemical variations within podiform chromitite deposits in the Shetland Ophiolite: Implications for petrogenesis and PGE concentration. *Econ. Geol.* **2015**, *110*, 187–208. [CrossRef]

26. Konstantopoulou, G.; Economou-Eliopoulos, M. Distribution of Platinum-group Elements and Gold in the Vourinos Chromitite Ores, Greece. *Econ. Geol.* **1991**, *86*, 1672–1682. [CrossRef]

27. Economou-Eliopoulos, M.; Parry, S.J.; Christidis, G. Platinum-group element (PGE) content of chromite ores from the Othrys ophiolite complex, Greece. In *Mineral Deposits: Research and Exploration. Where Do They Meet?* Papunen, H., Ed.; Balkema: Rotterdam, The Netherlands, 1997; pp. 414–441.

28. Economou-Eliopoulos, M.; Vacondios, I. Geochemistry of chromitites and host rocks from the Pindos ophiolite complex, northwestern Greece. *Chem. Geol.* **1995**, *122*, 99–108. [CrossRef]

29. Economou-Eliopoulos, M.; Sambanis, G.; Karkanas, P. Trace element distribution in chromitites from the Pindos ophiolite complex, Greece: Implications for the chromite exploration. In *Mineral Deposits*; Stanley, Ed.; Balkema: Rotterdam, The Netherlands, 1999; pp. 713–716.

30. Economou-Eliopoulos, M. On the origin of the PGE-enrichment in chromitites associated with ophiolite complexes: The case of Skyros island, Greece. In *9th SGA Meeting*; Andrew, C.J., Borg, G., Eds.; Digging Deeper: Dublin, Ireland, 2007; pp. 1611–1614.

31. Economou-Eliopoulos, M. Platinum-group element distribution in chromite ores from ophiolite complexes: Implications for their exploration. *Ore Geol. Rev.* **1996**, *11*, 363–381. [CrossRef]

32. Zhelyaskova-Panayiotova, M.; Economou-Eliopoulos, M. Platinum-group element (PGE) and gold concentrations in oxide and sulfide mineralizations from ultmmafic rocks of Bulgaria. *Ann. Univ. Sofia Geol. Congr.* **1994**, *86*, 196–218.

33. Rassios, A.; Dilek, Y. Rotational deformation in the Jurassic Meohellenic ophiolites, Greece, and its tectonic significance. *Lithos* **2009**, *108*, 192–206. [CrossRef]

34. Economou, M.; Dimou, E.; Economou, G.; Migiros, G.; Vacondios, I.; Grivas, E.; Rassios, A.; Dabitzias, S. *Chromite Deposits of Greece: Athens*; Theophrastus Publications: Athens, Greece, 1986; pp. 129–159.

35. Kapsiotis, A.; Grammatikopoulos, T.A.; Tsikouras, B.; Hatzipanagiotou, K. Platinum-group mineral characterization in concentrates from high-grade PGE Al-rich chromitites of Korydallos area in the Pindos ophiolite complex (NW Greece). *SGS Miner. Serv.* **2009**, *60*, 178–191. [CrossRef]

36. Bonev, N.; Moritz, R.; Borisova, M.; Filipov, P. Therma–Volvi–Gomati complex of the Serbo-Macedonian Massif, northern Greece: A Middle Triassic continental margin ophiolite of Neotethyan origin. *J. Geol. Soc.* **2018**. [CrossRef]

37. Melcher, F.; Grum, W.; Simon, G.; Thalhammer, T.V.; Stumpfl, E. Petrogenesis of the ophiolitic giant chromite deposit of Kempirsai, Kazakhstan: A study of solid and fluid inclusions in chromite. *J. Petrol.* **1997**, *10*, 1419–1458. [CrossRef]

38. Barnes, S.-L.; Naldrett, A.J.; Gorton, M.P. The origin of the fractionation of the platinum-group elements in terrestrial magmas. *Chem. Geol.* **1985**, *53*, 203–323. [CrossRef]

39. Economou-Eliopoulos, M. Apatite and Mn, Zn, Co-enriched chromite in Ni-laterites of northern Greece and their genetic significance. *J. Geochem. Exp.* **2003**, *80*, 41–54. [CrossRef]

40. Sack, R.O.; Ghiorso, M.S. Chromian spinels as petrogenetic indicators: Thermodynamic and petrological applications. *Am. Mineral.* **1991**, *76*, 827–847.

41. Pagé, P.; Barnes, S.-J. Using trace elements in chromites to constrain the origin of podiform chromitites in the Thetford Mines Ophiolite, Québec, Canada. *Econ. Geol.* **2009**, *104*, 997–1018. [CrossRef]

42. Uysal, I.; Sadiklar, M.; Tarkian, M.; Karsli, O.; Aydin, F. Mineralogy and composition of the chromitites and their platinum-group minerals from Ortaca (Muğla-SW Turkey): Evidence for ophiolitic chromitite genesis. *Mineral. Petrol.* **2005**, *83*, 219–242. [CrossRef]

43. Leblanc, M.; Violette, J.F. Distribution of aluminum-rich and chromium-rich chromite pods in ophiolite peridotites. *Econ. Geol.* **1983**, *78*, 293–301. [CrossRef]

44. Zhou, M.F.; Sun, M.; Keays, R.R.; Kerrich, R.W. Controls on platinum-group elemental distributions of podiform chromitites: A case study of high-Cr and high-Al chromitites from Chinese orogenic belts. *Geochim. Cosmochim. Acta* **1998**, *62*, 677–688. [CrossRef]

45. Zaccarini, F.; Garuti, G.; Proenza, J.A.; Campos, L.; Thalhammer, O.A.R.; Aiglsperger, T.; Lewis, J.F. Chromite and platinum group elements mineralization in the Santa Elena ultramafic nappe (Costa Rica): Geodynamic implications. *Geol. Acta* **2011**, *9*, 407–423.

46. Colás, V.; Padrón-Navarta, J.A.; González-Jiménez, J.M.; Griffin, W.L.; Fanlo, I.; O'reilly, S.Y.; Gervilla, F.; Proenza, J.A.; Pearson, N.J.; Escayola, M.P. Compositional effects on the solubility of minor and trace elements in oxide spinel minerals: Insights from crystal-crystal partition coefficients in chromite exsolution. *Am. Mineral.* **2016**, *101*, 1360–1372. [CrossRef]

47. Barnes, S.J.; Roeder, P.L. The range of spinel compositions in terrestrial mafic and ultramafic rocks. *J. Petrol.* **2001**, *42*, 2279–2302. [CrossRef]

48. Andreozzi, G.B.; Princivalle, F.; Skogby, H.; Della Giusta, A. Cation ordering and structural variations with temperature in MgAl2O4 spinel: An X-ray single-crystal study. *Am. Mineral.* **2000**, *85*, 1164–1171. [CrossRef]

49. Ma, Y.; Liu, X. Kinetics and Thermodynamics of Mg-Al Disorder in $MgAl_2O_4$-spinel: A Review. *Molecules* **2019**, *24*, 1704. [CrossRef] [PubMed]

50. Mueller, R.F. Kinetics and thermodynamics of intracrystalline distribution. *Mineral. Soc. Am. Spec. Pap.* **1969**, *2*, 83–93.

51. Mueller, R.F. Model for order-disorder kinetics in certain quasi-binary crystals of continuously variable composition. *J. Phys. Chem. Solids* **1967**, *28*, 2239–2243. [CrossRef]

52. Martignago, F.; Dal Negro, A.; Carbonin, S. How Cr^{3+} and Fe^{3+} affect Mg-Al order disorder transformation at high temperature in natural spinels. *Phys. Chem. Miner.* **2003**, *30*, 401–408. [CrossRef]

Minerals **2019**, *9*, 623

53. Princivalle, F.; Martignago, F.; Dal Negro, A. Kinetics of cation ordering in natural Mg (Al, Cr3+)$_2$O$_4$ spinels. *Am. Mineral.* **2006**, *91*, 313–318. [CrossRef]
54. Lavina, B.; Reznitskii, L.Z.; Bosi, F. Crystal chemistry of some Mg, Cr, V normal spinels from Sludyanka (Lake Baikal, Russia): The influence of V3+ on structural stability. *Phys. Chem. Miner.* **2003**, *30*, 599–605. [CrossRef]

Article

Trace Element Distribution in Magnetite Separates of Varying Origin: Genetic and Exploration Significance

Demetrios G. Eliopoulos [1] and Maria Economou-Eliopoulos [2,*]

[1] Institute of Geology and Mineral Exploration (IGME), Sp. Loui 1, Olympic Village, GR-13677 Acharnai, Greece; eliopoulos@igme.gr
[2] Department of Geology and Geoenvironment, University of Athens, 15784 Athens, Greece
*   Correspondence: econom@geol.uoa.gr

Received: 10 November 2019; Accepted: 2 December 2019; Published: 6 December 2019

Abstract: Magnetite is a widespread mineral, as disseminated or massive ore. Representative magnetite samples separated from various geotectonic settings and rock-types, such as calc-alkaline and ophiolitic rocks, porphyry-Cu deposit, skarn-type, ultramafic lavas, black coastal sands, and metamorphosed Fe–Ni-laterites deposits, were investigated using SEM/EDS and ICP-MS analysis. The aim of this study was to establish potential relationships between composition, physico/chemical conditions, magnetite origin, and exploration for ore deposits. Trace elements, hosted either in the magnetite structure or as inclusions and co-existing mineral, revealed differences between magnetite separates of magmatic and hydrothermal origin, and hydrothermal magnetite separates associated with calc-alkaline rocks and ophiolites. First data on magnetite separates from coastal sands of Kos Island indicate elevated rare earth elements (REEs), Ti, and V contents, linked probably back to an andesitic volcanic source, while magnetite separated from metamorphosed small Fe–Ni-laterites occurrences is REE-depleted compared to large laterite deposits. Although porphyry-Cu deposits have a common origin in a supra-subduction environment, platinum-group elements (PGEs) have not been found in many porphyry-Cu deposits. The trace element content and the presence of abundant magnetite separates provide valuable evidence for discrimination between porphyry-Cu–Au–Pd–Pt and those lacking precious metals. Thus, despite the potential re-distribution of trace elements, including REE and PGE in magnetite-bearing deposits, they may provide valuable evidence for their origin and exploration.

Keywords: magnetite separates; trace elements; exploration; ophiolites; skarn; porphyry-Cu; Fe–Ni-laterites

---

1. Introduction

Magnetite is a widespread mineral, as disseminated or massive ore, in various geotectonic settings and rock-types, such as magmatic, hydrothermal, ophiolites, calk-alkaline rocks (porphyry-Cu deposits, skarn type), laterites, and metamorphosed rocks/ores, all formed under a variety of conditions [1–13]. It has an inverse cubic (space group $Fd3m$) spinel-type structure with the general stoichiometry AB_2O_4, where A and B are tetrahedral (Fe^{3+}) and octahedral (Fe^{3+} and Fe^{2+}) coordination sites, respectively [14]. In addition, the geochemistry of magnetite separates is also affected by the co-precipitation of other minerals [5–9]. Trace elements hosted either in the structure of magnetite, such as cations which can substitute on the A (Mg, Fe^{2+}, Ni, Mn, Co, or Zn) and B (Al, Cr, Fe^{3+}, V, Mn or Ga) sites [14,15], and/or in mineral inclusions. They may result in distinctive trace element signatures in magnetite separates and can be used as a prospective tool for many types of ore deposits, and/or a variety of trace-element discrimination diagrams can classify magnetite separates from different types of ore deposits [4–13]. However, it has been proposed that the discrimination diagram (Ti + V) versus (Ca + Al + Mn) can be used with caution as a petrogenetic indicator for trace element data in the case of magnetite separates which have unusual composition and/or have been re-equilibrated. [16].

The present study is focused on massive and disseminated magnetite separates from various geotectonic settings covering a wide range of formation conditions: Hydrothermal magnetite associated with calc-alkaline rocks (skarn type, porphyry-Cu–Au deposit), ophiolite complexes (associated with sulphides or apatite), magmatic magnetite from the magmatic sequence of ophiolites and ultramafic lavas/wherlites, magnetite from metamorphosed laterites, and magnetite separates from black coastal sand. Major and minor/trace element contents, including rare earth elements (REEs), in these magnetite separate types are presented and are compared to literature data. The purpose of this application of the magnetite chemistry and texture features is to provide a potential relationship between compositional variation and physico/chemical conditions, the magnetite origin, and exploration of ore deposits.

2. Materials and Methods

2.1. Mineral Chemistry

Polished sections (two-tree samples from each type) from magnetite ores (Tables 1–4) were carbon- or gold-coated and examined by a scanning electron microscope (SEM) using energy-dispersive spectroscopy (EDS). The SEM images and EDS analyses were carried out at the University of Athens (NKUA), using a JEOL JSM 5600 scanning electron microscope (Tokyo, Japan), equipped with the ISIS 300 OXFORD automated energy dispersive analysis system (Oxfordshire, UK) with the following operating conditions: Accelerating voltage 20 kV, beam current 0.5 nA, time of measurement (dead time) 50 s, and beam diameter 1–2 μm. The following X-ray lines were used: FeKα, NiKα, CoKα, CuKα, CrKα, AlKα, TiKα, CaKα, SiKα, MnKα, and MgKα. Standards used were pure metals for the elements Cr, Mn, Ni, Co, Zn, V, and Ti, and Si, MgO for Mg, and Al_2O_3 for Al. Contents of Fe_2O_3 and FeO for spinels were calculated on the basis of the spinel stoichiometry. The H_2O for (Co, Mn, Ni)-hydroxides was calculated to obtain 2 (OH) *pfu*.

Whole Rock Analysis

The studied magnetite samples were massive and disseminated mineralizations, derived from large (weighing approximately 2 kg) samples, which were crushed and pulverized in an agate mortar. The samples of magnetite separates from the Skouries deposit are representative of increasing depth from 43 to 363 m (the depth for the sample skou8b was 280 m) [4]. The multi-stage water flotation was applied and magnetite separates were concentrated magnetically from the water suspension of the rock powder. This portion was furthermore pulverized in the agate mortar at a size fraction −75 μm, and magnetite separates were separated furthermore from silicates. Thus the magnetite separate concentrates with weak silicates and potential very fine inclusions of various minerals (SEM/EDS images; Table 1) were used for bulk rock analysis.

Major and minor/trace elements in magnetite separate samples and two samples from the extinct fumaroles co-existing with native sulphur from the volcanic environment in the Aegean Sea (Santorini and Nisyros islands) were determined at the SGS Global—Minerals Division Geochemistry Services Analytical Laboratories Ltd., Vancouver, BC, Canada. The samples were dissolved using sodium peroxide fusion, combined ICP-AES and ICP-MS (Package GE_ICP91A50). Detection limits of the method are provided (Table S1). On the basis of the quality control report provided by Analytical Labs, the results of the reference material analysis in comparison to expected values, and the results from multistage analysis of certain samples, showed accuracy and precision of the method in good agreement with international standard (<10%).

3. Geological Outline and Mineralogical Features

3.1. Hydrothermal Magnetite Separates

3.1.1. Massive Magnetite Separates of Skarn-Type (Lavrion, Plaka Mine)

The ancient (earlier than 1000 B.C.), famous mine of Lavrion in Attica (Greece) is associated with a granodiorite intrusion, of Upper Miocene age, within the metamorphic -Cycladic Crystalline belt, which belongs to the carbonate replacement deposits (CRDs) [17]. It is well known for the exploitation of mixed massive sulphide mineralization composed by sphalerite, pyrite and galena (B.P.G), especially the production of silver [17–20]. Early hornfels formation was followed by the multistage development of a skarn deposit at the area of Plaka, around the granodiorite body, located into two stratigraphic horizons, near the contacts between the Kaessariani schist with lower and upper marble (Figure 1) [2].

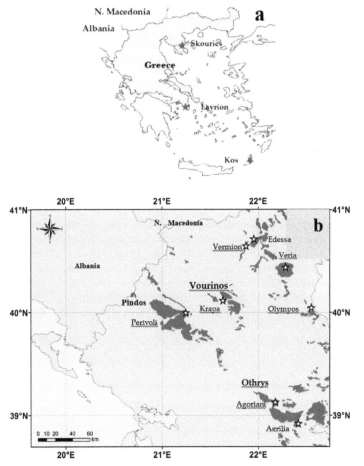

Figure 1. Sketch map showing the location of the studied magnetite separates hosted in calc-alkaline rocks (**a**). Ophiolitic rocks and metamorphosed Ni-laterite deposits (**b**).

Texture characteristics of the magnetite mineralization show multiple stages of the mineralization and that early magnetite and magnetite–hematite representing endoskarns (stage I) was followed by pyrite, pyrrhotite, and Fe-rich sphalerite, Ag-rich galena, arsenopyrite, forming exoskarns (stage II) [2,16] high-Fe sphalerite, chalcopyrite, and galena forming (Figure 2) [17–19]. Magnetite occurs as euhedral or anhedral crystals, homogeneous or having abundant porosity and solid inclusions, mostly silicate minerals. It is almost pure, having very low Si, Al, and Mn contents, while Ti and V are lower than the detection limit (Table 1).

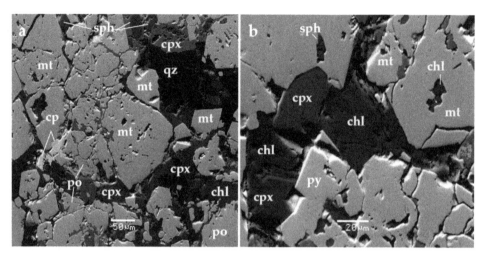

Figure 2. Back-scattered electron (BSE) images showing homogeneous magnetite separates (mt) in sharp contact with magnetite, having abundant porosity and solid inclusions, mostly silicate minerals, associated with clinopyroxene, quartz, chlorite, pyrrhotite, chalcopyrite, and sphlerite (**a**); magnetite is associated with silicates, sphalerite, and pyrite (**b**). Abbreviations: mt = magnetite; po = pyrrhotite; cp = chalcopyrite; py = pyrite; sph = sphalerite; cpx = clinopyroxene; qtz = quartz; chl = chlorite.

3.1.2. Disseminated Magnetite in Porphyry-Cu Deposit (Skouries)

The Skouries porphyry-Cu deposit is related to tertiary subvolcanic–porphyritic stocks, which belong to the Vertiskos Formation of the Serbo-Macedonian massif (SMM), Greece, and are controlled by deep fracture systems [21]. The typical alteration types of the porphyry-Cu intrusions are more or less presented in the Skouries intrusion, due to the repeated overprinting and intense silicification, with potassic being the predominant alteration type. Two mineral assemblages occurring as veinlets and disseminations can be distinguished: (1) Magnetite, reaching up to 10 vol.% (average 6 vol.%)—bornite–chalcopyrite, linked to pervasive potassic and propylitic alteration type, in the central parts of the deposit; and (2) chalcopyrite–pyrite, dominant at the peripheral parts of the deposit [4]. Chalcopyrite, and to a lesser extent bornite, contain occasional exsolutions of clausthalite (PbSe) (Figure 3b). There is a frequent association of magnetite separates and Cu-minerals (bornite and chalcopyrite) with inclusions of thorite, U-bearing thorite, rare earth element (REE)-enriched silicates of the epidote-group (allanite), Ce- and Th-rich monazites, and zircons identified as accessory minerals (Figure 3b–d; Table 1).

Table 1. Representative electron SEM/EDS analyses of magnetite in various rock-types (Figures 2–5, 7 and 8). Symbol n.d. = no detected.

Deposit type	CALC-ALKALINE							OPHIOLITES					PLACER
	Skarn-type		Hydrothermal			Hydrothermal		Hydrothermal		Magmatic			
Location	Lavrion	Plaka	Skouries			Othrys	Edessa	Pindos	Vourinos	Vourinos	Othrys	Othrys	KOS
sample	Plaka P.L.mt1	Plaka P.L.mt2	SK43	SG-6	Skou8	Agoriani O.Ag.t	Samari EdK2	Perivoli Pl.Mt	Krapa VKMt	Krapa VKMt1	Agrilia 15Agr2	Agrilia O.Agr34	KOS1
wt.%													
SiO_2	1.7	2.6	0.9	0.7	0.4	0.5	2.3	0.3	0.5	0.9	0.3	0.7	0.5
Al_2O_3	0.5	0.9	n.d	n.d.	n.d.	0.6	0.2	n.d.	1.7	2.1	11.1	7.5	1.5
Cr_2O_3	n.d.	n.d.	2.1	0.9	n.d.	n.d.	0.1	n.d.	n.d.	n.d.	3.9	1.5	0.2
V_2O_5	n.d.	n.d.	0.7	0.5	n.d.	n.d.	n.d.	n.d.	1.4	1.8	0.7	0.6	0.5
Fe_2O_3	67.5	67.7	66.7	67.6	67.9	68.1	68.1	69.7	55.1	55.4	45.2	51.4	57.1
FeO	30.4	30.4	30.6	30.8	30.7	30.1	29.9	30.2	33.5	33.3	32.3	31.1	34.5
MgO	n.d.	0.3	n.d.	n.d.	0.5	0.9	n.d.	0.2	n.d.	0.9	3.1	3.5	n.d.
MnO	0.5	0.3	n.d.	n.d.	n.d.	n.d.	0.2	0.2	0.5	0.6	0.2	0.4	1.2
CoO	n.d.	n.d.	n.d.	n.d.	n.d.	n.d.	n.d.	0.2	n.d.	n.d.	0.2	n.d.	n.d.
TiO_2	n.d.	n.d.	n.d.	n.d.	0.6	n.d.	n.d.	n.d.	6.9	4.4	3.6	4.6	5.1
Total	100.6	101.5	101	99.6	99.8	99.3	100.8	100.6	99.6	99.4	100.9	100.7	100.6

Table 2. Representative electron SEM/EDS analyses of fine-grained minerals associated with magnetite (Figure 3) from the Skouries porphyry-Cu deposit. Symbol n.d. = no detected.

wt.%	Monazite	Thorite	U-Thorite	Na, Ce-epidote	Allanite	Allanite
SiO_2	1.1	18.8	15.7	38.3	32.1	35.4
Al_2O_3	n.d.	n.d.	n.d.	27.2	16.2	14.1
FeO	n.d.	n.d.	0.6	8.3	14.9	17.4
CaO	0.8	n.d.	n.d.	22.2	12.4	13.6
P_2O_5	30.3	n.d.	n.d.	n.d.	n.d.	n.d.
ZrO_2	n.d.	n.d.	n.d.	n.d.	n.d.	n.d.
La_2O_3	17.2	n.d.	n.d.	0.9	5.1	3.9
Ce_2O_3	31.6	n.d.	n.d.	2.2	14.5	12.5
Nd_2O_3	10.5	n.d.	n.d.	n.d.	4.7	2.2
Pr_2O_3	2.8	n.d.	n.d.	n.d.	n.d.	n.d.
ThO_2	5.1	79.4	70.8	n.d.	n.d.	n.d.
UO_3	n.d.	n.d.	11.7	n.d.	n.d.	n.d.
Total	99.4	98.2	98.1	99.1	99.9	99.1

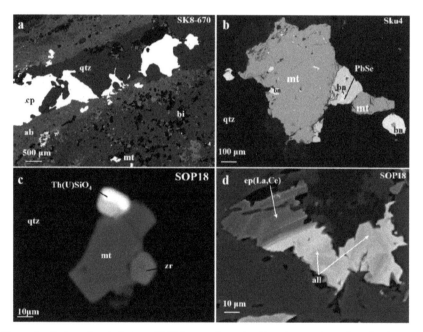

Figure 3. Back-scattered electron (BSE) images from the Skouries porphyry deposit, showing textural relationships between co-existing minerals. Quartz–chalcopyrite vein crosscutting biotite and magnetite-rich potassic altered porphyry (**a**); a close intergrowth of disseminated Ti–magnetite, chalcopyrite, bornite, and clausthalite (PbSe) (**b**); magnetite associated with thorite and zircon hosted by quartz (**c**); rare earth element (REE)-enriched epidote overgrown by allanite (**d**). Their composition is given (Table 2). Abbreviations: qtz = quartz; ab = albite; bi = biotite; mt = magnetite; cp = chalcopyrite; bn = bornite; PbSe = clausthalite; zr = zircon; ep = epidote; all = allanite.

3.1.3. Massive Magnetite Associated with Fe–Cu–Ni–Co Sulphides in Ophiolites (Pindos)

Sulfide mineralization in the Pindos ophiolite complex is located in the Smolicas Mountains, near the Aspropotamos dismembered ophiolite unit. This sequence, belonging to the ophiolitic unit of the Pindos complex, includes: (1) An intrusive section composed of dunites, lherzolites, olivine-websterites, olivine-gabbros, anorthosite gabbros, gabbros, and occasionally gabbronorites; and (2) a volcanic and subvolcanic sequence composed mainly by basalts and basaltic andesite pillow lavas ranging from high- to low-Ti affinity [22,23]. At the Perivoli (Tsoumes Hill) of the Aspropotamos unit, within gabbro close to its tectonic contact with serpentinized harzburgite, small irregular to lens-like occurrences (4 m × 1.5 m) of massive sulfide mineralization of Fe–Cu–Ni–Co type have been described [24]. Magnetite, forming often a network texture, is associated with sulphides, either as massive ore with inclusions of sulfides (chalcopyrite, pyrite, and pyrrhotite), or as individual grains dispersed within sulfide ore. A salient feature of magnetite is a network texture with occasionally deformed crystals (Figure 4). The composition of magnetite separates is characterized by very low Al, Ti, Cr, Mn, Ni, and V contents (Table 1 and Table S1).

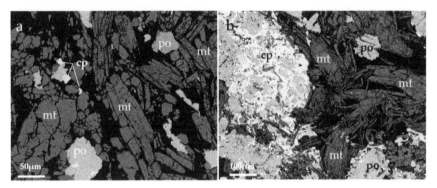

Figure 4. Back-scattered electron (BSE) images showing elongated curved magnetite associated with sulphides (**a**,**b**) from the Perivoli area, Pindos ophiolite complex. Abbreviations as in Figure 3.

3.1.4. Massive Magnetite in Mafic Ophiolitic Rocks Edessa (Samari)

The isolated ophiolite masses of Eastern Edessa belong to the Axios zone (Almopias subzone) [25–27]. At the area of Samari, Edessa, with dominant rock types diabase, gabbros, and serpentinized peridotites, ore bodies consisting of magnetite separates and sulphides have been described [27]. Magnetite occurs (1) as disseminated euhedral and/or subhedral crystals, along with sulphides (pyrite, chalcopyrite) within gabbros and diabase, accompanied by epidote, chlorite, quartz, sphene and tremolite; and (2) as massive ore hosted within gabbros, close to its contact with peridotite, with a gangue silicate mass dominant by greenalite, while sulphide veinlet crosscutting the magnetite separate ore are common (Figure 5) [27]. The magnetite is pure, characterized by the presence of only negligible contents of other elements (Table 1).

Figure 5. Back-scattered electron (BSE) images showing massive magnetite associated with little amounts of pyrite in a matrix of chlorite (greenalite) (**a**) and inclusions of chlorite in magnetite (**b**) from the Samari area (E. Edessa). Abbreviations as in Figures 2 and 3.

3.1.5. Massive Magnetite Associated with Apatite in Ophiolites

Major Mesozoic ophiolite complexes in Greece, including the Othris, Pindos, and Vourinos suites, were all obducted from the Middle-to-Late Jurassic and have been interpreted as parts of the same oceanic slab, although fore-arc and back-arc ridges have been tectonically fragmented and separated by overlapping sediments [28,29]. The Othrys ophiolite complex consists of a stack of thrust sheets in which overlapping stratigraphic successions, consisting mainly of harzburgites, lherzolites, gabbros, mafic dikes, and pillow lavas, have been recognized, while small irregularly shaped bodies of gabbro and gabbroic dikes or veins intrude the plagioclase lherzolite of the mantle sequence [30]. Irregular to lens-like occurrences (0.5 m × 1 m) of massive pure magnetite associated with large (up to 3 cm long) well-formed crystals of chlor–hydroxyapatite apatite, and lesser amounts of serpentine, tremolite, and Ni silicates (nepouite, pimelite), and by Ni sulfides (pentlandite, violarite, heazlewoodite) have been described in the Agoriani area of the Othrys ophiolite complex, central Greece [31]. The analysis of magnetite separates showed very little Al, Mg, Ni, Ti, and REE content (Table S1).

3.2. Magmatic Magnetite Separates

3.2.1. Disseminated Magnetite in Ultramafic Lavas and Wherlites

The Agrilia formation, Othrys ophiolite complex, about 6 km Northwestern of Lamia, is an unusual ultramafic lava characterized by high Mg content (31–33 wt.% MgO) [32]. Olivines from the Agrilia ultramafic lavas display high forsterite (Fo) contents, and calculated parental magma (in equilibrium with $Fo_{90.5}$) approximately 17 wt.% MgO [33]. Disseminated chromite occurs within groundmass (devitrified glass) and as inclusions in olivine and clinopyroxene [32–35]. The chromite spinel within groundmass is commonly rimmed by Ti–magnetite with sharp contacts (Figure 6a). Chromite is Cr-rich, exhibiting a limited range of variation (Table 3). Amphibole (tschermakite after [34], rhönite, chromite, and Ti–magnetite all associated with groundmass (Figure 6)). Rhönite in association with Ti–magnetite within devitrified glass (Table 3), and inter-grown with amphibole, may be of genetic significance [33,34]. The above ultramafic lavas are associated with medium-to-coarse grained wehrlite. They are mainly composed by olivine, clinopyroxene, hornblende, and in lesser amounts, spinel, orthopyroxene, phlogopite, and magnetites.

Table 3. Representative electron SEM/EDS analyses of rhönite (Rho) and amphibole (amph) within devitrified glass (Figure 6) from the Agrilia ultramafic lavas [35]. Symbol n.d. = no detected.

Host				Devitrified Glass (d.G.)				
	Rhonite (Rho)			d.G.			Amphibole (amph)	
SiO_2	26.3	26.6	25.8	33.1	45.6	39.8	40.4	55.2
MgO	8.8	8.7	8.8	14.5	6.5	9.2	10.6	11.1
Al_2O_3	15.7	14.9	15.2	12.1	9.9	16.5	17.5	10.6
Cr_2O_3	n.d.	n.d.	n.d.	0.8	1.1	n.d.	n.d.	1.3
CaO	10.7	10.6	10.6	4.3	3.1	10.1	9.6	10.4
TiO_2	4.9	3.9	4.7	1.6	0.2	1.9	1.2	0.2
FeO	28.9	29.8	30.3	27.1	21.8	17.7	17.1	6.2
MnO	0.2	0.2	0.3	0.2	0.2	n.d.	0.3	n.d.
Na_2O	0.5	0.5	1.3	1.6	6.9	2.5	1.7	2.7
K_2O	n.d.	n.d.	n.d.	0.2	0.9	0.3	0.2	0.3
NiO	n.d.	n.d.	n.d.	n.d.	n.d.	n.d.	n.d.	n.d.
Cl	n.d.	n.d.	n.d.	n.d.	n.d.	0.2	0.2	n.d.
Total	96.0	95.2	97.0	95.5	96.2	98.2	98.8	98.0

3.2.2. Disseminated Magnetite Separates in Norite Gabbros (Central Vourinos, Krapa Hills)

The Vourinos ophiolite complex located in NW Greece constitutes a complete but tectonically disrupted ophiolite sequence. Two magmatic series have been preserved at the central part of the complex, which constitute the earlier Krapa sequence, consisting of magnetite-bearing gabbro-norite and leuco-gabbro-norite (minor magnetite-bearing) and the younger Asprokambo sequence [36]. The Fe–Ti mineralization occurs as disseminated grains in a proportion of 5–20 vol.% between coarse-grained clinopyroxene and plagioclase, while sulphides (pyrrhotite, pyrite, chalcopyrite) are present in lesser amounts. Ti–magnetite occurs as interstitial to the major phases as small disseminated euhedral and irregular grains (Figure 7). Ti-bearing magnetite separates exhibits low Si, Mg, and high Al (1.7–2.3 wt.% Al_2O_3), Ti (3.5–6.9 wt.% TiO_2), and V (1.4–2.1 wt.% V_2O_5 contents) (Table 1).

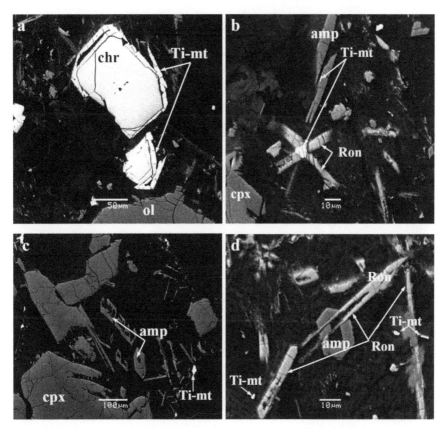

Figure 6. Back-scattered electron (BSE) images from ultramafic lavas of the Agrilia Formation, showing the association of Ti–magnetite (Ti–Mt) with chromite (chr) (**a**), microcrystals of rhönite (Ron) (**b**) associated, and randomly oriented amphibole (Amp) (**c,d**).

3.3. Magnetite in Coastal Black Sand

The volcanic rocks of the South Aegean arc form a chain from the Saronikos Gulf in the west, to Kos, Nisyros, and the east, through the Milos and Santorini islands [37]. Multiple eruptions of calc-alkaline to tholeiitic composition and Plio-Pleistocene age are located from the Santorini toward the Nisyros–Kos volcanic rocks (Figure 1a) [37–39]. Those volcanic rocks are part of the volcanic system related to the northward subduction of the last remnant of the oceanic crust of the African plate beneath the southern edge of the active margin of the European plate [38–41]. Magnetite and other heavy minerals have been found in all volcanic types in small quantities (1–2 modal %).

The investigated coastal black sand samples (black sea sand) collected from the Northern part of Kos Island revealed the presence of abundant Ti–magnetite with inclusions of apatite, ilmenites, Fe–Mn-oxides and monazite [(Ce, La, Th, Nd, Y) PO_4] (Figure 8). Feldspars and quartz are common too.

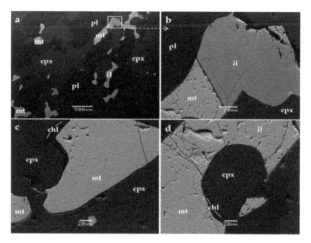

Figure 7. Back-scattered electron (BSE) images from the gabbro-norite sequence of Central Vourinos (Krapa area) showing disseminated Ti–magnetite as small euhedral and irregular phase associated with ilmenite (il) interstitially to the major phases of clinopyroxene (cpx) anf plagioclase (pl) (**a–d**).

Figure 8. Back-scattered electron (BSE) images of coastal black sand collected from the northern part of Kos Island, showing that Ti–magnetite is the dominant oxide, while apatite (apat) forms inclusions of approximately 100-μm-long grains (**a–c**); Fe–Mn-oxides (**b**), monazite [(Ce, La, Th, Nd, Y) PO_4 and (Ce, La, Nd, Pr, Sm) PO_4] are common (**a,d**).

The sample from the Nisyros caldera (neighboring Kos Island) comes from whitish extinct fumaroles co-existing with native sulphur, that is dominant by opal (Figure 9). The analyzed sample of the formation from the Nisyros caldera is opal–CT and quartz [41], which it is associated with abundant native sulphur (Figure 9).

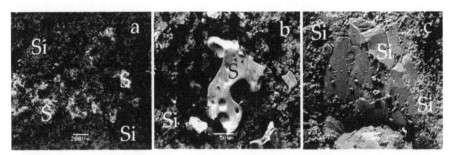

Figure 9. Back-scattered electron (BSE) images (unpolished sections) of extinct fumarole formation from the hydrothermal crater of the Nisyros caldera, showing native sulphur (**a**,**b**) and the association of native sulphur with whitish Si material (opal) (**c**), having low REE content (Table S1).

3.4. Magnetite Separates from Metamorphosed Fe–Ni-Laterites (Olympos, Edessa, and Vermion)

The dismembered ophiolite masses of the Olympos, Edessa, and Vermio mountains (Upper Jurassic–Lower Cretaceous), composed mainly by serpentinized harzburgite and, to a lesser extentm crustal magmatic rocks (pyroxenites and gabbros), outcrop along the eastern margin of the Pelagonian massif. They are considered to have been derived from the Almopias subzone of the Vardar zone and were overthrust westward onto the Pelagonian massif [22]. They are mostly composed of a mantle sequence (harzburgites) and crustal magmatic rocks, both characterized by supra-subduction zone (SSZ) features [23]. Small laterite bodies are located in the Olympos, Edessa, and E. Vermio Mountains, at the contact of serpentinized harzburgites with Upper Cretaceous limestones or within the serpentinites themselves. Due to intense tectonism, the Fe–Ni-laterite occurrences are often entirely enclosed within serpentinized harzburgites [1,25]. Apart from fine-grained magnetite separates in the matrix of the Fe–Ni ore, magnetite separates are associated with chromite, as the product of a gradual alteration of chromite or epitaxial as rims on chromite (Figure 10). A salient feature of zoned chromite grains from the Olympos, Edessa, and Vermion laterite occurrences is a Cr, Al, and Mg gradual decrease as Fe increases outward from the core; Mn, Zn, and Co increases gradually outwards, attaining greatest values at the periphery of chromite cores and in the Fe–chromite, up to 13.0, 2.1, and 4.1 wt.%, respectively, and drops off to negligible values in magnetite separates (Table 4). At the area of Vermio, there is abundant garnet (grossularite) and calcite, while Ni is mainly hosted in chlorite, serpentine, and theophrastite containing 80 wt.% NiO (Figure 10; Table 4) [42]. A (Co, Mn, Ni)-hydroxide with a wide compositional range occurs in a spatial association with theophrastite (Figure 11; Table 4) [43]. Several SEM-EDS analyses of hydroxides can be classified into two groups (Figure 11): Group A includes phase mineral dominated by cobalt, and group B those dominated by Mn, towards the peripheral parts of the concentrated development (Figure 10).

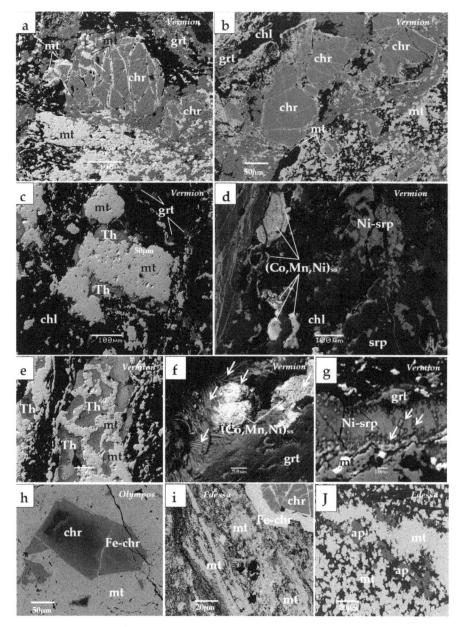

Figure 10. Back-scattered electron (BSE) images of metamorphosed Ni-laterites from northern Greece. Magnetite (mt) surrounding chromite (chr) grains and occurring along cracks (**a,b,e,f**); various morphological forms and textural relationship between magnetite–theophrastite (Th) and garnet (grt) (**c,f,g**); magnetite crosscutting veinlets (**e,i,j**); magnetite–(Co–Mn–Ni)(OH)$_{ss}$ showing successive thin layers, composed by fine fibrous crystals (**d**); microtextures resembling fossilized microorganisms coated by (Co–Mn–Ni)(OH)$_{ss}$ (white arrows) (**f**) or isolated and aggregates of bacterio-morphic Fe–Ni–serpentine (white arrows) within a veinlet of Ni–serpentine (Ni–srp) (**g**); chromite–Fe–chromite (**h**); chromite–magnetite (**i**); magnetite–apatite (ap) (**j**).

Table 4. Representative electron SEM/EDS analyses of chromite cores, magnetite (mt) and Fe–chromite (Fe–chr), Theophrastite and (Co, Mn, Ni)-hydroxides from Ni-laterites of the Vermion mountain (Northern Greece; Figure 10). Symbol n.d. = no detected.

Location	Olympos			Edessa						E. Vermion		
	Paliambela					Nissi	Mavrolivado			Stoumari		
Mineral	Core Chromite	Fe-chr	Rim mt	Core Chromite	Fe-chr	Rim mt	Core Chromite	Fe-chr	Rim mt	Theophastit	(Co,Mn,Ni)-hydroxides	
sample	L1	L1	L1	Ed10	Ed10	EdP10	V.AK	V.AK	V.AK			
wt.%												
SiO_2	0.2	n.d.	0.4	n.d.	n.d.	0.5	0.3	0.7	1.2	n.d.	n.d.	n.d.
Al_2O_3	8.2	8.6	0.4	9.7	6.5	0.2	24.4	7.7	0.5	n.d.	n.d.	n.d.
Cr_2O_3	62.1	53.7	5.9	60.2	58.1	1.6	43.3	42.3	1.6	n.d.	n.d.	n.d.
Fe_2O_3	1.8	3.5	62.2	2.7	3.3	67.7	2.2	17.6	66.5	n.d.	n.d.	n.d.
FeO	13.9	27.6	28.4	15.5	8.3	29.3	17.3	26.6	28.9	1.0	n.d.	n.d.
MgO	11.3	0.6	0.4	11.2	3.5	0.3	11.1	2.2	0.3	n.d.	13.9	19.2
MnO	1.9	4.1	0.5	1.6	13.1	n.d.	n.d.	2.8	0.3	n.d.	n.d.	n.d.
ZnO	n.d	0.4	n.d.	n.d.	2.1	n.d.	0.5	0.7	0.3	n.d.	25.0	n.d.
CoO	n.d.	n.d.	0.4	n.d.	4.1	n.d.	n.d.	n.d.	n.d.	n.d.	38.0	46.9
TiO_2	n.d.	n.d.	n.d.	n.d.	n.d.	n.d.	n.d.	n.d.	n.d.	n.d.	n.d.	n.d.
NiO	n.d	n.d.	1.6	n.d.	n.d.	1.4	n.d.	0.2	n.d.	79.2	17.4	13.6
Total	99.4	98.5	100.2	100.9	99.0	101	99.1	100.8	99.6	80.2	80.4	79.7
Cr/(Cr + Al)	0.83	0.81	0.81	0.78	0.86	0.87	0.54	0.79	0.94	79.44		
Mg/(Mg + Fe^{2+})	0.55	0.03	0.13	0.47	0.17	0.01	0.51	0.12	0.01			

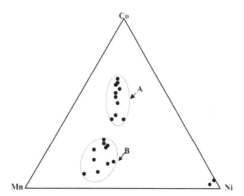

Figure 11. Triangular plot of Co, Mn, and Ni (atom%) showing the compositional variation of the (Co, Mn, Ni)solid solutions [43].

The (Co, Mn, Ni)solid-solution phases are associated with theophrastite, pyrochroite, Ni–serpentine (garnierite), garnet (grossularite), and magnetite, all having a common origin, subsequently of the strong late tectonic evolution, overprinting earlier deformation events (Figure 10c–g). They occur as successive thin (a few to tens of μm layers, composed by fine fibrous crystals up to a few μm (Figure 10d), microtextures resembling fossilized microorganisms coated by (Co–Mn–Ni) (OH)ss (white arrows) (Figure 10f), or isolated and aggregates of bacterio-morphic Fe–Ni–serpentine (white arrows) within a veinlet of Ni–serpentine (Ni–srp) (Figure 10g).

4. Geochemical Characteristics of Magnetite Separates

Although a good separation of magnetite from very fine inter-grown silicates and/or sulphides (Figures 2–10) was not achieved, it is clear that the highest Cr, Mn, and Co contents were recorded in Fe–Ni-laterite occurrences of northern Greece, ranging from 1.1 wt.% Cr in Vermio to 9.0 wt.% Cr in Olympos, from 0.38 wt.% Mn in Vermio to 0.69 wt.% Mn in Olympos, and from 460 ppm Co in Olympos to 1060 ppm Co in the Edessa laterites (Table S1).

The highest Ti and V contents were determined in disseminated magnetite separates from the magmatic sequence (norite gabbros) of the Vourinos (Krapa) ophiolite complex (2.56 wt.% Ti, 6330 ppm V), the Agrilia ultramafic rocks from the Othrys complex (7.48 wt.% Ti, 1760 ppm V), and the coastal black sands (or placer deposits) from Kos Island. In addition, magnetite separates from the Agrilia ultramafic rocks is richer in Al, Mg, Ti, and Cr compared to that from the Krapa norite-gabbros. Magnetite separates from the Skouries deposit shows relatively significant contents of Ni (up to 640 ppm Ni) and Cr (up to 1060 ppm Cr) (Table S1).

Hydrothermal magnetite separates from the ancient Lavrion (Plaka) mine, that is of skarn-type, and the Skouries porphyry deposits, that is disseminated, both associated with calc-alkaline rocks differ in terms of the higher Ba, Bi, Cr, Ni, Co, V, Sr, Sc, Rb, ΣREE, Y, Hf, Th, U, and Zr contents in the latter (Table S1). Gallium is higher in magnetite separates associated with cal-alkaline rocks (38–59 ppm) compared to that associated with ophiolites (2–21 ppm Ga) (Table S1). The highest Ga contents were recorded in magnetite separates from the Skouries porphyry-Cu deposit (40–59 ppm), skarn-type from Lavrion mine (38–39 ppm), and coastal sand from Kos Island (42 ppm), all related with calc-alkaline rocks. A potential order of the average Ga content in magnetite separates, from low to high is from Ni-laterites, hydrothermal (ophiolites) associated with sulphides, ultramafic lavas (ophiolites), hydrothermal (ophiolites) associated with apatite, magmatic (norite-gabbros), skarn-type (calc-alkaline), coastal sands (calc-alkaline volcanic), and porphyry-Cu deposits (hydrothermal calc-alkaline), resulting into a discrimination using a plot of (V + Ti) versus Ga content (Figure 12).

The rare earth element (REE) content of the majority of hydrothermal magnetite separates analyzed, such as those from the Lavrion mine (skarn-type), the Skouries porphyry-Cu deposit, the Pindos

(Perivoli) and Edessa (Samari) is generally low (a few decades ppm and REEs exhibit similar trends. Although the REE content, as well as actinides (Th, U) and Zr, in the porphyry (Skouries) deposit are higher compared to those in skarn-type (Lavrion), the REE patterns are similar in terms of the negative Eu, Tm, and Ho anomalies, but the former display occasionally positive Eu anomalies as well, and slight negative anomalies in Sm and Tm (Figure 13a,b). Moreover, considering the magnetite separate samples from the Skouries deposit, it seems likely that there is a small increase of the REE_{total} with the increasing depth from 43 to 363 m. In addition, the magnetite separate sample from the lowest depth (43 m) exhibits a positive Eu anomaly, in contrast to the samples from deeper parts of the deposit (Figure 13; Table S1). The hydrothermal magnetite separates associated with apatite from the Othrys (Agoriani) complex show higher REE content (54 and 80 ppm), that is, increasing with the presence of apatite, reaching values to more than 1300 ppm REE in apatite [31].

Calculations of Ce, Eu, Pr, and Gd anomalies were defined [44] by equations such as: Ce/Ce* = (2Ce/Ce chondrite)/(La/La chondrite + Pr/Pr chondrite) and Eu/Eu* = (2Eu/Eu chondrite)/(Sm/Sm chondrite + Gd/Gd chondrite). Assuming that values > 1 and < 1 are called positive and negative anomalies, respectively, chondrite normalized REE patterns for Ni-laterites from the Vermio mountains showed negative Ce, Eu anomalies, which is in agreement with the Ce/Ce* (0.56–0.24) and Eu/Eu* (0.40–0.24) values (Table S1). The highest REE contents were recorded in Ti–magnetite separates from the ultramafic lavas (130 ppm) and from the coastal sands, Kos Island (220 ppm). Sulphur occurs to be impregnated with particles of rocks, including opal–CT, quartz, alunite, anhydrite, and kaolinite (Figure 9) [41]. There is enrichment for light rare earth elements (LREE) from La to Sm and decrease for the heavier REE in magnetite separates compared to andesitic volcanic rock from Santorini Island (Figure 13f), and enrichment for all REE compared to the recent formation, associated with abundant native sulphur in Nisyros Island (Figure 9a,b). A distinct negative Eu anomaly is a common feature of the magnetite separates from the coastal sand of Kos and the andesitic rock from Santorini Island.

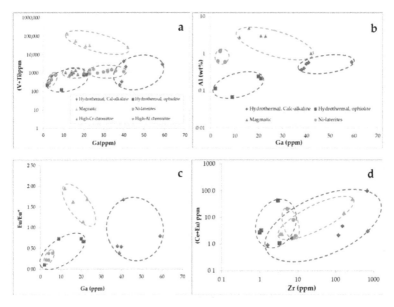

Figure 12. *Cont.*

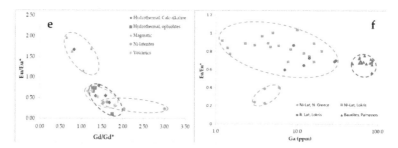

Figure 12. Plots of (V + Ti) and Al versus Ga contents (**a**,**b**), Eu/Eu* ratio versus Ga content (**c**); (Ce + Eu) versus Zr content (**d**), Eu/Eu* versus Gd/Gd* ratios (**e**), all for magnetite separates (data from Table S1) and plot of the Eu/Eu* ratio versus Ga content for Fe–Ni-laterites from Northern and Central Greece (Ni-lat), bauxite latererite (B-lat), and bauxites from Parnassos-Giona zone (**f**). Data from Table S1 and literature [45,46].

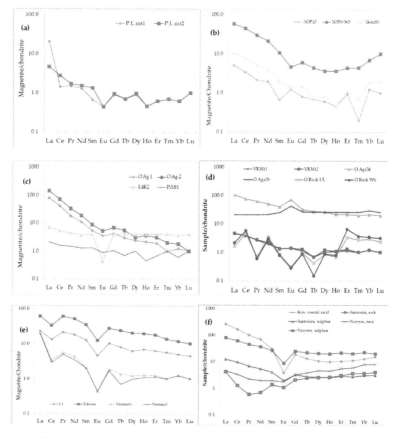

Figure 13. Chondrite normalized diagrams for magnetite separates (**a**–**f**) and volcanic rock (**d**,**f**). Data from Table S1 and [47] for ultramafic lavas (sample labeled as Rock UL, *n* = 3, and wehrlite (Wh), *n* = 7) from the Othrys ophiolite complex. Chondrite values from McDonough and Sun [48].

Moreover, the same trace elements were analyzed in two samples representing extinct fumaroles of andesitic and rhyolite composition, co-existing with native sulphur from the Aegean active volcanic arc, collected from Santorini (andesite) and Nisyros Island (Figures 9 and 13f). There is enrichment of LREE from La to Sm and decrease of the heavier REE in magnetite separates from costal sands (Kos), that is comparable to andesitic material from Santorini Island (Figure 13f). In addition, they exhibit elevated REE content compared to the extinct fumarole from Nisyros Island (Table S1; Figure 12f). Furthermore, REE patterns of both extinct fumarole material and native sulphur from Nisyros differ in terms of the low LREE content, the relative HREE enrichment compared to magnetite separates and andesitic material from Santorini and the negative Ce, Pr, and Eu anomalies (Figure 13f; Table S1).

Relatively high REE contents were recorded in Ti–magnetite separates from the ultramafic lavas from the Agrilia (99 ppm), while in wehrlite it is only 23 ppm. However, both are clearly higher compared to ultramafic lavas and wehrlite rocks (literature data), ranging from 14 to 17 ppm REE_{total}, respectively [47]. Chondrite normalized REE for magnetite separates from ultramafic lavas and wehrlite display sub-parallel patterns, with slight LREE enrichment in ultramafic lava, and an inverse relationship in terms of heavy rare earth elements (HREE) (Figure 13d). The Eu content in Ti–magnetite separates from the ultramafic lavas of Agrilia (Othrys) is enriched, showing a positive anomaly; in contrast, it is depleted in the magmatic sequence (norite gabbros) of the Vourinos (Krapa) complex (Figure 13d). In addition, the REE patterns for ultramafic lavas and wehrlites (literature data [47]) show a slight HREE enrichment for wehrlites compared to ultramafic lavas too (Figure 13d), while both rock types exhibit clear positive Ce, Nd, Gd, and Er anomalies and negative Pr, Eu, and Tb anomalies (Figure 13d).

In summary, the analyzed samples of magnetite, both massive ore and separates, covering a wide range of associated rock type, reveal wide compositional variations (Table S1 and Table 5). Although much compositional overlaps were recorded, a potential classification of magnetite separates is: (a) Magmatic: (a1) Associated with ophiolites and (a2) associated with calc-alkaline volcanic rocks (coastal sands); (b) hydrothermal: (b1) Associated with ophiolites and (b2) associated with calc-alkaline rocks (skarn-type and porphyry-Cu–Au deposits); and (c) epigenetic origin of magnetite separates in laterites (Figures 2–13). Average values show the highest ΣREE contents in disseminated magnetite separates from ultramafic lavas (Othrys) and coastal black sands (Kos Island) (Table 5).

Available data from previous publications have shown that an enrichment in Cr and Ni is a common feature of porphyry-Cu–Au–Pd–Pt, such as the Skouries, Elatsite, and other porphyry deposits worldwide, characterized by significant Pd and Pt contents, in contrast to the porphyry-Cu ± Mo deposits of Russia and Mongolia with lower Pd–Pt content (Figure 14) [4,12–14,49].

Table 5. Summary of the certain trace element contents (ICP-MS analyses) in magnetite separates and/or host ores.

	Location	Rock-Type	Texture	ΣREE	Ce/Ce*	Eu/Eu*	Pr/Pr*	Gd/Gd*	Ga	V	Ti	Cr	Ni	Reference
MAGMATIC														
Ophiolites	Vourinos (Krapa)	Norite gabbro	Disseminated	6	1.06	1.43	0.92	1.14	22	6520	2.6	1070	100	Present study
	Othrys	Ultramafic lava	Disseminated	132	0.9	1.97	0.98	0.65	12	1820	13.4	1600	160	Present study
Calc-alkaline	Kos	Coastal sand	Placer	224	0.92	0.32	0.85	1.81	42	1480	2.5	1150	11	Present study
HYDROTHERMAL														
Ophiolites	Edessa (Samari)	Peridotite	Massive	10	0.94	0.11	0.95	1.75	2	34	0.02	560	56	Present study
	Othrys (Agoriani)	Association with apatite	Massive	76	0.82	0.70	0.71	1.28	20	640	0.02	1100	280	Present study
	Pindos (Perivoli)	Irregular	Massive	3.1	0.90	0.74	1.02	1.34	9	80	<0.01	12	6	Present study
Calc-alkaline	Skouries	Porphyry-Cu	Disseminated	26	0.98	1.01	0.88	1.20	46	760	0.25	630	340	Present study
	Lavrion (Plaka)	Skarn	Massive	3	0.81	0.48	0.94	1.64	3.9	44	0.02	160	30	Present study
METAMORPHOSED	Olympos, Vermio, Edessa	Ni-laterites (n = 4)	Major, matrix	37	0.41	0.32	1.45	0.93	4.2	280	380	33,980	9800	Present study
Fe-Ni-laterite	Lokris	Ni-laterites (n = 13)	Rare, matrix	120	1.02	0.85	0.96	1.14	7.6	75	2900	1900	2200	[46]
Bauxite laterite	Lokris	Bauxitic laterites (n = 6)	Rare, matrix	350	1.01	0.71	0.97	1.2	17	340	7200	3500	1400	[46]
Bauxite	Parnassos-Giona	Bauxites (n = 17)	Rare, matrix	470	2.64	0.69	0.58	0.8	70	460	16,000	820	150	[45]

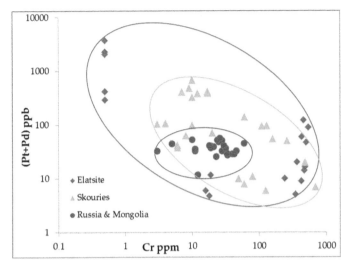

Figure 14. Plot of the (Pt + Pd) versus Cr content in porphyry-Cu–Au–Pd–Pt and porphyry-Cu–Mo deposits. Data from literature [12,14,49].

5. Discussion

The trace element content in magnetite separates, coupled with mineral chemistry and texture relationships, may provide evidence on (1) their parent magma or hydrothermal ore forming system, (2) partition coefficient of trace elements into magnetite separates, and (3) physico/chemical conditions (temperature, fO_2) during deposition and/or re-deposition of magnetite separates, and can be used as a prospective tool for many types of ore deposits [5–10,50]. The recorded wide variations in the trace element contents of massive and concentrates of disseminated magnetite separates from Greece may be used to discriminate between magmatic and hydrothermal magnetite separates, associated either with calc-alkaline rocks and ophiolites, between porphyry-Cu + platinum-group elements (PGEs) and porphyry-Cu lacking precious metals deposits, and between laterite types.

5.1. Trace Element Distribution in Magnetite Structure and Associated Minerals

The geochemistry of magnetite separates is affected by a variety of cations accommodated in the spinel-type structure [3,14,15] and the co-precipitation of other minerals [5–10]. The high-resolution studies of minerals applying TEM techniques provide a new and very important way to elucidate valuable information on the physico/chemical processes and mechanisms during formation of ore deposits and post-depositional changes of ore minerals [51]. The investigation of hydrothermal magnetite separates from the Los Colorados iron oxide–apatite deposit in Chile showed three generations of magnetite, which vary significantly in terms of their Si, Al, Ca, Mg, Ti, Mn, Na, and K minor and trace element contents, in contrast to the V content (remarkably constant), and concluded a complex incorporation of trace elements during growth of magnetite [51]. Although, there is a discrepancy between experimentally determined values and those derived from natural occurrences, the partition coefficients for spinel–liquid pairs ($D^{sp/lq}$), for high field strength elements, such as Ga, V, Zr, Zn, and Co, are extremely low [50,52]. Considering the presented data (Table S1), potential components of sulphide (Cu, Zn) and/or silicate (Si, Mg, K) inclusions in magnetite separates are used with caution. Selected elements, such as V, Ti, and Ga, in individual types of magnetite separates are plotted, showing different fields for magmatic and hydrothermal magnetite, and furthermore, between hydrothermal magnetite associated with ophiolites and calc-alkaline rocks (Figure 12).

Hydrothermal apatite from the Othrys (Agoriani) ophiolite complex contain over 1300 ppm ΣREE [31], whereas the associated magnetite separates contain less than 100 ppm ΣREE, due probably to incomplete separation of apatite (low P and Ca contents) (Table S1) [31]. Similarly, magnetite separates from the coastal black sand from Kos, containing inclusions of apatite (Figure 8b,c), exhibit relatively high (over 220 ppm ΣREE) (Table S1) and monazite of varying composition [(Ce, La, Th, Nd, Y) PO_4 and (Ce, La, Nd, Pr, Sm)PO_4] (Figure 8a,d). Magnetite separates from black coastal sands can be linked back to a volcanic source, and may be a potential REE source [41]. On a global scale, in many placer deposits, such as the river placers in Malaysia, paleo-placers in South Africa, Elliot Lake, Ontario, and beach sand deposits in India, monazite is an important REE mineral and many monazite ore deposits are economically important REE resources, although small contents of Th and U make monazite a restricted mineral in India [53]. Also, hydroxyl–bastnaesites and a (La, Nd, Y) (CO_3) F member of the bastnaesite group, which are among the most economically important as well, have been identified at the lowest part of karst-type laterites in Greece, close to their contact with carbonates [54–57].

The first data on magnetite separates from black coastal sand from Kos Island are compared with potential parental rocks of the neighboring active Nisyros Island and Santorini, all belonging to the Aegean volcanic arc, showing REE, Ti, V, Cr, Mn, Zn, and Ga contents in magnetite separates are higher than those in the volcanic rocks of basaltic composition from Santorini Island [50] and Nisyros Island (Table S1). The similarity between the REE patterns, including the clear Eu negative anomaly of magnetite separates from coastal black sand of Kos Island and that of andesitic material from Santorini Island (Figure 13f), may provide evidence of an andesitic source of the placer magnetite separates in Kos. Also, the composition of native sulfur has been used to qualitatively estimate elemental abundances in low-temperature fumarole gases. At low temperatures (<250 °C), oxygen in the atmosphere interacts with H_2S, resulting to precipitation of native sulphur [58].

5.2. Re-Mobilization and Re-Deposition of Magnetite

Magnetite separates as a late magmatic or early hydrothermal phase in both calc-alkaline systems and ophiolite complexes may be dissolved and re-precipitated, due to changes in temperature, salinity, pressure, and fO_2 [51,59–61]. An increase in temperature would enhance the solubility and, consequently, the undersaturation of iron in the fluids [61], leading to dissolution of the primary magnetite separates and formation of the secondary varieties. Thus, primary magnetite may have been re-mobilized during late brittle deformation and motion along the detachments contacts [60]. In general, textural and compositional data for magnetite from nine iron skarn deposits in Canada, Romania, and China have shown that most samples have been re-equilibrated by dissolution and re-precipitation and/or recrystallization [16,51]. These authors concluded that the applicability and reliability of trace element discrimination diagrams may be used with caution where magnetite has been subjected to multiple re-equilibration processes, and that detailed textural characterization must be undertaken in order to evaluate the effects of hydrothermal overprinting events and other re-equilibration processes on the trace element compositions of magnetite.

A peculiarity of the above Fe–Ni-laterite occurrences is the relatively high Cr, Mn, Co, and Ni content in Fe–chromite zones of spinels. The epigenetic mobilization of Fe, Mn, Co, and Ni and their re-deposition is suggested by a significant Mn, Zn, and Co enrichment in Fe–chromite (Table 1) and the abundance of Ni–serpentine (Figure 10; Table 4). The substitution for Mg^{2+} and Fe^{2+} by Mn, Zn, and Co in the chromite lattice resulted to a gradual transformation of chromite to Fe–chromite, while in the magnetite, that is considered to be the final product of this alteration process [1,62] and the above elements drop to negligible amounts (Table 4). Calculations for spinel structures [14,15,63,64] have shown that in normal spinels, there is a tendency for the electrostatic energy to favor the larger ions ($Fe^{2+} > Mg^{2+}$ and $Fe^{3+} > Al^{3+}$) into the tetrahedral site. Thus, the resulting small expansion of the tetrahedral and octahedral sites hampers the entry of other trace elements in the magnetite separate lattice at low temperatures [65,66]. Besides, the re-mobilization of Fe and re-deposition of magnetite at low temperature is indicated by its association with (Ni, Fe, Co, Mn)-hydroxides. Specifically, the texture relationships between magnetite, theophastite [$Ni(OH)_2$], and (Co, Fe, Mn, Ni) $(OH)_2$ (Figure 10a–h) suggest a common mineral forming system, along shear zones, subsequently of the strong late tectonic evolution, which overprints earlier deformation events, probably at low temperature, approximately 100 °C [67] and alkaline conditions, as it is suggested by the presence of calcite and the high Ca content in the magnetite separates concentrates (Table S1). Moreover, the association of the (Co, Mn, Ni) $(OH)_2$ solid-solution phases with microtextures resembling fossilized microorganisms coated by (Co–Mn–Ni) $(OH)_2$ (Figure 10f, white arrows) and the occurrence of isolated and aggregates of bacterio-morphic Fe–Ni–serpentine (Figure 10g) suggest the potential effect of microorganisms in the Vermio mineral-forming processes. Currently, much research interest has focused on a consortium of microorganisms of varying morphological forms, producing enzymes, which are considered to be a powerful factor to catalyze redox reactions, and act as nucleation sites for the precipitation of secondary minerals [68–70]. In general, the presence of organic matter and microorganism traces in highly tectonized zones of the metamorphosed Fe–Ni-laterites may suggest direct involvement of microorganisms, and have created appropriate conditions for metal bio-leaching and bio-mineralization. In general karst-type Fe–Ni-laterites, where dominant Fe minerals are goethite ± hematite [54,55,71], may display REE contents ranging from a few decades to thousands of more than 5000 ppm (Table S1) [54], depending on the site of the Ni-laterite samples along vertical profiles and the post-deposition processes affected their composition. The epigenetic mobilization of REE, along with Fe, Mn, Ni, Co, under reducing and acidic conditions and, subsequently, re-deposition under alkaline conditions at the lowest parts of contacts between laterites of Central Greece (Lokris) and carbonate rocks has been well established as well [54–56,71]. Thus, the potential depletion of Ga and Al, and elevated Mn, Co, Zn, and Fe ± REE contents has been recorded in the above large Fe–Ni-laterite deposits, but in small Fe–Ni-laterite occurrences in Northern Greece, strongly transformed during post-magmatic metamorphism, the dominant Fe mineral is depleted magnetite, due probably to their small size, facilitating potential changes during sub-solidus reactions. Therefore, the dominance of magnetite in Fe–Ni-laterite ores is not an encouraging factor for the exploration of Ni sources of laterite-type.

5.3. Implications of the Trace Element Distribution for the Origin of Magnetite Mineralization

Despite the potential re-distribution of trace elements in magnetite separate-bearing deposits, disseminated magnetite separates from mafic–ultramafic magmatic rocks of ophiolite complexes exhibit much higher Ti and V contents compared to hydrothermal magnetite, and they are comparable to magnetite separates from coastal sands derived from the weathering of calc-alkaline volcanic rocks (Table 5). The massive hydrothermal magnetite separates of skarn-type differs from the disseminated magnetite separates from porphyry deposits, in terms of the lower Ba, Bi, Cr, Ni, Co, V, Sr, Sc, Rb, ΣREE, Y, Hf, Th, U, and Zr contents (Table S1). Also, Ga is higher in magnetite separates associated with calc-alkaline rocks (38–59 ppm) compared to that related to ophiolites (2–21 ppm Ga) (Table S1). The highest Ga contents were recorded in magnetite separates from the Skouries porphyry-Cu deposit (40–59 ppm), skarn-type from Lavrion mine (38–39 ppm), and coastal sand from Kos Island (42 ppm), all related with calc-alkaline rocks. A potential order of the average Ga content in magnetite separates concentrates, from low to high, is from Ni-laterites, hydrothermal (ophiolites) associated with sulphides, ultramafic lavas (ophiolites), hydrothermal (ophiolites) associated with apatite, magmatic (norite gabbros), skarn-type (calc-alkaline), coastal sands (calc-alkaline volcanic), and porphyry-Cu deposits (hydrothermal calc-alkaline), resulting in a discrimination using a plot of (V + Ti) versus Ga content (Figure 12).

Magnetite separates associated with Fe–Cu–Ni–Co sulphides with dominant minerals pyrrhotite, chalcopyrite, and pentlandite exhibit relatively high Cu and Zn, due to incomplete separation from the associated sulphides, but it is REE-, Ti-, and V-depleted. Although an originally primary magmatic origin for these sulphides is not precluded, characteristics of the highly transformed pure magnetite separates (Table 1 and Table S1) at Pindos may indicate that potential original magmatic features have been overprinted by a low-level hydrothermal circulation process.

The plot of the Eu/Eu* ratio versus Ga content shows a less negative anomaly Eu/Eu* (average 0.85) for large Fe–Ni-laterite deposits [56,72,73], bauxite laterites and bauxites (63 and 0.69, respectively) [74], and the most negative Eu/Eu* anomaly (average 0.32), for the magnetite separates from Fe–Ni-laterites of Northern Greece (Figure 12e; Table 5). Assuming that the presence of Eu anomaly, caused by the ability of Eu to exist in either Eu^{2+} or Eu^{3+} states, the Eu/Eu* anomalies seem to be effected by physico/chemical conditions, such as a redox potential [74]. Diagenetic re-mobilization of Eu is possible under conditions of reduction to Eu^{2+} at low oxidation potential, as is suggested by the presence of organic matter [73–76]. In addition, the increasing order of the average Ga content from the Fe–Ni-laterites to bauxite laterites and bauxites (Table 5; Figure 12e) is in agreement with the preference of Ga to high-Al ores. The average Ga content for karst bauxite deposits is 58 ppm with a range from less than 10 to 812 ppm, because the close geochemical affinity of Ga to Al enables Ga to substitute easily in rock-forming alumino-silicates [77]. Such a preference Ga with high-Al chromitites, as well and the structural incorporation of Ga into chromite lattice, is evidenced by the progressive and linear increase of Al or decreasing Cr/(Cr + Al) atomic ratio [78,79].

5.4. Implications of Magnetite Separates for Exploration of Precious Metals in Porphyry-Cu Deposits

Alkaline or K-rich calc-alkaline porphyry deposits worldwide represent significant gold resources, owing to their large sizes. They are derived from hydrous calc-alkaline magmas and are genetically linked to magmatic and hydrothermal processes. Despite their common origin in a supra-subduction environment, platinum-group elements (PGEs) have been recorded only in certain porphyry-Cu deposits. During the last decades, significant Pd and Pt contents were described in the Cordillera of British Columbia, in the Balkan–Karpathian system, including the Skouries deposit, in the Philippines and elsewhere [4,12,13,80,81]. The estimated Pd, Pt, and Au potential for porphyry deposits, combined with the association of merenskyite [(Pd, Pt) (Te, Bi)$_2$)] with major Cu minerals (bornite and chalcopyrite), are considered to be encouraging economic factors for Pd and Pt, as by-products, the Au being the main precious metal product [13,80,81].

It has been suggested that first-stage arc magmas tend to generate Cu-rich, relatively Au-poor porphyry systems, while relatively Au-rich residue is left in the lower crust and lithospheric mantle after first-stage arc magmas [82]. Subsequently, a second stage of melting, due probably to lithospheric thickening, thermal rebound, mantle lithosphere delamination, or lithospheric extension, may generate magmas with relatively high Au/Cu ratios [82–84] and PGE enrichments [84–87]. Critical requirements for a significant base/precious metal potential in porphyry Cu + Au + Pd ± Pt deposits are considered to be the geotectonic environment, controlling the precious/base metal endowment in the parent magma (metasomatized asthenospheric mantle wedge) [86,87]. It is indicated by significant Cr, Co, Ni, and Re contents, the oxidized nature of parent magmas, that facilitate the capacity for transporting sufficient Au and PGE, and the degree of evolution of the mineralized system, as is exemplified by the enrichment of (Pd + Pt) versus (Os + Ir + Ru) and LREE versus HREE [4,86,87]. Arc magmas with high potential to produce hydrothermal systems with ideal chemistry for transporting precious metals (Cu + Au + Pd ± Pt deposits) are characterized by fO_2 more than two log units above fayalite-magnetite-quartz (FMQ), where fO_2 is oxygen fugacity and FMQ is the fayalite–magnetite–quartz oxygen buffer [88]. Since the oxidized nature of parent magma is connected with the ability to produce a magmatic–hydrothermal system with ideal chemistry that facilitates the capacity for transporting sufficient precious metals, abundant magnetite (reaching up to 10 vol.% in the Skouries deposit) linked to pervasive potassic and propylitic alteration type [13], in contrast to "reduced" porphyry Cu–Au deposits, lacking primary hematite, magnetite separates, and sulphate minerals [89], is considered to be a characteristic feature of the Pd-bearing porphyry-Cu deposits. Thus, the abundance of magnetite separates is a characteristic feature of all porphyry-Cu–Au–Pd deposits, as indicated by unexpected high-grade Cu–(Pd + Pt) (up to 6 ppm) mineralization at the Elatsite porphyry deposit, which is found in a spatial association with the Chelopech epithermal deposit (Bulgaria) and the Skouries porphyry deposit, Greece [4,12]. Disseminated hydrothermal magnetite separate is often associated with fine (less than 10 μm) crystals of U-thorite and zircon within veinlets of quartz (Figure 3c), which is consistent with the elevated Th, U, and Zr contents (Table S1).

The enrichment in Cr, Co ± Ni is a common feature, recorded in the Skouries, Elatsite, and other porphyry deposits of the Balkan Peninsula, characterized by significant Pd and Pt contents, in contrast to the porphyry-Cu ± Mo deposits of Russia and Mongolia with lower (Pd + Pt) content (less than 25 ppm Cr [49,90]) supports the origin of their parent magma from an enriched mantle source [91,92]. In addition, a salient feature is the evolved geochemical signature of the Skouries and Elatsite porphyry-Cu deposits, as indicated by the negative trend between (Pd + Pt) and Cr contents (Figure 14). The application of stable isotope analyses of oxygen ($\delta^{18}O$ = 4.33–9.45‰) and hydrogen (δD = −110‰ to −73‰) on quartz veins from various drill holes and depths may provide evidence to constrain the origin of fluids trapped in quartz veins [85]. Although any systematic variation of the isotopic data with depth is uncertain, it seems likely that the samples with higher Cr contents correspond to lower $\delta^{18}O$ values, while those with lower Cr contents correspond to higher $\delta^{18}O$ values. This isotope variation, coupled with the higher (Pd + Pt) content Pd/Pt ratios corresponding to higher $\delta^{18}O$ values and lower Cr contents, may point toward the more evolved mineralizing system (Figure 14) [86].

6. Conclusions

Texture, mineral chemistry and geochemical data on massive and disseminated magnetite separates from a wide range of geotectonic settings and rock types led us to the following conclusions:

- Magnetite separates of hydrothermal origin associated with calc-alkaline rocks can be distinguished from that associated with ophiolites in terms of their higher Ga content, while magnetite separates of magmatic origin are characterized by the highest (V + Ti) content.
- Hydrothermal magnetite separates from skarn-type (Lavrion mine) differs compared to the disseminated separates from porphyry-Cu deposit (Skouries) in terms of the higher Ba, Bi, Cr, Ni, Co, V, Sr, Sc, Rb, ΣREE, Y, Hf, Th, U, and Zr contents in the latter.
- Although an initial primary magmatic origin for the Fe–Cu–Ni–Co sulphides in the Pindos ophiolite is not precluded, geochemical characteristics (REE-, Ti-, and V-depleted) of the highly transformed magnetite separates may indicate that potential original magmatic features have been overprinted by low-level hydrothermal circulation processes.
- Similarity between the REE patterns, including the clear Eu negative anomaly of magnetite separates from coastal black sand of Kos Island, may provide evidence of an andesitic source.
- A major requirement controlling the Pd and Pt potential of porphyry-Cu deposits is probably the oxidized nature of parent magmas that facilitate the capacity for transporting sufficient Pd and Pt, and the crystallization of abundant magnetite.
- Abundance of magnetite in porphyry-type deposits may be used to discriminate between porphyry-Cu-Au-Pd-Pt and porphyry-Cu-Mo deposits lacking precious metals, providing evidence for the exploration of precious metals in porphyry systems.

Supplementary Materials: The following are available online at http://www.mdpi.com/2075-163X/9/12/759/s1, Table S1: Major and trace element contents in representative magnetite separates from various geotectonic settings (Greece).

Author Contributions: D.G.E. collected the samples, provided the field information, and performed all types of analyses. Both authors (D.G.E. and M.E.-E.) contributed to the elaboration and interpretation of the data. The writing of the paper was completed by M.E.-E.

Funding: This research was funded by the National and Kapodistrian University of Athens (NKUA) (Grant No. KE_11730).

Acknowledgments: Many thanks are expressed to the two anonymous reviewers for their constructive criticism and suggestions on an earlier draft of the manuscript. Vassilis Skounakis, University of Athens, is thanked for his assistance with the SEM/EDS analyses.

Conflicts of Interest: The authors declare no conflicts of interest.

References

1. Paraskevopoulos, G.M.; Economou, M.I. Zoned Mn-rich chromite from podiform type chromite ore in serpentinites of northern Greece. *Am. Miner.* **1981**, *66*, 1013–1019.

2. Economou, M.; Skounakis, S.; Papathanasiou, C. Magnetite deposits of skarn type from the Plaka area of Laurium. *Greece* **1981**, *40*, 241–252.

3. Lindsley, D.H. Experimental studies of oxide minerals: Reviews Mineral. *Geochemistry* **1991**, *25*, 69–106.

4. Eliopoulos, D.G.; Economou-Eliopoulos, M. Platinum-group element and gold contents in the Skouries porphyry-copper deposit, Chalkidiki Peninsula, northern Greece. *Econ. Geol.* **1991**, *86*, 740–749. [CrossRef]

5. Dupuis, C.; Beaudoin, G. Discriminant diagrams for iron oxide trace element fingerprinting of mineral deposit types. *Miner. Depos.* **2011**, *46*, 319–335. [CrossRef]

6. Dare, S.A.S.; Barnes, S.-J.; Beaudoin, G. Variation in trace element content of magnetite crystallized from a fractionating sulfide liquid, Sudbury, Canada: Implications for provenance discrimination. *Geochim. Cosmoch. Acta* **2012**, *88*, 27–50. [CrossRef]

7. Nadoll, P.; Mauk, J.L.; Hayes, T.S.; Koenig, A.E.; Box, S.E. Geochemistry of magnetite from hydrothermal ore deposits and host rocks of the Mesoproterozoic Belt Supergroup, United States. *Econ. Geol.* **2012**, *107*, 1275–1292. [CrossRef]

8. Nadoll, P.; Angerer, T.; Mauk, J.L.; French, D.; Walshe, J. The chemistry of hydrothermal magnetite: A review. *Ore. Geol. Rev.* **2014**, *61*, 1–32. [CrossRef]

9. Dare, S.A.; Barnes, S.-J.; Beaudoin, G.; Muric, J.; Boutroy, E.; Potvin-Doucet, C. Trace elements in magnetite as petrogenetic indicators. *Miner. Depos.* **2012**, *49*, 785–796. [CrossRef]

10. Rapp, J.F.; Klemme, S.; Butler, I.B.; Harley, S.L. Extremely high solubility of rutile in chloride and fluoride-bearing metamorphic fluids: An experimental investigation. *Geology* **2010**, *38*, 323–326. [CrossRef]

11. Economou-Eliopoulos, M. Apatite and Mn, Zn, Co-enriched chromite in Ni-laterites of northern Greece and their genetic significance. *J. Geochem. Exp.* **2003**, *80*, 41–54. [CrossRef]

12. Augé, T.; Petrunov, R.; Bailly, L. On the mineralization of the PGE mineralization in the Elastite porphyry Cu–Au deposit, Bulgaria: Comparison with the Baula-Nuasahi Complex, India, and other alkaline PGE-rich porphyries. *Can. Miner.* **2005**, *43*, 1355–1372. [CrossRef]

13. Economou-Eliopoulos, M. Platinum-Group Element Potential of Porphyry Deposits. In *Mineralogical Association of Canada Short Course 35*; Mineralogical Association of Canada: Quebec, QC, Canada, 2005; pp. 203–245.

14. Navrotsky, A.; Kleppa, O.J. The thermodynamics of cation distributions in simple spinels. *J. Inorg. Nucl. Chem.* **1967**, *29*, 2701–2714. [CrossRef]

15. O'Neill, H.C.; Navrotsky, A. Simple spinels: Crystallographic parameters, cation radii, lattice energies, and cation distribution. *Am. Miner.* **1983**, *68*, 181–194.

16. Hu, H.; Lentz, D.; Li, J.W.; McCarron, T.; Zhao, X.F.; Hall, D. Reequilibration processes in magnetite from iron skarn deposits. *Econ. Geol.* **2015**, *110*, 1–8. [CrossRef]

17. Marinos, G.P.; Petrascheck, W.E. Laurium. *Geol. Geophys. Res.* **1956**, *4*, 1–247.

18. Skarpelis, N.; Tsikouris, B.; Pe-Piper, G. The Miocene igneous rocks in the Basal Unit of Lavrion (SE Attica, Greece): Petrology and geodynamic implications. *Geol. Mag.* **2007**, *145*, 1–15. [CrossRef]

19. Voudouris, P.; Melfos, V.; Spry, P.; Bonsall, T.; Tarkian, M.; Economou-Eliopoulos, M. Mineralogical and fluid inclusion constraints on the evolution of the Plaka intrusion-related ore system, Lavrion, Greece. *Miner. Petrol.* **2008**, *93*, 79–110. [CrossRef]

20. Ducoux, M.; Branquet, Y.; Jolivet, L.; Arbaret, L.; Grasemann, B.; Rabillard, A.; Gumiaux, C.; Drufin, S. Synkinematic skarns and fluid drainage along detachments: The West Cycladic Detachment System on Serifos Island (Cyclades, Greece) and its related mineralization. *Tectonophysics* **2017**, *695*, 1–26. [CrossRef]

21. Kockel, F.; Mollat, H.; Gundlach, H. Hydrothermally altered and (copper) mineralizedp orphyritic intrusionsin the ServomacedoniaMn assif s (Greece). *Miner. Depos.* **1975**, *10*, 195–204. [CrossRef]

22. Pearce, J.A.; Lippard, S.J.; Roberts, S. Characteristics and tectonic significance of supra-subduction zone ophiolites. *Geol. Soc. Lond. Spec. Pub.* **1984**, *16*, 77–94. [CrossRef]

23. Saccani, E.; Photiades, A.; Santato, A.; Zeda, O. New evidence for supra-subduction zone ophiolites in the Vardar Zone from the Vermion Massif (northern Greece): Implication for the tectono-magmatic evolution of the Vardar oceanic basin. *Ofioliti* **2008**, *33*, 17–37.

24. Economou-Eliopoulos, M.; Eliopoulos, D.; Chryssoulis, S. A comparison of high-Au massive sulfide ores hosted in ophiolite complexes of the Balkan Peninsula with modern analogues: Genetic significance. *Ore Geol. Rev.* **2008**, *33*, 81–100. [CrossRef]

25. Paraskevopoulos, G.M.; Economou, M.I. Genesis of magnetite ore occurrences by metasomatism of chromite ores in Greece. *Neues Jahrb. Fur Miner. Abh.* **1980**, *140*, 29–53.

26. Mercier, J. Édute géologique des zones internes des Hellénides en Macédoine centrale (Grèce). Ire Thèse (1966). *Ann. Géol. Des. Pays Hell.* **1968**, *20*, 596.

27. Sideris, C.; Skounakis, S.; Economou, M. The ophiolite complex of Edessa area and the associated mineralization. International Symposium on metallogeny of mafic and ultramafic complexes. *I.G.C.P. 169 Athens* **1980**, *2*, 142–153.

28. Hynes, A.J. The Geology of Part of the Western Othrys Mountains, Greece. Ph.D. Thesis, University of Cambridge, Cambridge, UK, 1972.

29. Rassios, A.; Dilek, Y. Rotational deformation in the Jurassic Mesohellenic Ophiolites, Greece, and its tectonic significance. *Lithos* **2009**, *108*, 207–223. [CrossRef]

30. Rassios, A.; Konstantopoulou, G. Emplacement tectonism and the position of chrome ores in the Mega Isoma peridotites, SW Othris, Greece. *Bull. Geol. Soc. Greece* **1993**, *28*, 463–474.

31. Mitsis, I.; Economou-Eliopoulos, M. Occurrence of apatite with magnetite in an ophiolite complex (Othrys), Greece. *Am. Miner.* **2001**, *86*, 1143–1150. [CrossRef]

32. Paraskevopoulos, G.; Economou, M. Komatiite-type ultramafic lavas from the Agrilia Formation, Othrys ophiolite complex, Greece. *Ofioliti* **1986**, *11*, 293–304.

33. Economou-Eliopoulos, M.; Paraskevopoulos, G. Platinum -group elements and gold in komatiitic rocks from the Agrilia Formation, Othrys ophiolite complex, Greece. *Chem. Geol.* **1989**, *77*, 149–158. [CrossRef]

34. Koutsovitis, P.; Magganas, A.; Ntaflos, T. Rift and intra-oceanic subduction signatures in the Western Tethys during the Triassic: The case of ultramafic lavas as part of an unusual ultramafic–mafic–felsic suite in Othris, Greece. *Lithos* **2012**, *144*, 177–193. [CrossRef]

35. Baziotis, I.; Economou-Eliopoulos, M.; Asimow, P.D. Ultramafic lavas and high-Mg basaltic dykes from the Othris ophiolite complex, Greece. *Lithos* **2017**, *288*, 231–247. [CrossRef]

36. Rassios, A. Geology and Evolution of the Vourinos Complex, Northern Greece. Ph.D. Thesis, University of California (Davis), Davis, CA, USA, 1981.

37. Francalanci, L.; Varekamp, J.C.; Vougioukalakis, G.; Defant, M.J.; Innocenti, F.; Manetti, P. Crystal retention, fractionation and crustal assimilation in aconvecting magma chamber Nisyros Volcano, Greece. *Bull. Volcanol.* **1995**, *56*, 601–620. [CrossRef]

38. Papanikolaou, D. Geotectonic evolution of the Aegean. *Bull. Geol. Soc. Greece* **1993**, *27*, 33–48.

39. Royden, L.H.; Papanikolaou, D.J. Slab segmentation and late Cenozoicdisruption of the Hellenic arc. *Geochem. Geophys. Geosyst.* **2011**, *12*, Q03010. [CrossRef]

40. Mitropoulos, P.; Tarney, J.; Saunders, A.D.; Marsh, N.G. Petrogenesis of Cenozoic rocks from the aegean island arc. *J. Volcanol. Geotherm. Res.* **1987**, *32*, 177–193. [CrossRef]

41. Tzifas, I.T.; Misaelides, P.; Godelitsas, A.; Gamaletsos, P.N.; Nomikou, P.; Karydas, A.G.; Kantarelou, V.; Papadopoulos, A. Geochemistry of coastal sands of Eastern Mediterranean: The case of Nisyros volcanic materials. *Chem. Der Erde-Geochem.* **2017**, *77*, 487–501. [CrossRef]

42. Marcopoulos, T.; Economou, M. Theophrastite, $Ni(OH)_2$, a new mineral from Northern Greece. *Am. Miner.* **1981**, *66*, 1020–1021.

43. Economou-Eliopoulos, M.; Eliopoulos, D. *A New Solid Solution [(Co,Mn,Nn)(OH)₂], in the Vermion Mt (Greece) and its Genetic Significance for the Mineral Group of Hydroxides*; McLaughlin, E.D., Braux, L.A., Eds.; Chemical Mineralogy, Smelting and Metallization; Nova Science Publishers: New York, NY, USA, 2009; pp. 1–18.

44. Taylor, S.R.; McLennan, S.M. *The Continental Crust: Its Composition and Evolution*; Blackwell Scientific: Oxford, UK, 1985.

45. Gamaletsos, P. Mineralogy and Geochemistry of Bauxites from Parnassos-Ghiona Mines and the Impact on the Origin of the Deposits. Ph.D. Thesis, University of Athens, Athens, Greece, 2014.

46. Kalatha, S. Metallogenesis of Bauxite Laterites and Fe-Ni-laterites-Enrichnment in Rare Earth Elements. Ph.D. Thesis, University of Athens, Athens, Greece, 2017.

47. Barth, M.; Gluhak, T. Geochemistry and tectonic setting of mafic rocks from the Othris Ophiolite, Greece. *Contrib. Miner. Petrol.* **2009**, *157*, 23–40. [CrossRef]

48. McDonough, W.F.; Sun, S.S. Chemical evolution of the mantle. The composition of the Earth. *Chem. Geol.* **1995**, *120*, 223–253. [CrossRef]

49. Sotnikov, V.I.; Berzina, A.N.; Economou-Eliopoulos, M.; Eliopoulos, D.G. Palladium, platinum and gold distribution in porphyry Cu _ Mo deposits of Russia and Mongolia. *Ore Geol. Rev.* **2001**, *18*, 95–111. [CrossRef]

50. Schock, H.H. Distribution of rare-earth and other trace elements in magnetite separates. *Chem. Geol.* **1979**, *26*, 119–133. [CrossRef]

51. Deditius, A.P.; Reich, M.; Simon, A.C.; Suvorova, A.; Knipping, J.; Roberts, M.P.; Rubanov, S.; Dodd, A.; Saunders, M. Nanogeochemistry of hydrothermal magnetite. *Contrib. Miner. Petrol.* **2018**, *173*, 46. [CrossRef]

52. Horn, I.; Foley, S.F.; Jackson, S.E.; Jenner, G.A. Experimental determined partitioning of high field strength and selected transition elements between spinel and basaltic melt. *Chem. Geol.* **1994**, *117*, 193–318. [CrossRef]

53. Balaram, V. Rare earth elements: A review of applications, occurrence, exploration, analysis, recycling, and environmental impact. *Geosci. Front.* **2019**, *10*, 1285–1303. [CrossRef]

54. Maksimovic, Z.; Skarpelis, N.; Panto, G. Mineralogy and geochemistry of the rare earth elements in the karstic nickel deposit of Lokris area, Greece. *Acta Geol. Hung.* **1993**, *36*, 331–342.

55. Economou-Eliopoulos, M.; Eliopoulos, D.G.; Apostolikas, A.; Maglaras, K. Precious and rare earth element distribution in Ni-laterite deposits from Lokris area, Central Greece. In Proceedings of the Fourth Biennial SGA Meeting, Turku, Finland, 11–13 August 1997; pp. 411–413.

56. Kalatha, S.; Perraki, M.; Economou-Eliopoulos, M.; Mitsis, I. On the origin of bastnaesite-(La,Nd,Y) in the Nissi (Patitira) bauxite laterite deposit, Lokris, Greece. *Minerals* **2017**, *7*, 45. [CrossRef]

57. Goodenough, K.M.; Schilling, J.; Jonsson, E.; Kalvig, P.; Charles, N.; Tuduri, J.; Deady, E.A.; Sadeghi, M.; Schiellerup, H.; Muller, A.; et al. Europe's rare earth element resource potential: An overview of REE metallogenetic provinces and their geodynamic setting. *Ore Geol. Rev.* **2016**, *72*, 838–856. [CrossRef]

58. Colony, W.E.; Nordlie, B.E. Liquid sulfur at Volcan Azufre, Galapagos Islands. *Econ. Geol.* **1973**, *68*, 371–380. [CrossRef]

59. GustafsonL, B.; Hunt, J.P. The porphyry copper deposit at E1 Salvador, Chile. *Econ. Geol.* **1975**, *70*, 857–912. [CrossRef]

60. Hemley, J.J.; Hunt, J.P. Hydrothermal ore-forming processes in the light of studies in rock-buffered systems. II. Some general geologic applications. *Econ. Geol.* **1992**, *87*, 23–43.

61. Hemley, J.J.; Cygan, G.L.; Fein, J.B.; Robinson, G.R.; d'Angelo, W.M. Hydrothermal ore-forming processes in the light of studies in rock-buffered systems I.: Iron-copper-zinc-lead sulfide solubility relations. *Econ. Geol.* **1992**, *87*, 1–22. [CrossRef]

62. Bliss, N.W.; MacLean, W.H. The paragenesis of zoned chromite from central Manitoba. *Geochim. Cosmochim. Acta* **1975**, *39*, 973–990. [CrossRef]

63. Bosi, F.; Andreozzi, G.B.; Hålenius, U.; Skogby, H. Zn-O tetrahedral bond length variations in normal spinel oxides. *Am. Mineral.* **2011**, *96*, 594–598. [CrossRef]

64. Martignago, F.; Dal Negro, A.; Carbonin, S. How Cr^{3+} and Fe^{3+} affect Mg-Al order disorder transformation at high temperature in natural spinels. *Phys. Chem. Miner.* **2003**, *30*, 401–408. [CrossRef]

65. Gervilla, F.; Padrón-Navarta, J.A.; Kerestedjian, T.; Sergeeva, I.; González-Jiménez, J.M.; Fanlo, I. Formation of ferrian chromite in podiform chromitites from the Golyamo Kamenyane serpentinte, Eastern Rhodopes,SE Bulgaria: A two-stage process. *Contrib. Mineral. Petrol.* **2012**, *164*, 643–657. [CrossRef]

66. Colás, V.; González-Jiménez, J.M.; Griffin, W.L.; Fanlo, I.; Gervilla, F.; O'Reilly, S.Y.; Pearson, N.J.; Kerestedjian, T.; Proenza, J.A. Fingerprints of metamorphism in chromite: New insights from minor and trace elements. *Chem. Geol.* **2014**, *389*, 137–152. [CrossRef]

67. Glemser, O.; Einerhand, J. Uber hohere Nickelhydroxide. *Z. Anorg. Chem.* **1950**, *261*, 26–42. [CrossRef]

68. Baskar, S.; Baskar, R.; Kaushik, A. Role of microorganisms in weathering of the Konkan-Goa laterite formation. *Curr. Sci.* **2003**, *85*, 1129–1134.

69. Russell, M.J.; Hall, A.J.; Boyce, A.J.; Fallick, A.E. On hydrothermal convection systems and the emergence of life. *Econ. Geol.* **2005**, *100*, 418–438.

70. Southam, G.; Saunders, J. The geomicrobiology of ore deposits. *Econ. Geol.* **2005**, *100*, 1067–1084. [CrossRef]

71. Valeton, I.; Biermann, M.; Reche, R.; Rosenberg, F. Genesis of Nickel laterites and bauxites in Greece during the Jurassic and Cretaceous, and their relation to ultrabasic parent rocks. *Ore Geol. Rev.* **1987**, *2*, 359–404. [CrossRef]

72. Eliopoulos, D.; Economou-Eliopoulos, M. Geochemical and mineralogical characteristics of Fe–Ni and bauxite–laterite deposits of Greece. *Ore Geol. Rev.* **2000**, *16*, 41–58. [CrossRef]

73. Kalatha, S.; Economou-Eliopoulos, M. Framboidal pyrite and bacteriomorphic goethite at transitional zones between Fe–Ni–laterites and limestones: Evidence from Lokris, Greece. *Ore Geol. Rev.* **2015**, *65*, 413–425. [CrossRef]

74. Norton, S. Laterite and bauxite formation. *Econ. Geol.* **1973**, *63*, 353–361. [CrossRef]

75. Laskou, M.; Economou-Eliopoulos, M. The role of microorganisms on the mineralogical and geochemical characteristics of the Parnassos-Ghiona bauxite deposits, Greece. *J. Geochem. Explor.* **2007**, *93*, 67–77. [CrossRef]

76. Laskou, M.; Economou-Eliopoulos, M. Bio-mineralization and potential biogeochemical processes in bauxite deposits: Genetic and ore quality significance. *Miner. Petrol.* **2013**, *107*, 471–486. [CrossRef]

77. Schulte, R.F.; Foley, N.K. *Compilation of Gallium Resource Data for Bauxite Deposits*; U.S. Geological Survey: Reston, VA, USA, 2014.

78. Zhou, M.F.; Robinson, P.T.; Su, B.X.; Gao, J.F.; Li, J.W.; Yang, J.S.; Malpas, J. Compositions of chromite, associated minerals, and parental magmas of podiform chromite deposits: The role of slab contamination of asthenospheric melts in suprasubduction zone environments. *Gondwana Res.* **2014**, *26*, 262–283. [CrossRef]

79. Eliopoulos, I.P.D.; Eliopoulos, G.D. Factors Controlling the Gallium Preference in High-Al Chromitites. *Minerals* **2019**, *9*, 623. [CrossRef]

80. Tarkian, M.; Koopmann, G. Platinum-group minerals in the Santo Tomas II (Philex) porphyry copper–gold deposit, Luzon Island, Philippines. *Miner. Depos.* **1995**, *30*, 39–47. [CrossRef]

81. Thompson, J.F.H.; Lang, J.R.; Stanley, C.R. Platinum-group elements in alkaline porphyry deposits, British Columbia. *Explor. Min. B. C. Mines Branch Part B* **2001**, *2001*, 57–64.

82. Richards, J.P. Tectono-magmatic precursors for porphyry Cu–(Mo–Au) deposit formation. *Econ. Geol.* **2003**, *98*, 1515–1533. [CrossRef]

83. Richards, J.P. Postsubduction porphyry Cu–Au and epithermal Au deposits: Products of remelting of subduction-modified lithosphere. *Geology* **2009**, *37*, 247–250. [CrossRef]

84. Richards, J. Tectonic, magmatic, metallogenic evolution of the Tethyan orogeny: From subduction to collision. *Ore Geol. Rev.* **2015**, *70*, 323–345. [CrossRef]

85. Tarkian, M.; Hünken, U.; Tokmakchieva, M.; Bogdanov, K. Precious-metal distribution and fluid-inclusion petrography of the Elatsite porphyry copper deposit, Bulgaria. *Miner. Depos.* **2003**, *38*, 261–281. [CrossRef]

86. Eliopoulos, D.G.; Economou-Eliopoulos, M.; Zelyaskova-Panayiotova, M. Critical factors controlling Pd and Pt potential in porphyry Cu–Au deposits: Evidence from the Balkan Peninsula. *Geosciences* **2014**, *4*, 31–49. [CrossRef]

87. Holwell, D.; Fiorentini, M.; McDonald, I.; Lu, Y.; Giuliani, A.; Smith, D.; Keith, M.; Locmelis, M. A metasomatized lithospheric mantle control on the metallogenic signature of post-subduction magmatism. *Nat. Commun.* **2019**, *10*, 1–10. [CrossRef]

88. Mungall, J.E.; Andrews, D.R.A.; Cabri, L.J.; Sylvester, P.J.; Tubrett, M. Partitioning of Cu, Ni, Au, and platinum-group elements between monosulfide solid solution and sulfide melt under controlled oxygen and sulphur fugacities. *Geochim. Cosmochim. Acta* **2005**, *69*, 4349–4360. [CrossRef]

89. Sillitoe, R.H. Porphyry copper systems. *Econ. Geol.* **2010**, *105*, 3–41. [CrossRef]

90. Berzina, A.P.; Berzina, A.N.; Gimon, V.O. Paleozoic–Mesozoic porphyry Cu (Mo) and Mo (Cu) deposits within the southern margin of the Siberian Craton: Geochemistry, geochronology, and petrogenesis (a review). *Minerals* **2016**, *6*, 125. [CrossRef]

91. McInnes, B.I.A.; Cameron, E.M. Carbonated, alkaline hybridizing melts from a sub-arc environment: Mantle wedge samples from the Tabar-Lihir-Tanga-Feni arc, Papua New Guinea. *Earth Planet. Sci. Lett.* **1994**, *122*, 125–141. [CrossRef]

92. Moritz, R.; Kouzmanov, K.; Petrunov, R. Upper Cretaceous Cu–Au epithermal deposits of the Panagyurishte district, Srednogorie zone, Bulgaria. *Swiss Bull. Miner. Petrol.* **2004**, *84*, 79–99.

Article

Mineralogical and Geochemical Constraints on the Origin of Mafic–Ultramafic-Hosted Sulphides: The Pindos Ophiolite Complex

Demetrios G. Eliopoulos [1], Maria Economou-Eliopoulos [2,*], George Economou [1] and Vassilis Skounakis [2]

[1]   Institute of Geology and Mineral Exploration (IGME), Sp. Loui 1, Olympic Village,
      GR-13677 Acharnai, Greece; eliopoulos@igme.gr (D.G.E.); Georgeoik7@gmail.com (G.E.)
[2]   Department of Geology and Geoenvironment, University of Athens, 15784 Athens, Greece;
      vskoun@geol.uoa.gr
*   Correspondence: econom@geol.uoa.gr

Received: 16 April 2020; Accepted: 13 May 2020; Published: 18 May 2020

Abstract: Sulphide ores hosted in deeper parts of ophiolite complexes may be related to either primary magmatic processes or links to hydrothermal alteration and metal remobilization into hydrothermal systems. The Pindos ophiolite complex was selected for the present study because it hosts both Cyprus-type sulphides (Kondro Hill) and Fe–Cu–Co–Zn sulphides associated with magnetite (Perivoli-Tsoumes) within gabbro, close to its tectonic contact with serpentinized harzburgite, and thus offers the opportunity to delineate constraints controlling their origin. Massive Cyprus-type sulphides characterized by relatively high Zn, Se, Au, Mo, Hg, and Sb content are composed of pyrite, chalcopyrite, bornite, and in lesser amounts covellite, siegenite, sphalerite, selenide-clausthalite, telluride-melonite, and occasionally tennantite–tetrahedrite. Massive Fe–Cu–Co–Zn-type sulphides associated with magnetite occur in a matrix of calcite and an unknown (Fe,Mg) silicate, resembling Mg–hisingerite within a deformed/metamorphosed ophiolite zone. The texture and mineralogical characteristics of this sulphide-magnetite ore suggest formation during a multistage evolution of the ophiolite complex. Sulphides (pyrrhotite, chalcopyrite, bornite, and sphalerite) associated with magnetite, at deeper parts of the Pindos (Tsoumes), exhibit relatively high Cu/(Cu + Ni) and Pt/(Pt + Pd), and low Ni/Co ratios, suggesting either no magmatic origin or a complete transformation of a preexisting magmatic assemblages. Differences recorded in the geochemical characteristics, such as higher Zn, Se, Mo, Au, Ag, Hg, and Sb and lower Ni contents in the Pindos compared to the Othrys sulphides, may reflect inheritance of a primary magmatic signature.

Keywords: sulphides; ophiolites; ultramafic; selenium; gold; Pindos

---

1. Introduction

Traditionally, the sulphide mineralization associated with ophiolite complexes is that of Cyprus-type volcanogenic massive sulphide (VMS) deposits. They may be derived from the interaction of evolved seawater with mafic country rocks, under greenschist facies metamorphic conditions and subsequent precipitation on and near the seafloor, when ore-forming fluids are mixed with cold seawater [1–5]. They are associated with basaltic volcanic rocks and are important sources of base and trace metals (Co, Sn, Se, Mn, Cd, In, Bi, Te, Ga, and Ge) [1]. Massive sulphide deposits have been described in the Main Uralian Fault Zone (Ivanovka and Ishkinino deposits), southern Urals; they are mafic–ultramafic-hosted VMS deposits and show mineralogical, compositional, and textural analogies with present-day counterparts on ultramafic-rich substrates [6]. Recently, an unusual association of magnetite with sulphides of Cyprus-type VMS deposit was described in Ortaklar, hosted in the

Koçali Complex, Turkey, which is part of the Tethyan Metallogenetic Belt [7]. Additionally, the largest magnetite deposit in a series of apatite and sulphide-free magnetite orebodies hosted in serpentinites of Cogne ophiolites, in the Western Alps, Italy, is characterized by typical hydrothermal compositions [8].

Although the Fe–Cu–Ni–Co-sulphide mineralization was initially considered an unusual type in ophiolite complexes, several occurrences have been located, like those in pyroxenite cumulates of the Oregon ophiolite [9], in dunites associated with chromitites of the Acoje ophiolite, Philippines [10], in layered gabbros of the Oman ophiolite [11], in dunites of the upper mantle–crust transition zone of the Bulqiza (Ceruja, Krasta), Albania ophiolite [12], Shetland (Unst), UK ophiolite [13], and the Moa-Baracoa ophiolitic massif (Cuba) [14]. On the basis of magmatic texture features and steep positive chondrite-normalized Platinum-Group Elements (PGE) patterns, sulphide mineralization of that type has been interpreted as reflecting the immiscible segregation of sulphide melts [9,13,14]. Moreover, the occurrence of Fe–Ni–Cu±Zn-sulphide mineralization (with dominant minerals pyrrhotite, chalcopyrite, and minor pentlandite) in mantle serpentinized peridotites and mafic to ultramafic rocks of ophiolite complexes of Limassol, Cyprus, Othrys (Eretria) in Greece, Pindos (Tsoumes) ophiolite and elsewhere has been the topic of research for extensive studies [15–20]. On the other hand, texture and geochemical characteristics, including PGE contents, and a very low partition coefficient for Ni and Fe between olivine and sulphides are inconsistent with sulphides having an equilibrium with Ni-rich host rocks at magmatic temperature [16,18]. Although the initial magmatic origin is not precluded, present characteristics of the highly transformed ore at the Eretria (Othrys) area may indicate that the magmatic features have been lost or that metals were released from the host rocks by a low-level hydrothermal circulation process [16]. Fe–Cu–Zn–Co–Ni mineralization is also reported in seafloor VMS deposits from modern oceans (as well as in their possible analogues on several ophiolites on land, e.g., Urals) indicating that these deposits can be formed by purely hydrothermal processes [19].

Despite the extensive literature data on a diverse array of sulphide mineralizations, sulphide ores hosted in mafic–ultramafic ophiolitic rocks are characterized by structure as well as mineralogical and geochemical features, suggesting either magmatic origin or links to serpentinization processes and metal remobilization from primary minerals into hydrothermal systems. The present study is focused on some new SEM/EDS and geochemical data on Cyprus-type and Fe–Cu–Co–Zn-type sulphides hosted in deeper parts of the Pindos ophiolite complex, aiming to improve our understanding of the factors controlling trace element incorporation into sulphide minerals and their origin.

2. Materials and Methods

2.1. Mineral Chemistry

Polished sections (20 samples) from sulphide ores were carbon-coated and examined by a scanning electron microscope (SEM) using energy-dispersive spectroscopy (EDS). The SEM images and EDS analyses were carried out at the University of Athens (NKUA, Athens, Greece), using a JEOL JSM 5600 scanning electron microscope (Tokyo, Japan), equipped with the ISIS 300 OXFORD automated energy-dispersive analysis system (Oxford, UK) under the following operating conditions: accelerating voltage 20 kV, beam current 0.5 nA, time of measurement (dead time) 50 s, and beam diameter 1–2 μm. The following X-ray lines were used: FeKα, NiKα, CoKα, CuKα, CrKα, AlKα, TiKα, CaKα, SiKα, MnKα, and MgKα. Standards used were pure metals for the elements Cr, Mn, Mo, Ni, Co, Zn, V, and Ti, as well as Si and MgO for Mg and Al_2O_3 for Al.

2.2. Whole Rock Analysis

The studied sulphide samples were massive and disseminated mineralizations, derived from large (weighing approximately 2 kg) samples, which is necessary to obtain statistical representative trace element distribution in sulphide ores. They were crushed and pulverized in an agate mortar. Major and minor/trace elements were determined at the SGS Global—Minerals Division Geochemistry Services Analytical Laboratories Ltd., Vancouver, BC, Canada. The samples were dissolved using sodium peroxide fusion, combined Inductively Coupled Plasma and Atomic Emission Spectrometry, ICP-AES and Mass Spectrometry, ICP-MS (Package GE_ICP91A50). On the basis of the quality control report provided by Analytical Labs, the results of the reference material analysis in comparison to expected values, and the results from the multistage analysis of certain samples, showed an accuracy and a precision of the method in good agreement with the international standard (<10%).

3. A Brief Outline of Characteristics for the Studied Sulphides

The Pindos ophiolite complex, of Middle to Upper Jurassic age, is located in the northern-western part of Greece (49° N, 21° E), lies tectonically over Eocene flysch of the Pindos zone, and contains a spectrum of lavas from Mid-Ocean Ridge, MOR basalts through island arc tholeiites (IATs) to boninite series volcanics (BSVs) [21–23]. Two tectonically distinct ophiolitic units can be distinguished: (a) the upper unit (Dramala Complex), including mantle harzburgites, and (b) a lower unit, including volcanic and subvolcanic sequences at the Aspropotamos Complex (Figure 1). This complex consists of a structurally dismembered sequence of ultramafic and mafic cumulate ophiolitic rocks, including gabbros, which is locally underlain by sheets of serpentinite [21,22]. Sulphide mineralization in the Pindos ophiolite complex is located near the Aspropotamos dismembered ophiolite unit, belonging to the lower ophiolitic unit of the complex and includes the following.

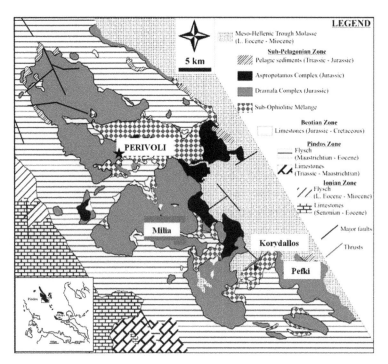

Figure 1. Simplified geological map of the southern part of the Pindos ophiolite, showing the Perivoli (Tsoumes–Kondro) area (modified after [22,24]).

A volcanic and subvolcanic sequence composed mainly of basalts and basaltic andesite pillow lavas ranging from high to low Ti affinity [20–24]. The different magmatic groups may have been derived from different mantle sources and/or various degrees of partial melting [20–24]. Massive Cyprus-type sulphide ore occurrences in the form of small lenses (maximum 4 × 40 m), are located in an abandoned mine, at the Kondro Hill, very close to the village of Perivoli (Figure 1). The estimated ore potential is about 10,000 tons with an average 6.6 wt.% Cu and 9.4 wt.% Zn [25]. They occur on the top of diabase (massive or pillow lavas) and are directly overlain by metalliferous (Fe-Mn-oxide-bearing sediments). Due to the tectonic disruption of the Aspropotamos unit, the spatial association between massive and stockwork disseminated mineralization is unclear.

Several sub-vertical veins of quartz with veinlets and a brecciated pipe-shaped diabase dike (stockwork ore zone) have been described in the Neropriona area of the Aspropotamos unit, Kondro Hill, with disseminations of pyrite + chalcopyrite, and dominant mineral-altered plagioclase and clinopyroxene, penninite, kaolinite, quartz, epidote, and calcite [20,25–27]. Small irregular to lens-like occurrences (4 × 1.5 m) of massive Fe–Cu–Zn–Co-type sulphide mineralization associated with magnetite are exposed at the Perivoli (Tsoumes) Hill (Figure 1). These are hosted within gabbro, close to its contact with serpentinized harzburgite [20,25], consisting of pyrrhotite, pyrite, chalcopyrite, and sphalerite associated with magnetite. The contact between ore and hosting rock is not sharp, appearing as irregular nets of veinlets. Rounded fragments of highly altered rock and massive fragments of sulphide ore are broadly parallel to the shear plane of a thrust fault.

4. Mineralogical Features

4.1. Cyprus-Type Sulphides

The massive ore is mainly composed of pyrite, chalcopyrite, bornite, and in lesser amounts covellite, siegenite, sphalerite, and clausthalite, while pyrrhotite is lacking (Figure 2). Chalcopyrite, bornite, and sphalerite occur in at least two different generations. Pyrite grains vary from euhedral to subhedral and rarely framboidal. Textural relationships indicate that early pyrite, commonly occurring as large crystals but often exhibiting dissolution, is extensively penetrated and replaced by fine-grained chalcopyrite, bornite, and sphalerite in a matrix of quartz (Figure 2). Copper-bearing sphalerite, with up to 3.6 wt.% Fe, 4.2 wt.% Cu, and 1.7 wt.% Bi, occurs within pyrite crystals and/or cements minor chalcopyrite and pyrite (Figure 2b,e,f; Table 1). Pyrite is extensively replaced by intergrowths between chalcopyrite or bornite and Fe-poor sphalerite (Figure 2b,d; Figure 3) and occasionally contains Co (Table 1). Fine-grained intergrowths of framboidal or colloform pyrite-bornite, occurs in a matrix of quartz (Figure 2g,h). Fine-grained chalcopyrite or bornite are often found in cross-cutting veins, hosting selenides (mainly clausthalite, PbSe) (Table 1), the telluride mineral melonite ($NiTe_2$), gold, galena, and barite [26]. Furthermore, present investigation reveals the formation of aggregates of secondary minerals, occurring as characteristic crusts on bornite surfaces (Figure 2i–l). These minerals are present-day grown minerals, on the surface of polished sections of sulphide ore, exposed to air, under room conditions (20–25 °C) and moderate air humidity (atmospheric water). Gold, as inclusions in chalcopyrite reaching a maximum size of 20 μm with Ag contents of up to 9 wt.%, is a rare component of the ores [26]. Additionally, we observed the presence of submicroscopic gold, i.e., <1 μm and thus invisible under an optical microscope, in grains of pyrite, chalcopyrite, and bornite, that increases with decreasing crystal size, reaching contents up to 7.7 ppm Au in pyrite, 8.8 ppm Au in very fine intergrowths between pyrite and sphalerite, and 17.3 ppm Au in fine intergrowths between pyrite and bornite [20].

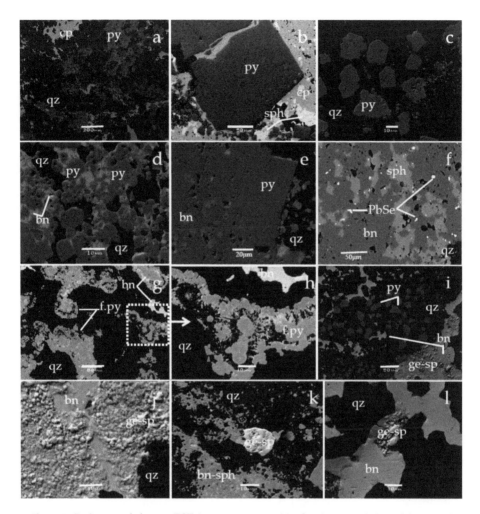

Figure 2. Backscattered electron (BSE) images representative of various morphological forms and textures of pyrite associated with chalcopyrite and sphalerite in a matrix of quartz (**a–d**); euhedral pyrite displaying erosion and replacement by chalcopyrite–sphalerite intergrowths (**b**); a close intergrowth between pyrite and bornite (**d,e**); selenides (clausthalite) as inclusions within bornite (**f**); fine-grained intergrowths of framboidal or colloform pyrite-bornite, in a matrix of quartz (**g,h**); replacement of bornite by neo-formed intergrowths of Cu minerals (**i–l**), which are Cu-enriched and Fe-depleted (Table 1). Scale bar: 200 μm (**a**); 50 μm (**b,f,g,i**); 20 μm (**e**); 10 μm (**c,d,h,j,k,l**). Abbreviations: py = pyrite; cp = chalcopyrite; sph = sphalerite; bn = bornite; PbSe = clausthalite; qz = quartz; f.py = framboidal forms of pyrite with tiny inclusions of Cu minerals; ge-sp = intergrowths of neo-formed Cu minerals with composition corresponding to geerite and spionkopite.

Table 1. Representative SEM/EDS analyses of minerals from the Pindos sulphide ores. (n.d.: Lower than detection limit).

	Sulphides of Cyprus Type									Kondro		
	Pyrite			Sphalerite			Chalcopyrite				Bornite	
Mineral wt%	1	2	3	4	5	6	7	8	9	10	11	12
Fe	44.6	43.7	42.8	3.6	4.2	1.8	28.7	29.9	29.2	11	11.7	11.9
Cu	n.d.	0.5	2.1	4.2	1.2	0.4	33.6	34.0	35.3	60.6	60	62.2
Zn	n.d.	n.d.	n.d.	57.7	62.7	63.3	1.6	n.d.	n.d.	n.d.	n.d.	n.d.
Co	n.d.	1.2	n.d.	n.d.	n.d.	n.d.	n.d.	n.d.	n.d.	1.2	n.d.	n.d.
Bi	1.2	n.d.	n.d.	1.7	1.4	1.5	0.8	1.2	n.d.	n.d.	n.d.	n.d.
Se	n.d.	n.d.	n.d.	n.d.	n.d.	n.d.	n.d.	n.d.	1.1	n.d.	n.d.	0.6
S	54.1	53.9	53.9	32.8	33.3	32.9	35.3	35.4	32.9	27.1	27.8	25.3
Total	99.9	99.3	99.3	100	100.5	99.9	100	100.4	99.8	99.9	99.5	100

	Epigenetic Cu-minerals (Figures 2 and 3)							Selenides-claousthalite			Tellurides-Melonite	
Mineral wt%	13	14	15	16	17	18	19	20	21	22	23	24
Fe	3.8	3.2	2.2	0.8	3.2	1.4	0.6	3.2	2.2	2.1	0.9	1.2
Cu	70.8	70.6	71.1	82.4	73.2	73.2	76.2	4.6	1.9	2.3	1.5	0.9
Zn	n.d.	4.6	1.8	n.d.	n.d.	n.d.	0.9	n.d.	n.d.	n.d.	n.d.	n.d.
Se	n.d.	n.d.	n.d.	n.d.	n.d.	n.d.	n.d.	21.6	25.1	24.8	1.4	0.5
Pb	n.d.	n.d.	n.d.	n.d.	n.d.	n.d.	n.d.	70.5	71	69.8	n.d.	n.d.
Co	0.3	n.d.	n.d.	n.d.	n.d.	n.d.	n.d.	n.d.	n.d.	n.d.	n.d.	n.d.
Ni	n.d.	n.d.	n.d.	n.d.	n.d.	n.d.	n.d.	n.d.	n.d.	n.d.	1.9	3.8
Te	n.d.	n.d.	n.d.	n.d.	n.d.	n.d.	n.d.	n.d.	n.d.	n.d.	13.8	12.4
S	24.5	21.6	23.9	26.2	24.2	25	23	n.d.	n.d.	n.d.	80.4	80.9
Total	99.4	100	99	99.4	100.6	99.6	100.7	99.9	100.2	99	99.9	99.7

Sulphides associated with Magnetite — Tsoumes

	Pyrrhotite			Pyrite		Chalcopyrite			Sphalerite	
wt%	25	26	27	28	29	30	31	32	33	34
Fe	61.0	58.9	59.7	44.9	44.5	32.0	30.6	30.4	7.5	6.4
Cu	n.d.	n.d.	n.d.	n.d.	n.d.	31.4	33.8	33.6	n.d.	n.d.
Zn	n.d.	0.8	n.d.	n.d.	n.d.	2.3	n.d.	n.d.	57.8	58.6
Bi	n.d.	1.4	n.d.	n.d.	n.d.	n.d.	n.d.	n.d.	1.4	1.9
S	39.4	39.8	40	55.3	54.2	34.3	35.9	10	33.1	33.2
Total	100.4	100.6	99.7	100.2	99.7	100	100.3	99.6	99.8	100.1

1 = $Fe_{1.2}Bi_{0.02}Cu_{0.05}S_{6.8}$
2 = $Fe_{1.5}Co_{0.05}Cu_{0.03}S_{6.7}$
3 = $Fe_{5.0}Cu_{0.14}S_{6.8}$
4 = $Zn_{1.3}Fe_{0.3}Bi_{0.04}Cu_{0.05}S_{5.0}$
5 = $Zn_{4.7}Fe_{0.2}Bi_{0.05}Cu_{0.05}S_{5.0}$
6 = $Zn_{4.8}Fe_{0.2}Bi_{0.04}S_{5.0}$
7 = $Cu_{2.4}Fe_{2.4}Zn_{0.02}S_{5.1}$
8 = $Cu_{2.5}Fe_{4.00.02}S_{5.1}$
9 = $Cu_{2.5}Fe_{4.00.04}S_{5.1}$
10 = $Cu_{0.8}Fe_{1.0}Bi_{0.03}S_{4.3}$
11 = $Cu_{0.5}Fe_{1.0}S_{4.2}$
12 = $Cu_{0.9}Fe_{1.1}Se_{0.04}S_{4.0}$
13 = $Cu_{5.8}Fe_{0.3}S_{4.2}$
14 = $Cu_{5.8}Fe_{0.3}Zn_{0.4}S_{3.5}$
15 = $Cu_{5.8}Fe_{0.2}Zn_{0.2}S_{3.9}$
16 = $Cu_{6.1}Fe_{0.6}S_{3.8}$
17 = $Cu_{5.9}Fe_{0.3}Co_{0.3}S_{3.9}$
18 = $Cu_{5.9}Fe_{0.12}S_{4.0}$
19 = $Cu_{6.2}Fe_{0.6}Zn_{0.07}S_{3.7}$
20 = $Pb_{4.6}Se_{0.7}Fe_{0.8}Cu_{1.0}$
21 = $Pb_{4.7}Se_{4.4}Fe_{0.5}Cu_{0.4}$
22 = $Pb_{4.6}Se_{4.6}Fe_{0.5}Cu_{0.5}$
23 = $Te_{6.97}Ni_{2.6}Fe_{0.2}Cu_{0.2}Se_{0.2}$
24 = $Te_{7.17}Ni_{2.4}Fe_{0.3}Cu_{0.6}Se_{0.07}$
25 = $Fe_{7}S_{8.1}$
26 = $Fe_{6.6}Zn_{0.05}Bi_{0.03}S_{5.3}$
27 = $Fe_{7}S_{6.4}$
28 = $Fe_{3.5}S_{6.8}$
29 = $Fe_{3.5}S_{6.8}$
30 = $Fe_{2.5}Cu_{2.3}Zn_{1.6}S_{4.9}$
31 = $Fe_{2.5}Cu_{2.4}S_{5.1}$
32 = $Fe_{2.5}Cu_{2.4}S_{5.1}$
33 = $Zn_{4.3}Fe_{0.6}Bi_{0.03}S_{5.0}$
34 = $Zn_{4.4}Fe_{0.6}Bi_{0.04}S_{5.0}$

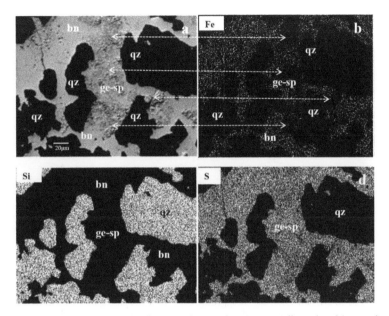

Figure 3. Backscattered electron (BSE)images showing bornite, partially replaced by neo-formed intergrowths of epigenetic minerals resembling geerite or spionkopite, in a matrix of quartz (**a**) and the single scanning for Fe (**b**), Si (**c**), and S (**d**). White arrows indicate the Fe depletion in the present-day formed intergrowths of high Cu minerals. Symbols, as in Figure 2. Scale bar for b, c, d as in 3a (20 μm)

4.2. Breccia Pipe

Disseminated pyrite, minor chalcopyrite, and sphalerite occur mostly in vesicles filled by quartz, kaolinite, chlorite, and epidote, within brecciated pipe-form diabase (a discharge pathway) underlying the Kondro massive ore [27]. Samples (*n* = 10) of pyrite separates from the diabase breccia have shown a limited range for δ^{34}S values from +1.0 to +1.5‰ [27].

4.3. Massive Fe–Cu–Co–Zn-Type Sulphides Associated with Magnetite

The sulphide ore is mainly composed of pyrrhotite, while pyrite, chalcopyrite, sphalerite, malachite, and azurite are present in lesser amounts. Pure magnetite, often forming a network texture, is associated with sulphides, either as massive ore with inclusions of sulphides (chalcopyrite, pyrite, and pyrrhotite), or as individual grains dispersed within sulphide ore (Figure 4). A characteristic feature of magnetite is its elongated and curved form and its textural relationship with the sulphides. Pyrrhotite, which is the most abundant sulphide, is followed by chalcopyrite and sphalerite, all showing an irregular contact with the magnetite, that often occurs surrounding sulphides (Figure 4a–c). Additionally, a salient feature is the occurrence of an (Fe/Mg) phyllosilicate associated with sphalerite, chalcopyrite, and magnetite (Figure 4d–f; Figure 5).

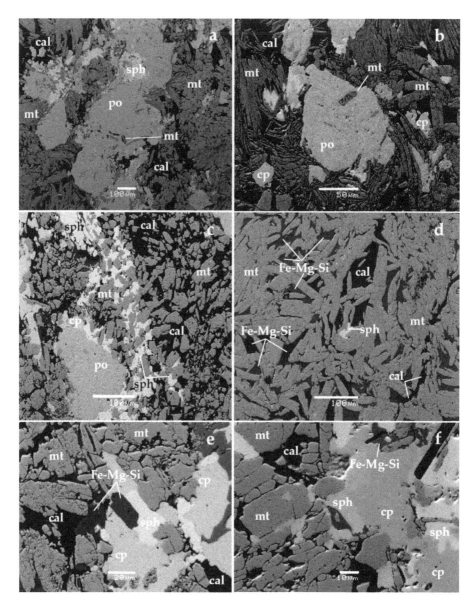

Figure 4. Backscattered electron (BSE) images showing intergrowths between pyrrhotite, sphalerite, and magnetite in a matrix of calcite (**a**); elongated curved crystals of magnetite and inclusions within pyrrhotite (**b**); intergrowths between pyrrhotite, sphalerite, chalcopyrite, and magnetite in a matrix of calcite (**c**); magnetite associated with sphalerite in a matrix of calcite and (Fe,Mg) silicate (**d**); sphalerite adjacent to chalcopyrite with inclusions of (Fe,Mg) silicate (**e**); transitional contact between chalcopyrite and sphalerite and their intergrowths with magnetite (**f**) Scale bar: 100 μm (**a,c,d**); 50 μm (**b**); 20 μm (**e**); 10 μm (**f**). Symbols: cal = calcite; Fe-Mg-Si = (Fe,Mg) silicate, and as in Figure 2.

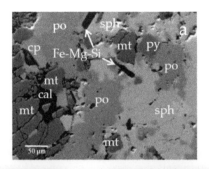

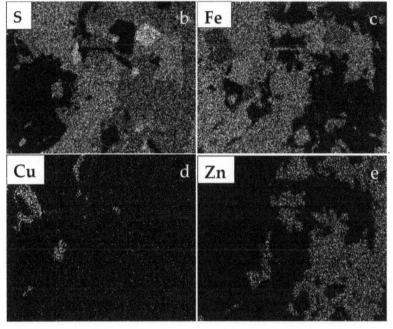

Figure 5. Backscattered electron (BSE) images showing intergrowths of pyrrhotite, pyrite, chalcopyrite, and sphalerite in a matrix of calcite and an unknown (Fe–Mg) silicate (**a**), Table 2; single scanning for S (**b**), Fe (**c**), Cu(**d**), and Zn (**e**). Scale bar for b, c, d as in 5a (50 μm). Symbols, as in Figures 2 and 4.

Table 2. Representative SEM/EDS analyses of (Fe,Mg) silicates from the Tsoumes massive sulphides.

wt %	Tsoumes (Fe,Mg)-Silicate	Laramie Complex [28] Hisingerite		
SiO$_2$	35	33.2	42.7	37.8
Al$_2$O$_3$	0.9	1.1	1.2	0.03
Cr$_2$O$_3$	n.d.	n.d.	0.03	0.04
Fe$_2$O$_{3t}$	42.3	43.2	35.83	46.94
MnO	n.d.	n.d.	0.15	0.49
MgO	11.9	12.1	7.46	2.15
NiO	n.d.	n.d.	0.05	0.01
CaO	n.d.	n.d.	0.81	0.39
Total	90.1	89.6	88.22	87.85

5. Geochemical Characteristics of Sulphides

Massive sulphide ores from the Kondro Hill exhibit uncommon high contents in Au (up to 3.6 ppm), Ag (up to 56 ppm), Se (up to 1900 ppm), Co (up to 2200 ppm), Mo (up to 370 ppm), Hg (up to 280 ppm), Sb (up to 10 ppm), and As up to 150 ppm, which are much higher than those in the Fe–Cu–Zn–Co-type sulphide hosted in ultramafic parts of the complex, along a shearing zone, close to a contact with gabbros (Table 3), as well as within brecciated pipe-form diabase (a discharge pathway) underlying the Kondro massive ore [20]. Major and minor elements, such as Fe, Cu, and Zn, are hosted in sulphides (pyrrhotite and pyrite, chalcopyrite, bornite, and epigenetic high Cu minerals, and sphalerite, respectively). Magnetite and Fe silicates (Tables 1 and 2), selenides, tellurides, gold, galena, and barite are occasionally present, but Mo-bearing minerals were not identified.

Major and trace elements in massive sulphide ores of Cyprus- and Fe–Cu–Co–Zn-type sulphides from the Pindos ophiolite complex, along with those from comparable ophiolites such as the Othrys and Troodos ophiolite complexes [29–33], are plotted in Figure 6. Although there are overlapping fields, it seems likely that the Pindos sulphides can be distinguished by their higher Zn (Figure 6a,c) and Co (Figure 6b) contents (Figure 6). The highest Se contents were recorded in the Pindos and the Apliki ores (Cyprus) accompanied by Cu and Au contents (Figure 6d,e). A positive trend is also clear between Au and As (Figure 6f).

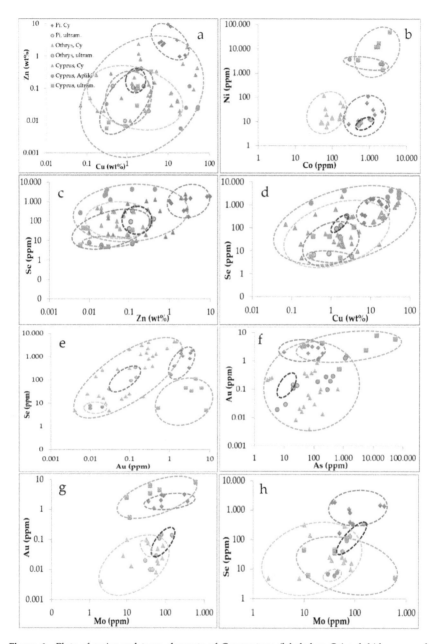

Figure 6. Plots of major and trace elements of Cyprus-type (labeled as Cy) sulphide ores and Fe-Cu ± Co ± Zn sulphides hosted in ultramafic rocks (labeled as ultram.) from the Pindos (Pi), Othrys, and Troodos ophiolite complexes. Although there are overlapping fields, the ores from the Pindos can be distinguished for the presence of ores with higher Zn (**a,c**) and Co (**b**) contents in Cyprus-type ores. The highest Se contents are recorded in the Pindos and the Apliki ores (Cyprus) accompanied by Cu and Au contents (**d,e**). A positive trend is also clear between Au and As (**f**). Data: Table 3; [3,7,16,18,20,29,30].

Table 3. Geochemical characteristics of sulphide ores hosted in the Pindos, Othrys, and Troodos ophiolites. Data: present study [20].

Location	Description	wt %	wt %	wt %	ppm											
		Fe	Cu	Zn	Co	Ni	Mo	Se	As	Au	Ag	Hg	Sb	Cu/(Cu + Ni)	Ni/Co	Pt/(Pt + Pd)
Pindos																
	Kondro Massive ore	26.9	6.9	2.9	1400	20	50	170	140	1.1	56	40	9.5	>0.99	0.014	—
		21.2	6.1	2.6	1250	32	80	400	150	1.3	34	70	7.1	>0.99	0.026	—
		26.5	16.4	3.2	600	12	310	1500	90	3.3	38	170	8.1	>0.99	0.02	—
		22.5	10.1	9.8	500	11	40	1900	64	3.6	32	280	10	>0.99	0.022	—
		11.9	25.4	1.1	2200	27	80	850	40	3.2	39	60	2.1	>0.99	0.012	—
		20.8	7.2	2.2	910	110	91	350	140	3	35	48	6.6	>0.99	0.12	—
		9.5	18.9	1	870	60	76	1100	10	2.1	40	128	1.3	>0.99	0.07	—
		26.4	11.6	8.3	280	8	36	1900	64	1.9	38	280	10	>0.99	0.028	—
		23.2	8.1	2.5	1000	13	370	1400	200	1.9	48	168	1.4	>0.99	0.013	—
Aspropotamos																
	Disseminated Diabase breccia	7.1	0.006	0.007	27	<5	<5	<5	25	0.018	<1	<1	0.3	—	—	0.95
		6.9	0.007	0.009	15	<5	<5	<5	25	0.025	<1	<1	0.6	—	—	0.97
	Tsoumes	49.8	1.8	0.4	600	9	80	130	20	0.15	6	10	1	>0.99	0.015	0.12
		33.4	2.5	0.12	1100	12	150	350	27	0.14	7	30	0.8	>0.99	0.01	0.87
		31.2	1.4	0.11	540	8	70	95	20	0.11	8	11	0.9	>0.99	0.015	0.22
		30.3	1.7	0.1	520	7	58	41	11	0.03	5	16	0.4	>0.99	0.013	0.87
Othrys																
	Eretria	39.4	3.3	0.02	2300	2400	40	6	2	0.01	<1	1	0.2	0.93	1.04	0.99
		35.9	0.44	0.12	400	3700	30	7	<5	0.02	<1	0.3	0.2	0.49	12.33	0.61
		37.4	0.89	0.01	2000	2500	46	8	<5	0.01	<1	0.6	0.2	0.78	1.25	0.86
Cyprus																
	Limassol	38.8	1.8	0.12	1500	12,000	40	35	47	3.5	2.1	2	18	0.6	8	—
		43.5	1.74	0.03	1800	17,100	610	5	12,800	8.2	8.2	<1	73	0.5	9.5	—
		43.8	3.8	0.34	2300	1400	70	110	190	4.5	4.5	<1	10	0.96	0.61	0.95
		38.7	0.91	0.007	3600	48,500	40	45	54,800	6.2	0.91	2	94	0.66	13.5	0.5
		54.4	0.35	0.009	1700	12,500	10	44	5600	2.5	0.35	1.5	13	0.21	7.35	0.61
		13.2	0.31	0.006	500	3500	12	6	730	0.53	0.31	1	6.6	0.53	7	

Symbol __: Pt contents < 10 ppb.

6. Discussion

6.1. A Comparison between Magmatic Sulphides and Those Hosted in Mafic–Ultramafic Ophiolites

It has been well established that the formation of magmatic deposits is related to the segregation of sulphide melts by immiscibility from basaltic magmas, which are able to collect precious metals (PGE, Ag, Au) as well as other chalcophile elements (Se, Te, Bi, Pb, As, Sb) because of their high partition coefficient ($D_{sulphide\ melt/silicate\ melt}$) [34,35]. The compositions of the magmatic ores may be mainly controlled by (a) the metal abundances of the mantle source and the degree of partial melting [36,37], (b) the degree of fractional crystallization and potential crustal contamination during magma ascent [38], and (c) interactions with magmatic-hydrothermal or metamorphic-hydrothermal fluids [39–41]. Compositional data obtained from a large number of magmatic sulphide deposits [42] show clear differences from the Fe–Cu ± Co ± Zn ± Ni-sulphide mineralization hosted in ophiolite complexes.

The unusual Fe–Cu–Ni–Co type of sulphides hosted in mafic–ultramafic ophiolitic rocks is characterized by varying structural, mineralogical, and geochemical features, which are not of magmatic origin. Sulphides hosted in the magmatic sequence of ophiolite complexes [9–14] have been interpreted as reflecting the immiscible segregation of sulphide melts, because they exhibit magmatic texture features and steep positive chondrite-normalized PGE patterns [9–14]. However, the Pindos (Tsoumes) Fe–Cu–Co–Zn-sulphides, consisting mainly of pyrrhotite, chalcopyrite, bornite, and sphalerite and hosted within gabbro, close to its tectonic contact with serpentinized harzburgite, differ compared to the Othrys (Eretria) sulphides, which are located at the peripheries of podiform chromite bodies hosted in serpentinized harzburgite and consist of pyrrhotite, chalcopyrite, and minor Co-pentlandite. Furthermore, the Pindos massive sulphides differ from those in the serpentinized rocks of the Limassol Forest (Troodos), which are composed dominantly of troilite, maucherite, pentlandite, chalcopyrite, bornite, vallerite, magnetite, minor sphalerite, graphite, molybdenite, and gold. In addition, the sulphides in the Limassol Forest contain much higher Au and As, up to 8 and 62 ppm, respectively [15,18], compared to the Pindos sulphides (Table 3). Despite the above differences between the Pindos, Othrys, and Limassol sulphides, they are all characterized by higher Cu/(Cu + Ni) ratios (>0.99, 0.5–0.93, and 0.2–0.96, respectively) compared to most magmatic deposits, having Cu/(Cu + Ni) ratios ranging from 0.05 to 0.14 [42]. In addition, the Ni/Co ratio in magmatic deposits typically ranges from 15 to 50 [42], whereas the range of Ni/Co ratio is 1.0–12 for the Eretria, 0.6–14 for the Limassol, and 0.01–0.02 for the Pindos (Tsoumes) samples (Table 3).

Model calculations have shown that the relatively high Co tenor and low Ni/Co cannot be explained by an earlier phase of fractional crystallization or sulphide segregation [43]. Thus, the higher Ni and Co contents in sulphide occurrences associated with chromitite bodies in the Othrys (Eretria) peridotites may suggest re-mobilization of Fe, Co, and Ni during hydrothermal alteration of peridotite in the presence of aqueous H_2S, and precipitation of Ni–Co–Fe sulphides [44,45]. The presence of graphite-like material in chromitites associated with sulphides from the Othrys complex, along shear zones that served as fluid pathways through the chromitites [46,47], may support the mobilization and re-precipitation of Fe–Ni–Co sulphides. Furthermore, the occurrence of phosphides such as Ni–V–Co phosphide [48], $Mo_3Ni_2P_{1+x}$ [49], NiVP [50], and the associated sulphide V_7S_8 [51] in chromitite concentrates from the Othrys ophiolite are consistent with extremely low fO_2 (reducing environment) during serpentinization [51] and re-precipitation of sulphides.

Intergrowths between sulphides and magnetite often forming curved crystals, which reflect a simultaneous deposition, coupled with the occurrence of calcite and an unknown (Fe,Mg) silicate (Figure 4d–f) resembling Mg–hisingerite [52] may provide evidence for the conditions of the Pindos sulphide deposition. Such a (Fe,Mg) silicate is unusual compared to the common presence of Mg-enriched serpentine in other ophiolite complexes. It seems to be comparable with the secondary phyllosilicates described in altered ferroan metaperidotite ("Oxide Body"), from the Laramie Complex (Laramie city, WY, USA), and serpentinites containing the Si-free minerals, such as brucite and NiFe alloy (awaruite). Such secondary phyllosilicates contain approximately equal amounts of end-members of the serpentine [$(Mg,Fe^{2+})_3Si_2O_5(OH)_4$] and hisingerite [$Fe^{3+}_2Si_2O_5(OH)_4.nH_2O$] [28]. The substitution of Fe^{3+} ions into the serpentine structure is crystallographically favorable because of the smaller ionic radius of Fe^{3+} compared to that of Mg^{2+} [53]. Moreover, based on a thermodynamic model for hydrothermal alteration in the Fe-silicate system, it has been shown that the formation of serpentine–hisingerite solid solutions after primary olivine may occur at elevated $a_{SiO2(aq)}$ and low $a_{H2(aq)}$ at low temperatures (about 200 °C) [28]. In addition, it has been suggested that H_2 production is associated with Fe(III) incorporation into serpentine (or magnetite) [54–56].

The association of magnetite with sulphides from the Pindos (Tsoumes) resembles an unusual association of magnetite with sulphides of Cyprus-type ophiolite-hosted VMS deposit in Ortaklar, located in the Koçali Complex, Turkey [7]. They are similar in terms of the deposition order, as suggested by the observed textural relationships: Fe-sulphide (pyrrhotite or pyrite) → chalcopyrite → sphalerite and subsequently magnetite (Figure 4; [7]) and the Cu, Zn, Pb, and Au-Ag contents [7]. Such temporal relationships among the primary ore minerals have been attributed to the evolution of ore-forming fluids. Specifically, increasing oxygen fugacity (fO_2) and pH would deplete sulphide (H_2S) and facilitate the magnetite precipitation in the hydrothermal fluids [7]. In addition, the Pindos sulphides exhibit similarities with the Cogne magnetite deposit (Western Alps, Italy), which is the largest in a series of apatite and sulphide-free magnetite orebodies that are hosted in serpentinites belonging to western Alpine ophiolitic units [8]. The authors applying thermodynamic modelling of fluid–rock interactions concluded that fractionation processes such as phase separation were critical to generate hydrothermal fluids capable of precipitating large amounts of magnetite in various types of ultramafic host rocks [8]. Although variable textures described in the large Cogne magnetite deposit differ from those in the Pindos magnetite ore, the trace element content of magnetite from the Cogne deposit, characterized by high Mg and Mn and low Cr, Ti, and V [8], is comparable to those in the Pindos magnetite separates [57]. Additionally, Fe–Cu–Zn–Co–Ni mineralization has been reported in seafloor VMS deposits from modern oceans, as well as in their potential analogues on several ophiolite complexes, as exemplified in the Urals, supporting the origin of such deposits by hydrothermal processes [19]. Moreover, hydrothermal products including Cu–Zn–(Co)-rich massive sulphides, hosted in ultramafic rocks at the Rainbow (Mid-Atlantic Ridge), exhibit structure, mineralogy, and bulk rock chemistry similar to those found in mafic volcanic-hosted massive sulphide deposits [58].

In general, the main factors controlling metal associations in seafloor massive sulphide (SMS) deposits may be the temperature of deposition, seafloor spreading rate and r/w ratio, and zone refining [59]. The authors emphasized the significance of the final depositional conditions and evolution of mound and vent structures rather than the original geochemistry of the hydrothermal fluid; the composition of the substrate may become relevant in subseafloor mineralization, where sulphides are precipitated by the reaction of ascending hydrothermal fluids with substrate host rocks [59]. Individual deposits may show a mixture of geochemical signatures, which may be related to mafic and ultramafic rocks [19,58]. Assuming that the leaching of elements from substrate rocks is influenced by the structure of the oceanic lithosphere and by the nature of the hydrothermal convection (spreading rate), some specific geochemical features, such as Au enrichment, Au/Ag and Co/Ni ratios, may be related to the nature of the substrate, the presence of a magmatic influx of volatiles and metals, the morphology of vent structures, the ridge spreading rate, or a combination of these factors [19,58,59]. Although the observed textural and mineralogical features (Figures 3–5) are inconsistent with an origin of the sulphides at magmatic temperatures, the recorded differences, such as the higher Zn, Se, Mo, Au, Ag, Hg, and Sb and lower Ni contents in the Pindos compared to the Othrys sulphides (Table 3; Figure 6), may reflect inheritance of a primary magmatic signature.

6.2. Genetic Significance of Trace Elements

The massive Cyprus-type sulphides from the Kondro Hill are characterized by elevated Zn, Co, Se, Au, As, Ag, Mo, and Sb content (Table 1) compared to those of the Othrys and most of the Troodos sulphide ores (Figure 6). Apart from the major elements, namely Fe, Cu, and Zn, hosted in pyrite, chalcopyrite, bornite, and sphalerite, Au occurs as submicroscopic particles (<1 μm) in grains of As-bearing pyrite, chalcopyrite, and bornite [20,26]. Selenium and Te are found as individual fine minerals, such as selenides (clausthalite) and tellurides (melonite) in Cu minerals (Figure 2f; [26]). Although Se can be easily hosted as a solid solution in high-temperature chalcopyrite [58], the presence of clausthalite in a late generation of fine-grained chalcopyrite–sphalerite intergrowths, penetrating into an earlier stage ore, may indicate re-distribution of Se. A late growth of clausthalite is also supported by the occurrence of clausthalite and tellurides filling cracks in pyrite and Cu minerals [26]. Furthermore, on the basis of thermodynamic calculations, it has been demonstrated that the presence of selenides in the oxidation zones of sulphide ores of Uralian VMS deposits is related to their stability under oxidizing conditions [60]. Molybdenite or other visible Mo minerals in the Pindos and Othrys sulphides have not yet been reported. It has been established that Mo displays siderophile, chalcophile, and lithophile behavior, depending on the composition of the system (including fO_2 and fS_2), temperature, and pressure [61]. Further research is required to define the potential presence of invisible Mo minerals (less than 1 μ) in the Pindos sulphide-magnetite ores.

6.3. Stability of Sulphides

A salient feature of the sulphide minerals is a varying stability. The occurrence of euhedral pyrite crystals, in contrast to microcrystalline unhedral Cu and Zn sulphides (Figure 2), may indicate that pyrite was more stable during subsequent modification of the orebody. Although early large crystals of pyrite may be replaced by chalcopyrite or bornite and Fe-poor sphalerite intergrowths (Figure 2), the formation of cruciform aggregates of secondary minerals occurs only on bornite surfaces (Figure 2i–l), probably reflecting a difference in their stability. The preferential leaching of Cu and Zn sulphide phases and the neo-formation of high Cu sulphides on bornite in contrast to neighboring pyrite may be the result of a preferred dissolution of Cu sulphides over pyrite, due to differing surface potentials [62]. It has been suggested that bornite with sulfur in excess (x-bornite) is stable at high temperature [62]. The authors of this study show that if the so-called sulfur-rich bornites are annealed at lower temperature, chalcopyrite or chalcopyrite and digenite exsolve, depending on the annealing temperature and composition. In addition to this exsolution, a new phase forms below approximately 140 °C, which is referred to as x-bornite, and it is a metastable phase. Although x-bornite is a metastable phase, the presented data (Figure 2i–l; Table 1) may confirm that x-bornite can remain for a long time in natural environments, and epigenetic minerals, with a stoichiometry resembling geerite or spionkopite [63,64], can be formed under environmental conditions in a short time.

7. Conclusions

The compilation of the mineralogical, geochemical, and mineral chemistry data from the sulphide occurrences hosted in the Pindos ophiolite complex and those from other ophiolites lead us to the following conclusions:

- Elevated contents of Au as invisible submicroscopic Au in pyrite and Cu minerals in the Pindos sulphides may reflect main collectors of Au at the time of the sulphide mineralization.
- The occurrence of clausthalite (PbSe) and fine-grained gold in chalcopytite–bornite–sphalerite intergrowths of a subsequent stage mineralization in the Pindos sulphides indicates their re-mobilization/re-deposition.
- Sulphides (pyrrhotite, chalcopyrite, bornite, and sphalerite) associated with magnetite, at deeper parts of the Pindos (Tsoumes), exhibit Cu/(Cu + Ni), Ni/Co, and Pt/(Pt + Pd) ratios, suggesting either no magmatic origin or a complete transformation of a preexisting magmatic assemblages.
- Textural features and the presence of the (Fe/Mg) phyllosilicate resembling Mg–hisingerite, and calcite in the matrix of the Pindos sulphides, suggest precipitation of the sulphide-magnetite ore at the deeper levels from a Fe-rich and alkaline ore-forming system.

- The preferential leaching of Fe and S and neo-formed high Cu sulphides on bornite, in contrast to neighboring pyrite, may be the result of a preferred dissolution of Cu sulphides over pyrite, confirming literature data on differing surface potentials between those sulphides.

- Assuming that trace elements in epigenetic minerals are derived from the decomposition of primary minerals, and coupled with the higher Zn, Se, Mo, Au, Ag, Hg, and Sb and lower Ni contents in the Pindos compared to the Othrys sulphides, this may reflect inheritance of a primary magmatic signature.

Author Contributions: D.G.E., M.E.-E., and G.E. collected the samples, provided the field information, and contributed to the conceptualization of the manuscript. V.S. performed the SEM/EDS analyses. M.E.-E. discussed the mineralogical and chemical data with the co-authors and carried out the original draft of the manuscript. All authors have read and agreed to the published version of the manuscript.

Funding: The National and Kapodistrian University of Athens (NKUA) is greatly acknowledged for the financial support (Grant No. KE_11730) of this work.

Acknowledgments: We thank the reviewers for the constructive criticism and suggestions on an earlier draft of the manuscript. In particular, the review of this work by the Academic Editor Paolo Nimis is greatly appreciated. Many thanks are due to our colleague Costas Mparlas for the donation of certain sulphide samples from his collection and valuable discussions.

Conflicts of Interest: The authors declare no conflict of interest.

References

1. Hannington, M.; Herzig, P.; Scott, S.; Thompson, G.; Rona, P. Comparative mineralogy and geochemistry of gold-bearing sulfide deposits on the mid-ocean ridges. *Mar. Geol.* **1991**, *101*, 217–248. [CrossRef]

2. Franklin, J.M.; Sangster, D.M.; Lydon, J.W. Volcanic-associated massive sulfide deposits. *Econ. Geol.* **1991**, *75*, 485–627.

3. Hannington, M.D.; Galley, A.; Gerzig, P.; Petersen, S. Comparison of the TAG mound and stockwork complex with Cyprus-type massive sulfide deposits. *Proc. Ocean Drill. Program* **1998**, *158*, 389–415.

4. Barrie, C.T.; Hannington, M.D. Classification of volcanic-associated massive sulfide deposits based on host-rock composition. In *Volcanic-Associated Massive Sulfide Deposits: Processes and Examples in Modern and Ancient Settings*; Society of Economic Geologists: Littleton, CO, USA, 1999; pp. 1–11.

5. Galley, A.; Hannington, M.; Jonasson, I. Volcanogenic massive sulphide deposits. *mineral deposits of Canada Spec. Publ.* **2007**, *5*, 141–161.

6. Nimis, P.; Zaykov, V.V.; Omenetto, P.; Melekestseva, I.Y.; Tesalina, S.G.; Orgeval, J.J. Peculiarities of some mafic–ultramafic- and ultramafic-hosted massive sulfide deposits from the Main Uralian Fault Zone, southern Urals. *Ore Geol. Rev.* **2008**, *33*, 49–69. [CrossRef]

7. Yıldırım, N.; Dönmez, C.; Kang, J.; Lee, I.; Pirajno, F.; Yıldırım, E.; Günay, K.; Seo, J.H.; Farquhar, J.; Chang, S.W. A magnetite-rich Cyprus-type VMS deposit in Ortaklar: A unique VMS style in the 1373 Tethyan metallogenic belt, Gaziantep, Turkey. *Ore Geol. Rev.* **2016**, *79*, 425–442. [CrossRef]

8. Toffolo, L.; Nimis, P.; Martin, S.; Tumiati, S.; Bach, W. The Cogne magnetite deposit (Western Alps, Italy): A Late Jurassic seafloor ultramafic-hosted hydrothermal system? *Ore Geol. Rev.* **2017**, *83*, 103–126. [CrossRef]

9. Foose, M.P. *The Setting of a Magmatic Sulfide Occurrence in a Dismembered Ophiolite, Southwest Oregon*; Distribution Branch, USA Geological Survey: Reston, VA, USA, 1985; p. 1626.

10. Bacuta, G.C.; Kay, R.W.; Gibbs, A.K.; Lipin, B.R. Platinum-group element abundance and distribution in chromite deposits of the Acoje Block. Zambales ophiolite Complex, Philippines. *J. Geochem. Explor.* **1990**, *37*, 113–145. [CrossRef]

11. Lachize, M.; Lorand, J.P.; Juteau, T. Calc-alkaline differentiation trend in the plutonic sequence of the Wadi Haymiliyah section, Haylayn Massif, Semail Ophiolite, Oman. *Lithos* **1996**, *38*, 207–232. [CrossRef]

12. Karaj, N. Reportition des platinoides chromites et sulphures dans le massif de Bulqiza, Albania. In *Incidence sur le Processus Metallogeniques dans les Ophiolites (These)*; Universite de Orleans: Orleans, France, 1992; p. 379.

13. Prichard, H.M.; Lord, R.A. A model to explain the occurrence of platinum- and palladium-rich 3065 ophiolite complexes. *J. Geol. Soc.* **1996**, *153*, 323–328. [CrossRef]

14. Proenza, J.A.; Gervilla, F.; Melgarejo, J.; Vera, O.; Alfonso, P.; Fallick, A. Genesis of sulfide-rich chromite ores by the interaction between chromitite and pegmatitic olivine–norite dikes in the Potosí Mine (Moa-Baracoa ophiolitic massif, eastern Cuba). *Miner. Depos.* **2001**, *36*, 658–669. [CrossRef]

15. Panayiotou, A. *Cu-Ni-Co-Fe Sulphide Mineralization, Limassol Forest, Cyprus*; Panayiotou, A., Ed.; Intern. Ophiolite Symposium: Nicosia, Cyprus, 1980; pp. 102–116.

16. Economou, M.; Naldrett, A.J. Sulfides associated with podiform bodies of chromite at Tsangli, Eretria, Greece. *Miner. Depos.* **1984**, *19*, 289–297. [CrossRef]

17. Thalhammer, O.; Stumpfl, E.F.; Panayiotou, A. Postmagmatic, hydrothermal origin of sulfide and arsenide mineralizations at Limassol Forest, Cyprus. *Miner. Depos.* **1986**, *21*, 95–105. [CrossRef]

18. Foose, M.P.; Economou, M.; Panayotou, A. Compositional and mineralogic constraints in the Limassol Forest portion of the Troodos ophiolite complex, Cyprus. *Miner. Depos.* **1985**, *20*, 234–240. [CrossRef]

19. Melekestzeva, I.Y.; Zaykov, V.V.; Nimis, P.; Tret'yakov, G.A.; Tessalina, S.G. Cu–(Ni–Co–Au)- bearing massive sulfide deposits associated with mafic–ultramafic rocks of the Main Urals Fault, South Urals: Geological structures, ore textural and mineralogical features, comparison with modern analogs. *Ore Geol. Rev.* **2013**, *52*, 18–36. [CrossRef]

20. Economou-Eliopoulos, M.; Eliopoulos, D.; Chryssoulis, S. A comparison of high-Au massive sulfide ores hosted in ophiolite complexes of the Balkan Peninsula with modern analogues: Genetic significance. *Ore Geol. Rev.* **2008**, *33*, 81–100. [CrossRef]

21. Kostopoulos, D.K. Geochemistry, Petrogenesis and Tectonic Setting of the Pindos Ophiolite, NW Greece. Ph.D. Thesis, Univ. of Newcastle, Newcastle, UK, 1989.

22. Jones, G.; Robertson, A.H.F. Tectono-stratigraphy and evolution of the Mesozoic Pindos ophiolite and related units, northwestern Greece. *J. Geol. Soc. Lond.* **1991**, *148*, 267–288. [CrossRef]

23. Pe-Piper, G.; Tsikouras, B.; Hatzipanagiotou, K. Evolution of boninites and island-arc tholeiites in the Pindos ophiolite, Greece. *Geol. Mag.* **2004**, *141*, 455–469. [CrossRef]

24. Kapsiotis, A.; Grammatikopoulos, T.; Tsikouras, B.; Hatzipanagiotou, K.; Zaccarini, F.; Garuti, G. Chromian spinel composition and Platinum-group element mineralogy of chromitites from the Milia area, Pindos ophiolite complex, Greece. *Can. Miner.* **2009**, *47*, 1037–1056. [CrossRef]

25. Skounakis, S.; Economou, M.; Sideris, C. The ophiolite complex of Smolicas and the associated Cu-sulfide deposits. In *Proceedings, International Symposium on the Metallogeny of Mafic and Ultramafic Complexes, UNESCO, GCP Project 169*; Augoustidis, S.S., Ed.; Theophrastus Publications S.A.: Athens, Greece, 1980; Volume 2, pp. 361–374.

26. Barlas, C.; Economou-Eliopoulos, M.; Skounakis, S. Selenium-bearing minerals in massive sulfide ore from the Pindos ophiolite complex. In *Mineral. Deposits at the Beginning of the 21st Century*; Piestrzyiski, A., Ed.; CRC Press: Rotterdam, The Netherlands, 2001; pp. 565–568.

27. Sideris, C.; Skounakis, S.; Laskou, M.; Economou, M. Brecciated pipeform diabase from the Pindos ophiolite complex. *Chem. Der Erde* **1984**, *43*, 189–195.

28. Tutolo, B.M.; Evans, B.W.; Kuehner, S.M. Serpentine–hisingerite solid solution in altered ferroan peridotite and olivine gabbro. *Minerals* **2019**, *9*, 47. [CrossRef]

29. Metsios, C. Metsios, C. Mobility of selenium—Environmental Impact. Master's Thesis, National University of Athens, Athens, Greece, 1999; 132p. (In Greek)

30. Constantinou, G. Metalogenesis associated with Troodos ophiolite. In Proceedings of the International Ophiolite Synposium, Nicosia, Cyprus, 1–8 April 1979; pp. 663–674.

31. Constantinou, G.; Govett, G.J.S. Genesis of sulphide deposits, ochre and umber of Cyprus. *Trans. Inst. Min. Met.* **1972**, *81*, B34–B46.

32. Rassios, A. Geology and copper mineralization of the Vrinena area, east Othris ophiolite, Greece. *Ofioliti* **1990**, *15*, 287–304.

33. Robertson, A.H.F.; Varnavas, S.P. The origin of hydrothermal metalliferous sediments associated with the early Mesozoic Othris and Pindos ophiolites, mainland Greece. *Sediment. Geol.* **1993**, *83*, 87–113. [CrossRef]

34. Naldrett, A. *Magmatic Sulfide Deposits—Geology, Geochemistry and Exploration*; Springer: Heidelberg, NY, USA, 2004; pp. 1–727.

35. Barnes, S.J.; Mungall, J.E.; Le Vaillant, M.; Godel, B.; Lesher, C.M.; Holwell, D.; Lightfoot, P.C.; Krivolutskaya, N.; Wei, B. Sulfide-silicate textures in magmatic Ni-Cu-PGE sulfide ore deposits: Disseminated and net-textured ores. *Am. Miner.* **2017**, *102*, 473–506. [CrossRef]

36. Naldrett, A.J.; Barnes, S.-J. The behaviour of platinum group elements during fractional crystallization and partial melting with special reference to the composition of magmatic sulfide ores. *Fortschr. Miner.* **1986**, *63*, 113–133.

37. Maier, W.D.; Barnes, S.J.; Campbell, I.H.; Fiorentini, M.L.; Peltonen, P.; Barnes, S.J.; Smithies, R.H. Progressive mixing of meteoritic veneer into the early Earth's deep mantle. *Nature* **2009**, *460*, 620–623. [CrossRef]

38. Lesher, C.M.; Burnham, O.M.; Keays, R.R.; Barnes, S.J.; Hulbert, L. Trace-element geochemistry and petrogenesis of barren and ore-associated komatiites. *Can. Miner.* **2001**, *39*, 673–696. [CrossRef]

39. Barnes, S.-J.; Prichard, H.M.; Cox, R.A.; Fisher, P.C.; Godel, B. The location of the chalcophile and siderophile elements in platinum-group element ore deposits (atextural, microbeam and whole rock geochemical study): Implications for the formation of the deposits. *Chem. Geol.* **2008**, *248*, 295–317. [CrossRef]

40. Hinchey, J.G.; Hattori, K.H. Magmatic mineralization and hydrothermal enrichment of the High Grade Zone at the Lac des Iles palladium mine, northern Ontario. *Can. Miner.* **2005**, *40*, 13–23. [CrossRef]

41. Su, S.G.; Lesher, C.M. Genesis of PGE mineralization in the Wengeqi mafic-ultramafic complex, Guyang County, Inner Mongolia, China. *Miner. Depos.* **2012**, *47*, 197–207. [CrossRef]

42. Naldrett, A.J. Nickel sulfide deposits: Classification, composition and genesis. *Econ. Geol* **1981**, *75*, 628–655.

43. Konnunaho, J.P.; Hanski, E.J.; Karinen, T.T.; Lahaye, Y.; Makkonen, H.V. The petrology and genesis of the Paleoproterozoic mafic intrusion-hosted Co–Cu–Ni deposit at Hietakero, NW Finnish Lapland. *Bull. Geol. Soc. Finl.* **2018**, *90*, 104–131. [CrossRef]

44. Shiga, Y. Behavior of iron, nickel, cobalt and sulfur during serpentinization, with reference to the Hayachine ultramafic rocks of the Kamaishi mining distric, northeastern Japan. *Can. Miner.* **1987**, *25*, 611–624.

45. Alt, J.C.; Shanks, W.C. Serpentinization of abyssal peridotites from the MARK area, Mid-Atlantic Ridge: Sulfur geochemistry and reaction modeling. *Geochim. Cosmochim. Acta* **2003**, *67*, 641–653. [CrossRef]

46. Etiope, G.; Ifandi, E.; Nazzari, M.; Procesi, M.; Tsikouras, B.; Ventura, G.; Steele, A.; Tardini, R.; Szatmari, P. Widespread abiotic methane in chromitites. *Sci. Rep.* **2018**, *8*, 8728. [CrossRef]

47. Economou-Eliopoulos, M.; Tsoupas, G.; Skounakis, V. Occurrence of graphite-like carbon in podiform chromitites of Greece and its genetic significance. *Minerals* **2019**, *9*, 152. [CrossRef]

48. Ifandi, E.; Zaccarini, Z.; Tsikouras, B.; Grammatikopoulos, T.; Garuti, G.; Karipi, S. First occurrences of Ni–V–Co phosphides in chromitite from the Agios Stefanos Mine, Othrys Ophiolite, Greece. *Ofioliti* **2018**, *43*, 131–145.

49. Zaccarini, F.; Bindi, L.; Ifandi, E.; Grammatikopoulos, T.; Stanley, C.; Garuti, G.; Mauro, D. Tsikourasite, Mo3Ni2P1 + x (x < 0.25), a new phosphide from the chromitite of the Othrys Ophiolite, Greece. *Minerals* **2019**, *9*, 248.

50. Bindi, L.; Zaccarini, F.; Ifandi, E.; Tsikouras, B.; Stanley, C.; Garuti, G.; Mauro, D. Grammatikopoulosite, NiVP, a new phosphide from the chromitite of the Othrys Ophiolite, Greece. *Minerals* **2020**, *10*, 131. [CrossRef]

51. Bindi, L.; Zaccarini, F.; Bonazzi, P.; Grammatikopoulos, T.; Tsikouras, B.; Stanley, C.; Garuti, G. Eliopoulosite, V7S8, a new sulfide from the podiform chromitite of the othrys ophiolite, Greece. *Minerals* **2020**, *10*, 245. [CrossRef]

52. Eggleton, R.A.; Tilley, D.B. Hisingerite: A ferric kaolin mineral with curved morphology. *Clays Clay Miner.* **1998**, *46*, 400–413. [CrossRef]

53. Wicks, F.J.; O'Hanley, D.S. Volume 19, hydrous phyllosilicates: Serpentine minerals: Structures and petrology. In *Reviews in Mineralogy*; BookCrafters: Chelsea, MI, USA, 1988; pp. 91–167.

54. Andreani, M.; Munoz, M.; Marcaillou, C.; Delacour, A. Mu XANES study of iron redox state in serpentine during oceanic serpentinization. *Lithos* **2013**, *178*, 70–83. [CrossRef]

55. Klein, F.; Bach, W.; Humphris, S.E.; Kahl, W.-A.; Jons, N.; Moskowitz, B.; Berquo, T.S. Magnetite in seafloor serpentinite–Some like it hot. *Geology* **2014**, *42*, 135–138. [CrossRef]

56. Bonnemains, D.; Carlut, J.; Escartı'n, J.; Me'vel, C.; Andreani, M.; Debret, B. Magnetic signatures of serpentinization at ophiolite complexes. *Geochem. Geophys. Geosyst.* **2016**, *17*, 2969–2986. [CrossRef]

57. Eliopoulos, D.; Economou-Eliopoulos, M. Trace element distribution in magnetite separates of varying origin: Genetic and exploration significance. *Minerals* **2019**, *9*, 759. [CrossRef]

58. Marques, A.F.A.; Barriga, F.; Scott, S.D. Sulfide mineralization in an ultramafic-rock hosted seafloor hydrothermal system: From serpentinization to the formation of Cu–Zn–(Co)-rich massive sulfides. *Mar. Geol.* **2007**, *245*, 20–39. [CrossRef]

59. Toffolo, L.; Nimis, P.; Tret'yakov, G.A.; Melekestseva, I.Y.; Beltenev, V.E. Seafloor massive sulfides from mid-ocean ridges: Exploring the causes of their geochemical variability with multivariate analysis. *Earth-Sci. Rev.* **2020**, in press. [CrossRef]

60. Belogub, E.V.; Ayupovaa, N.R.; Krivovichevb, V.G.; Novoselov, K.A.; Blinov, I.A.; Charykova, M.V. Se minerals in the continental and submarine oxidation zones of the South Urals volcanogenic-hosted massive sulfide deposits: A review. *Ore Geol. Rev.* **2020**, *122*, 103500. [CrossRef]

61. Fitton, J.G. Coupled molybdenum and niobium depletion in continental basalts. *Earth Planet. Sci. Lett.* **1995**, *136*, 715–721. [CrossRef]

62. Kullerud, G. The Cu–Fe–S system. In *Washington Year Book*; Carnegie Institution of Washington: Washington, DC, USA, 1964; Volume 63, pp. 200–202.

63. Goble, J.; Robinson, G. Geerite, $Cu_{1.60}S$, a new copper sulfide from Dekalb Township, New York. *Can. Miner.* **1980**, *18*, 519–523.

64. Goble, J.R. Copper sulfides from Alberta: Yarrowite Cu_9S_8 and Spionkopite $Cu_{39}S_{28}$. *Can. Miner.* **1980**, *18*, 511–518.

Article

A Comparative Study of Porphyry-Type Copper Deposit Mineralogies by Portable X-ray Fluorescence and Optical Petrography

Connor A. Gray and Adrian D. Van Rythoven *

Department of Environmental, Geographical, and Geological Sciences, Bloomsburg University of Pennsylvania, Bloomsburg, PA 17815, USA; connorgray@mymail.mines.edu
* Correspondence: avanrythov@bloomu.edu; Tel.: +1-570-389-3912

Received: 30 March 2020; Accepted: 8 May 2020; Published: 11 May 2020

Abstract: Porphyry-type deposits are crucial reserves of Cu and Mo. They are associated with large haloes of hydrothermal alteration that host particular mineral assemblages. Portable X-ray fluorescence analysis (pXRF) is an increasingly common tool used by mineral prospectors to make judgments in the field during mapping or core logging. A total of 31 samples from 13 porphyry copper deposits of the Western Cordillera were examined. Whole-rock composition was estimated over three points of analysis by pXRF. This approach attempts to capture the rapid and sometimes haphazard application of pXRF in mineral exploration. Modes determined by optical petrography were converted into bulk rock compositions and compared with those determined by pXRF. The elements S, Si, Ca, and K all were underestimated by optical mineralogy, and the elements Cu, Mo, Al, Fe, Mg, and Ti were overestimated by optical mineralogy when compared with pXRF results. Most of these porphyry samples occur in veined porphyritic quartz monzonite that is characteristic of these deposits. Sulfide and silicate vein stockworks are pervasive in most of the samples as well as dissemination of sulfides outwards from veinlets. Ore minerals present include chalcopyrite and molybdenite with lesser bornite. Chalcocite, digenite, and covellite are secondary. Potential sources of analytical bias are discussed.

Keywords: Western Cordillera; pXRF; North America; modal mineralogy; optical mineralogy; porphyry deposit; copper; molybdenum; major element; handheld X-ray fluorescence

1. Introduction

1.1. Porphyry-Type Deposits

Porphyry-type deposits, often termed just "porphyry deposits", represent one of the most studied geological systems on Earth, and the economic fortunes that entail from the proper extraction of the metals found in these deposits have pushed the academic and industrial communities to study them. This deposit model that came to popularity in the mid-late 1960s [1,2], was summarized by Sillitoe [3] and then later expanded upon [4–8]. Porphyry-type deposits represent 60% of world Cu production annually, and 65% of known world Cu resources [6]. These deposits also represent half of the world's Mo production and may also contain economic grades of Ag, Zn, W, Sn, Re, and Au [7]. Low-grade resources of Pt-group elements are also possible [9,10]. Tonnages commonly exceed one billion [8]. An example of a deposit of this scale is Bingham Canyon (Utah), a deposit analyzed in this study, at 2.6 Gt of ore [11]. These high tonnages are offset by lower grades of metals extracted, typically between 0.5–1.0 wt% Cu [7]. The deposits are typically found along convergent plate margins where the subduction of an oceanic plate below the adjacent continental or oceanic plate drives partial melting, thus creating arc magmatism [3]. Although similar types of deposits occur in rarer continental-continental orogens [12]. These magmatic arc systems allow metals found in greater concentrations deeper in the earth to rise buoyantly as magma and emplace themselves as intrusive bodies capable of metal concentration. These systems cool, and depressurization processes occur such that metals are highly concentrated in an immiscible hydrothermal fluid located at the top of such intrusive bodies. These metals are then injected into the surrounding rock by hydraulic fracturing forming large stockworks of cross-cutting veins that contain concentrated metals [8]. The particular samples analyzed in this study are from the Western Cordillera of the Americas. Deposits of this type are essential to the growth of economies in developing countries, such as Chile where copper mining has contributed an average of 10% of GDP over the past 20 years [13]. Deposits found along the Western Cordillera of North America are, or have been influential regarding the economies of the regions and countries these deposits are found in.

1.2. Portable X-ray Fluorescence

X-ray fluorescence analysis produces compositional data by exciting the sample in question with X-rays. Those X-rays will ionize the sample's atoms by knocking out core-shell electrons. Outer-shell electrons will then fall/relax into the subsequent core-shell vacancy. As this involves a loss of energy, a photon is emitted with energy equal to the difference in electron orbitals. The energy quantum of the photon is unique to the element in question and is also in the X-ray portion of the electromagnetic spectrum. Thus, the X-ray's energy is characteristic of the element and the number of photons with that quantum is proportional to the element's concentration in the sample. Engineering advances over the last few decades have allowed this laboratory-scale instrumentation to be miniaturized into a portable format. A variety of commercial portable and handheld X-ray fluorescence analyzers are now available [14]. Portable X-ray fluorescence (pXRF) is an increasingly common method for on-site and rapid materials analysis. For exploration and development of mineral deposits, it has particularly been applied to geochemical exploration [15–17], assessments of soil contamination [18], core logging [19], and mineralization type [20,21]. This is a method that requires little training, is portable, fast, fairly affordable, and can provide useful data. However, compared to other lab methods (including more sophisticated, lab-based, and non-portable XRF analyzers), analyses can be very susceptible to measurement bias, insensitivity to elements lighter than Mg or Al, and matrix effects [22,23]. Whereas "portable" can refer to a variety of XRF configurations, for this study it refers to handheld instruments, typically resembling a pistol.

As pXRF is often used as a tool by field or core logging geologists, it is typically one of the first analytical methods applied to samples. The absence of typical sample preparation in the field, the small sample point size, and human bias in selecting what points to analyze introduce additional uncertainties to the data. Although they may be collected under less than optimal analytical conditions, results from pXRF can influence decisions regarding geologic mapping, cross-sections, and selection of samples for more detailed assay/investigation. Another investigative method that can follow pXRF is optical petrography. In fact, data from pXRF may influence a geologist to select a sample for thin sectioning and petrographic examination. A geologist can later re-interpret and reconcile the pXRF data in the context of sample mineralogy. Likewise, pXRF data can be useful in interpreting the observed mineralogy in thin sections. The combined use of complementary but independent methods such as pXRF and optical petrography has the potential to augment the abilities of the investigating geologist in a fairly fast and economical manner.

In addition to its portability and ease of use, pXRF is typically non-destructive. Outcrop, drill core, grab samples, soil, and thin section billets (this study) can be analyzed with no crushing, grinding, fusion, dissolution, etc., that lab-based XRF or ICP-MS require. Similarly, polished thin sections that can be made from a modest amount of sample, are not expended when examined, preserve mineral textures, and can later be analyzed by other methods such as electron microscopy or other microbeam methods.

We present that pXRF used in a rapid fashion with no standardization, in the manner of many field applications, can still provide semi-quantitative data that are useful to answer geological questions in the field. We simulate these haphazard conditions using point analysis on thin-section billets of rock samples from porphyry-type deposits. Sources of analytical uncertainty arising from these conditions are examined and discussed. The results are compared with those of optical petrography, another common, albeit much older, standby in non-destructive geological sample characterization.

2. Materials and Methods

2.1. Sample Locations

The 31 hand samples used in this comparative study are from 13 different deposits found along the oceanic-continental subduction zone of the Western Cordillera of North America (Table 1, Figure 1). These are Cu ± Mo porphyry-type deposits and represent different portions within each deposit (for those deposits with more than one sample). The samples include specimens from the hypogene and supergene zones of these deposits. The zone of interest for this study was primarily the hypogene. These samples were collected and donated by Peter H. Kirwin and John R. Ray, formerly of the American Copper and Nickel Company.

Table 1. List of samples investigated in this study and their origins.

Sample	Locality	Location	Sample	Locality	Location
Be-1	Bethlehem	British Columbia	L-1	Lornex	British Columbia
Be-4	Bethlehem	British Columbia	Mo-2	Morenci	Arizona
Be-6	Bethlehem	British Columbia	Mo-4	Morenci	Arizona
Bi-2	Bingham Canyon	Utah	Mo-6	Morenci	Arizona
Bi-3	Bingham Canyon	Utah	Mo-8	Morenci	Arizona
Bu-2	Butte	Montana	OH-1	Orange Hill	Alaska
Bu-6	Butte	Montana	S-1	Sierrita	Arizona
Bu-7	Butte	Montana	S-2	Sierrita	Arizona
C-3	Cananea	Mexico	S-4	Sierrita	Arizona
C-4	Cananea	Mexico	SM-3	San Manuel	Arizona
CC-1	Copper Canyon	Nevada	SM-4	San Manuel	Arizona
F-1	Florence	Arizona	Y-3	Yerington	Nevada
F-2	Florence	Arizona			
F-3	Florence	Arizona			
IP-1	Ithaca Peak	Arizona			
IP-2	Ithaca Peak	Arizona			
IP-3	Ithaca Peak	Arizona			
IP-10	Ithaca Peak	Arizona			
IP-12	Ithaca Peak	Arizona			

2.2. Analytical Methods

In this comparative study two methods were used to determine the overall mineralogy of the samples. These methods include using pXRF giving elemental weight percent readings for each sample, as well as using polarized transmitted and reflected light microscopy to estimate mineral abundances and related textures. Limiting the study to these two methods allows the investigation to be conducted in an essentially non-destructive manner at minimal cost.

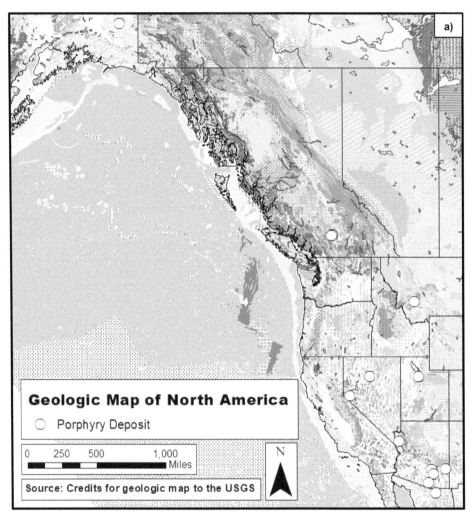

Figure 1. *Cont.*

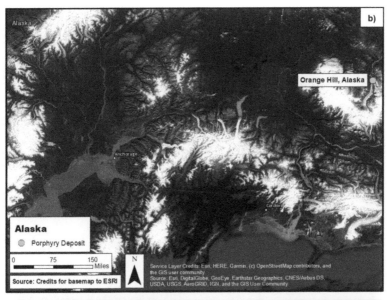

Figure 1. *Cont.*

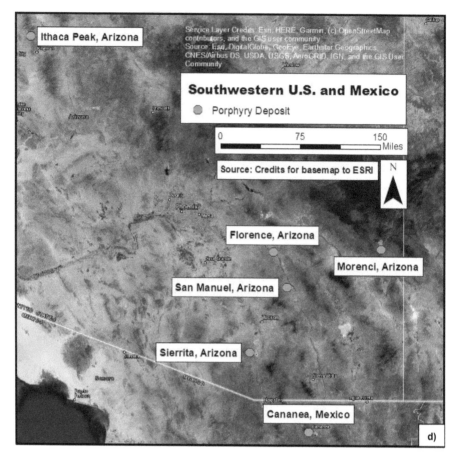

Figure 1. Maps of deposit locations in western North America that were sampled. (**a**) Overall geologic map of the Western Cordillera (map constructed in ArcGIS). (**b**) Satellite image of the Orange Hill deposit in Alaska. (**c**) Satellite image of the British Columbian, Nevadan, Montanan, and Utahan deposits. (**d**) Satellite image of Arizonan and Mexican deposits.

2.2.1. Optical Petrography

The 31 samples were selected from a larger collection on the basis of interesting mineralogy or textures visible in hand samples, geographic location, and budget limitations (for thin sectioning). They were cut into rectangular billets by a rock saw (e.g., Figure 2). The billets were then used to shave off 30 μm × 26 mm × 46 mm polished thin sections for further analysis under a petrographic microscope. The thin sectioned samples were then examined under a Leica DM2700 P microscope (Leica Microsystems Inc., Buffalo Grove, IL, USA) in transmitted and reflected light modes. In the first assessment of these thin sections, the textures were documented along with a qualitative listing of minerals present in each sample. More in-depth qualitative observations were then made at this point, and the genetic sequencings of minerals along with their characteristics were noted. In the second extensive assessment of these samples using petrographic methods, the overall mineral modes were determined. This assumed that the mineral abundances in terms of thin section area are representative of the volume% abundances (true modes) of the mineral.

Figure 2. Example photographs of two samples from this study. (**a**) Bu-2 (Butte, Montana) hand sample showing sulfide vein stockwork characteristic of porphyry-type deposits and alteration haloes surrounding the veins. (**b**) Thin section billet cut from the sample in (a) and impregnated with epoxy. (**c**) Be-6 (Bethlehem, British Columbia) hand sample showing sulfide vein stockwork characteristic of porphyry-type deposits and alteration adjacent to the veins. (**d**) Thin section billet cut from the sample in (c) and impregnated with epoxy. (**e**) IP-12 (Ithaca Peak, Utah) hand sample showing a high-grade vein of Cu-Fe sulfides. (**f**) Y-3 (Yerington, Nevada) somewhat brecciated hand sample showing cross-cutting veinlet stockwork formed in multiple generations of fractures. American quarter for scale (~2.43 cm in diameter). Locations to be cut for thin-sectioning are outlined in black marker.

2.2.2. Portable X-ray Fluorescence

When the overall modes were determined for each sample, the bulk elemental compositions were then determined by pXRF analysis of the flat face of the corresponding billet the thin section was cut from. A Thermo Scientific Niton^{TM} XL3t-950 GOLDD+ (Thermo Fisher Scientific, Waltham, MA, USA) was the pXRF analyzer used in this study. All analyses were done in Mining Cu/Zn mode. A 10-minute warmup and system check were conducted prior to every session of analyses. Analysis quality was checked for instrument drift using two manufacturer-provided standards: CRM 180-649 NIST 2709a and blank 180-647 99.995% SiO_2 blank that were checked using the pXRF prior to the study and after the study. These standards are supplied as fine homogenized powders. No significant deviation from known values (>10%) was noted. Samples were analyzed with pXRF for a 120-s interval, with 30 s each for the main (Mo, Zr, U, Rb, Th, Pb, Se, As, Hg, Nb, Cu, Ni, Co, Fe, Mn, Sr, Y), low (Sc, Ca, K, V, Ti, Cr), high (Ba, Cs, W, Te, Sb, Sn, Cd, Ag, Pd, La, Ce, Pr, Nd), and light (Al, P, Si, Cl, S, Mg) characteristic X-ray ranges of elements using a series of filters. This allowed for fairly precise readings of elemental weight percent (wt%) for each sample. Each sample was analyzed three times in three separate locations on the sectioned surface of the billet in an attempt to address inhomogeneities within these billets. The average elemental composition was determined using all three readings per sample. In the case of three very different analyses, a fourth location was analyzed on the billet.

2.3. Data Calculation

Modes from petrographic microscopy observations were converted into estimates of bulk elemental composition for each sample. The modes, taken as volumetric percentages, were converted into mineral mass percentages using established densities. These mineral mass proportions were then converted into elemental proportions. Density and compositional values were sourced from webmineral.com. Elements not analyzable by pXRF (mostly elements lighter than Mg) were left out of calculations in the modal abundance conversion to allow for simpler comparison. Estimates of solid solution members were made for plagioclase, chlorite, and muscovite. In terms of atoms per formula unit (apfu) it is assumed that plagioclase has 0.5 Na, 0.5 Ca, 1 Al, and 3 Si, chlorite has 3.75 Mg and 1.25 Fe, and biotite has 2.5 Mg and 0.5 Fe.

The whole-rock compositions calculated from the modes were then compared with the average whole-rock compositions determined by pXRF for purposes of comparison and to audit the accuracy of each method against the other.

3. Results

3.1. Hand Sample Examination

At the hand sample scale, prior to the thin sectioning of each sample, porphyry-type copper deposit characteristics were evident (e.g., Figure 2; see [24] for a review). Characteristic veining associated with hydraulic fracturing was seen in many samples as mineral assemblages of silicates and sulfides with rare carbonates. Sulfides associated with the hypogene system are typically dominated by chalcopyrite and pyrite. Silicates in this zone displayed less alteration than that of more near-surface samples, maintaining the dominantly quartz monzonitic mineralogy seen in a majority of hypogene samples although chalky sericitization of feldspars is common in the host rocks. In hand samples, characteristic zoning of alteration could be seen with some ambiguity but overall a grade from hypogene to supergene alteration was clear (c.f., [7]). The oxidation and hydration of hypogene sulfide minerals was a clear indicator of supergene alteration.

3.2. Optical Petrography

In thin sections, the observations made of the hand samples were expressed in clearer detail, and the overall modes observed in hand samples are consistent with those observed at a microscopic level (Table 2, Figure 3). The main sulfides present in hypogene samples show characteristic veins/veinlets of chalcopyrite, bornite, and/or pyrite (e.g., Figure 3g,f), along with sulfide grains disseminated distally from veinlets (e.g., Figure 3a,h). These veinlets also comprise hydrothermal quartz along with sericite and chlorite, and rarely calcite (e.g., Figure 3g). Molybdenite was the only Mo-bearing sulfide seen in thin section (e.g., Figure 3b,c,h, Table 2). Samples affected by some supergene fluid alteration have distinct mineralogies. The sulfide assemblages of the hypogene zone, though starting rich in Cu sulfides, quickly grade into zones of pyrite showing a lack of Cu concentration distally (e.g., Figure 3f). In the supergene samples, chalcocite, covellite, and/or digenite are observed as rims on partially replaced pyrite and/or chalcopyrite (Figure 3a). Specular hematite (possibly with magnetite in some cases) occurs in some samples and also appears to be hydrothermal. This is distinct from the late hematite after sulfides seen in some partially oxidized samples (see below).

Table 2. Mineral modes (in volume%) of samples. Mineral abbreviations as in Whitney and Evans [25].

Sample:	Be-1	Be-4	Be-6	Bi-2	Bi-3	Bu-2	Bu-6	Bu-7	C-3	C-4	CC-1	F-1	F-2	F-3	IP-1	IP-2
Ccp	0.3	1	1.5	1	1	0.5	–	1.5	–	–	1	4	0.8	2.5	–	0.1
Cv	–	–	–	0.3	–	0.2	–	0.3	–	2	–	–	–	–	–	1
Cct	–	–	–	1.5	0.1	–	0.5	–	–	–	–	–	–	–	0.5	0.5
Bn	–	1	0.2	0.1	1	–	–	–	–	–	–	–	–	–	–	–
Dg	–	–	–	–	0.3	–	–	–	–	–	–	–	–	–	–	–
Mol	–	–	–	0.2	0.1	–	–	0.5	0.5	–	–	–	–	–	–	–
Py	1	–	–	2	–	9	20	45	4	7	–	–	2	2	3	4
Chl	8	4.5	7	1	–	–	–	–	–	–	–	–	0.4	0.3	–	–
Rt	0.3	–	0.3	0.5	0.2	1	–	–	–	–	0.3	0.5	1	0.4	1	–
Hem	2	0.1	–	–	–	3	–	–	–	0.5	–	0.5	0.5	–	–	–
Cal	–	2	–	–	–	–	–	–	–	–	–	–	–	3	–	–
Qz	57.4	64.9	60	48.7	61.3	56.3	74.5	48.7	60.5	50	67	58	56.3	59.8	64.5	67.4
Or	–	–	–	–	–	–	3	–	–	–	–	–	10	–	–	–
Ms	27	25.5	22	42.7	25	30	2	4	35	40.5	23.7	34	12	32	31	27
Bt	–	0.5	–	1	10	–	–	–	–	–	–	8	3	10	–	–
Pl	3	–	8	–	–	–	–	–	–	–	–	–	–	5	–	–
Ap	1	0.5	1	1	1	–	–	–	–	–	–	–	2	–	–	–
Total	100.0	100.0	100.0	100.0	100.0	100.0	100.0	100.0	100.0	100.0	100.0	100.0	100.0	100.0	100.0	100.0

Sample:	IP-3	IP-10	IP-12	L-1	Mo-2	Mo-4	Mo-6	Mo-8	OH-1	S-1	S-2	S-4	SM-3	SM-4	Y-3
Ccp	0.1	0.05	22	6	1	–	2	1.5	1	1	2	0.5	3.5	4.5	2
Cv	0.1	0.1	–	–	–	–	–	–	–	–	–	–	–	–	–
Cct	–	0.2	–	–	–	–	–	–	–	–	–	–	–	–	–
Bn	–	–	10	–	–	–	–	–	–	–	–	–	–	–	–
Dg	–	–	–	–	–	–	–	–	–	–	–	–	–	–	–
Mol	0.3	–	–	0.1	–	–	–	–	–	–	0.2	0.3	–	–	–
Py	1	9	–	–	5	2.5	–	1.4	8	1.6	2	3	5	1	–
Chl	–	–	5	–	13	–	1.5	–	11	11	2	0.5	–	–	4
Rt	1	–	–	1	0.3	0.3	0.2	–	0.5	0.5	0.5	0.5	1.5	0.3	–
Hem	1	–	0.5	–	–	–	–	–	0.5	–	–	0.3	–	1	1.5
Cal	–	–	–	–	–	–	0.3	0.9	–	–	–	1	–	–	–
Qz	43.3	85.65	44.5	85.4	45	94.9	54.4	53.4	26.5	35	62.5	50.9	43	36.3	55.7
Or	16	–	1	–	–	–	–	7.1	–	8.9	4	3	–	–	5
Ms	10	5	12	8	34	2.3	34.5	33	15	10	25	36	37	26	25
Bt	24.5	–	1	–	–	–	2	2.5	3	–	–	–	4	25	–
Pl	1	–	4	–	–	–	5	–	35	32	1.3	4	6.3	4	6
Ap	1.7	–	–	0.5	1	–	–	–	–	–	0.5	–	0.7	0.7	0.5
Total	100.0	100.0	100.0	100.0	100.0	100.0	100.0	100.0	100.0	100.0	100.0	100.0	100.0	100.0	100.0

The host rocks vary depending on the particular deposit. Most are porphyritic hypabyssal/intrusive rocks that are quartz monzonitic/latitic, although granitic/rhyolitic and granodioritic/dacitic assemblages occur (c.f., [26]). Primary quartz is present in varying amounts but silicification is prevalent. Biotite, apatite, rare amphibole, and rare zircon are also present as primary minerals. In addition to silicification, potassic alteration, sericitization (mostly muscovite after plagioclase and orthoclase) and chloritization (of biotite) are prevalent (e.g., Figure 3e). Carbonation (of plagioclase and mafic minerals) occurs but is rare. In many samples, silicification and/or sericitization obliterated much of the primary phases and textures.

Supergene oxidation is evident in some samples by partial replacement of pyrite by masses of hematite ± goethite. These samples also showed some spots of decomposition of Cu-bearing sulfides as a greenish tinge or patina (i.e., chrysocolla-malachite) on hand samples, however, none of the areas thin sectioned contained these spots to any significance.

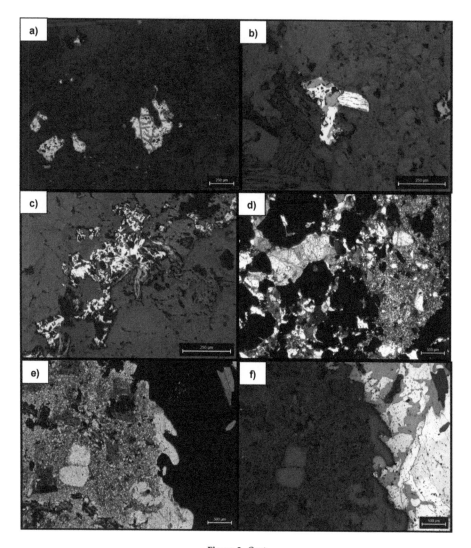

Figure 3. *Cont.*

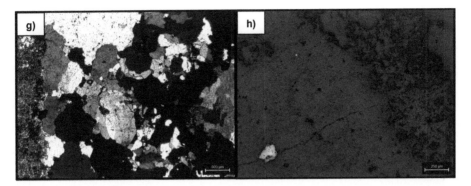

Figure 3. Example photomicrographs of two samples from this study. (**a**) Bi-2, chalcopyrite with alteration to chalcocite/covellite along crystal rims and fractures. Surrounding phases are quartz and sericite. RPPL (reflected plane-polarized light). (**b**) Bi-3, chalcopyrite, bornite, and molybdenite intergrown. Digenite alteration along the bornite and chalcopyrite crystal rims. Surrounding phases are biotite, quartz, and sericite. RPPL. (**c**) S-2, chalcopyrite and molybdenite in a quartz vein. RPPL (**d**) Bu-6, large pyrite crystals (some with possible zircon inclusions) in quartz + sericite + orthoclase. TXPL (transmitted cross-polarized light). (**e**) IP-12, vein of chalcopyrite + bornite + biotite in sericitized and chloritized porphyritic biotite quartz monzonite/latite. TPPL (transmitted plane-polarized light). (**f**) Same view as in (**e**), but in RPPL. (**g**) F-3, quartz + pyrite + chalcopyrite + calcite vein in sericitized monzonite host. (**h**) L-1, small grains of chalcopyrite (lower left) and molybdenite (upper right) in quartz and sericite. RPPL.

3.3. Estimation of Bulk Compositions from Modes

The estimated bulk rock compositions for these modes was calculated by converting the observed mineral abundances to weight proportions, and then elemental proportions based on idealized compositions of the minerals. Based on the observed minerals, only the elements Mg, Al, Si, P, S, K, Ca, Ti, Fe, Cu, and Mo were considered (Table 3).

Table 3. Bulk rock compositions (elemental wt%) for selected elements calculated from modes determined by optical petrography.

Sample	Be-1	Be-4	Be-6	Bi-2	Bi-3	Bu-2	Bu-6	Bu-7	C-3	C-4	CC-1	F-1	F-2	F-3	IP-1	IP-2
Mg	1.2	0.8	1.0	0.3	1.6	0.0	0.0	0.0	0.0	0.0	1.3	0.5	1.6	0.0	0.0	0.0
Al	6.3	5.8	6.1	8.7	5.8	5.7	0.6	0.6	7.2	7.9	5.5	7.2	4.6	6.6	6.4	5.5
Si	32.3	35.5	35.4	30.2	34.5	28.9	30.4	16.5	33.9	29.2	37.0	33.4	33.8	33.2	35.0	35.1
P	0.2	0.1	0.2	0.2	0.2	0.0	0.0	0.0	0.0	0.0	0.0	0.0	0.4	0.0	0.0	0.0
S	1.1	1.0	0.9	3.4	1.2	8.4	17.4	33.2	4.2	7.5	0.5	2.1	2.3	3.2	3.1	4.6
K	2.7	2.6	2.2	4.2	3.5	2.8	0.5	0.3	3.5	3.8	3.2	3.7	3.5	3.2	3.1	2.7
Ca	0.5	1.0	1.0	0.4	0.5	0.0	0.0	0.0	0.0	0.0	0.0	0.0	1.3	1.2	0.0	0.0
Ti	0.3	0.0	0.3	0.4	0.2	0.8	0.0	0.0	0.0	0.0	0.0	0.3	0.5	0.9	0.4	0.9
Fe	1.9	1.4	1.6	2.3	1.4	10.9	15.0	28.5	3.3	6.2	1.1	2.7	3.5	3.1	2.5	3.4
Cu	0.2	1.7	1.0	3.3	2.3	0.4	0.7	0.8	0.0	2.1	0.5	2.1	0.4	1.3	0.8	2.0
Mo	0.0	0.0	0.0	0.0	0.2	0.1	0.0	0.4	0.6	0.0	0.0	0.0	0.0	0.0	0.0	0.0
Balance	53.5	50.1	50.2	46.2	48.7	42.0	35.4	19.6	47.3	43.2	50.5	47.9	47.8	47.7	48.2	46.7

Sample	IP-3	IP-10	IP-12	L-1	Mo-2	Mo-4	Mo-6	Mo-8	OH-1	S-1	S-2	S-4	SM-3	SM-4	Y-3
Mg	3.7	0.0	0.8	0.0	1.9	0.0	0.5	0.4	2.0	1.6	0.3	0.1	0.6	3.7	0.6
Al	5.2	1.0	0.3	1.7	7.9	8.0	0.5	7.7	7.7	7.0	5.8	8.1	8.1	7.1	6.6
Si	30.7	37.8	20.9	40.0	28.2	43.7	34.0	33.4	26.3	31.8	34.9	32.0	28.4	26.6	34.0
P	0.3	0.0	0.0	0.1	0.2	0.0	0.0	0.0	0.0	0.0	0.1	0.0	0.1	0.1	0.1
S	1.3	8.6	13.9	3.3	5.2	2.5	1.1	2.2	7.9	2.1	3.2	3.4	6.4	3.2	1.1
K	5.4	0.5	1.2	0.8	3.3	0.2	3.7	4.5	1.7	2.2	3.0	3.9	3.9	4.8	3.2
Ca	0.8	0.0	0.2	0.2	0.4	0.0	0.5	0.4	2.4	2.3	0.3	0.7	0.7	0.6	0.7
Ti	0.9	0.0	0.0	0.0	0.9	0.3	0.3	0.2	0.0	0.5	0.5	0.5	0.4	1.3	0.3
Fe	3.9	7.4	11.5	2.8	6.0	2.2	2.4	2.1	8.9	3.1	2.8	3.2	5.8	5.7	3.4
Cu	0.2	0.5	19.7	3.2	0.5	0.0	1.1	0.8	0.5	0.5	1.0	0.3	1.7	2.2	1.1
Mo	0.3	0.0	0.0	0.1	0.0	0.0	0.0	0.0	0.0	0.0	0.2	0.4	0.0	0.0	0.0
Balance	47.2	44.3	31.5	47.8	45.5	50.6	49.7	48.5	42.7	48.8	47.8	47.6	43.7	44.8	49.1

3.4. Portable X-ray Fluorescence Analysis of Bulk Compositions

Analysis by pXRF provided compositional information in terms of elements Mg and heavier (Table 4). Overall composition across the samples varies significantly. Significant variations in Mg, Si, P, S, K, Ca, Ti, Fe, Cu, and Mo content occur. Other elements analyzed for, but are low in concentration (<0.5 wt%) and do not show significant variation are: Cl, V, Cr, Mn, Co, Ni, Zn, As, Se, Rb, Sr, Y, Zr, Nb, Ag, Cd, Sn, Sb, Ba, La, Ce, Pr, Nd, W, Au, Pb, Th, and U. Concentrations for V, Co, Se, Y, Sn, Sb, Au, and U were essentially non-detectable (less than approximately 0.001 wt%). Uncertainty in analyses is dominated by the variance between spots compared to the actual instrumental error with the possible exception of Mg, the element with the highest uncertainty in terms of measurement.

In terms of the economic metals, Cu grades vary from 0 to 14.8 wt% (mean: 0.8 wt%) and Mo grades vary from 0 to 0.3 wt% (mean 0.02 wt%). For the other elements, silicon is the most abundant and ranges from 23.2 to 53.8 wt% (mean: 37.4 wt%). Al is the second most consistently abundant element (0.1 to 8 wt%, mean: 4.9 wt%) followed by K (0 to 8.5 wt%, mean: 3.5 wt%). Elements that are high abundance in some samples (those with high sulfide mineral content) are Fe (0.5 to 20.8 wt%, mean: 3.6 wt%) and S (0.1 to 40 wt%, mean: 5.4 wt%). Less significant variations are Mg (0 to 3.5 wt%, mean: 0.6 wt%), P (0 to 0.4 wt%, mean 0.1 wt%), Ca (0 to 3.2 wt%, mean 0.8 wt%), and Ti (0 to 1.2 wt%, mean 0.3 wt%). The sum of chalcophile element concentrations: Fe, Cu, and Mo, correlate with S content (Figure 4).

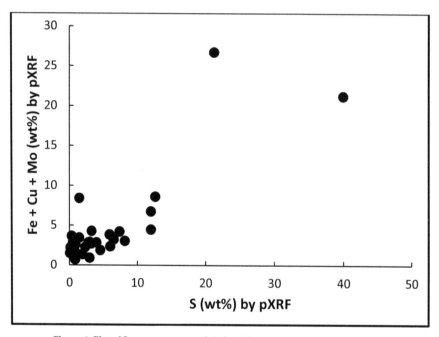

Figure 4. Plot of S content vs. sum of chalcophile elements determined by pXRF.

3.5. Comparison of Bulk Compositions from Both Methods

Calculated bulk rock composition (in terms of selected elements) based on estimated modes from optical petrography and the estimated bulk rock compositions based on pXRF analyses are compared on an element-by-element basis (Figure 5).

Table 4. Bulk rock compositions (elemental wt%) for selected elements determined by the average of point analyses for each thin section billet. Uncertainties are the averaged instrumental population standard deviation determined by the pXRF software: $(2\sigma)_{inst}$ and the sample standard deviation of the spot analyses across the billet: $(s)_{spot}$. Samples with more than three spots analyzed are Bu-7 (five spots), Bu-2 (four spots), and IP-3 (four spots).

Sample	Be-1	Be-4	Be-6	Bi-2	Bi-3	Bu-2	Bu-6	Bu-7	C-3	C-4	CC-1	F-1	F-2	F-3	IP-1	IP-2
Mg	0.558	0.422	0.921	0.610	0.776	0.332	0.000	0.000	0.000	0.328	0.559	0.141	0.851	0.151	0.000	0.000
Mg(2σ)inst	0.197	0.201	0.203	0.284	0.237	0.668	0.409	1.405	0.318	0.356	0.209	0.326	0.213	0.391	0.327	0.437
Mg(s)spot	0.306	0.081	0.470	0.107	0.088	0.423	0.000	0.000	0.000	0.294	0.247	0.244	0.488	0.262	0.000	0.000
Al	5.995	5.798	5.979	6.977	6.007	5.821	0.134	0.152	6.972	6.819	5.109	4.832	4.397	5.828	5.247	4.555
Al(2σ)inst	0.134	0.136	0.130	0.173	0.147	0.185	0.076	0.197	0.153	0.163	0.131	0.121	0.119	0.156	0.127	0.123
Al(s)spot	0.131	2.363	0.659	0.077	0.646	0.348	0.134	0.147	0.954	0.641	0.215	1.550	0.357	1.631	2.394	0.411
Si	36.175	41.352	32.875	33.531	37.622	34.213	49.225	23.243	39.127	36.880	40.978	38.249	40.287	40.416	41.148	36.948
Si(2σ)inst	0.213	0.207	0.215	0.201	0.209	0.208	0.182	0.137	0.220	0.198	0.217	0.210	0.232	0.196	0.208	0.187
Si(s)spot	1.925	7.025	1.674	1.190	0.584	7.114	2.716	20.339	2.561	2.991	1.034	2.019	3.793	6.274	5.434	5.187
P	0.085	0.110	0.201	0.372	0.150	0.071	0.015	0.000	0.000	0.020	0.000	0.000	0.085	0.000	0.080	0.000
P(2σ)inst	0.020	0.023	0.019	0.026	0.023	0.035	0.045	0.070	0.033	0.033	0.039	0.029	0.026	0.041	0.024	0.038
P(s)spot	0.047	0.018	0.045	0.100	0.079	0.057	0.027	0.000	0.000	0.034	0.000	0.000	0.075	0.000	0.012	0.000
S	0.136	0.581	0.472	2.778	0.564	12.577	8.156	39.920	1.933	6.521	0.056	0.903	0.885	3.250	4.570	6.029
S(2σ)inst	0.008	0.014	0.012	0.036	0.015	0.100	0.064	0.451	0.028	0.060	0.007	0.017	0.018	0.037	0.045	0.052
S(s)spot	0.155	0.625	0.127	0.470	0.187	12.607	0.753	23.920	0.798	2.363	0.006	0.425	0.596	1.270	3.526	5.011
K	2.768	3.145	1.272	8.525	7.306	3.381	0.030	0.028	4.740	3.737	5.275	4.364	1.939	6.701	3.242	3.886
K(2σ)inst	0.041	0.039	0.034	0.083	0.069	0.063	0.010	0.036	0.055	0.051	0.056	0.044	0.039	0.064	0.042	0.044
K(s)spot	0.484	1.162	0.140	0.160	0.781	0.502	0.005	0.043	0.493	0.246	0.651	3.827	1.012	1.825	1.416	0.881
Ca	1.410	1.282	1.086	0.137	0.231	0.021	0.009	0.014	0.054	0.080	0.460	0.599	1.185	0.341	0.000	0.028
Ca(2σ)inst	0.034	0.034	0.030	0.026	0.023	0.027	0.004	0.019	0.016	0.017	0.024	0.022	0.032	0.025	0.026	0.014
Ca(s)spot	0.529	0.276	0.077	0.029	0.088	0.024	0.000	0.019	0.006	0.010	0.135	0.247	0.168	0.210	0.000	0.007
Ti	0.262	0.113	0.448	0.414	0.228	0.336	0.000	0.000	0.429	0.201	0.087	0.070	0.594	0.042	0.068	0.034
Ti(2σ)inst	0.088	0.059	0.105	0.115	0.083	0.179	0.062	0.163	0.094	0.092	0.083	0.080	0.109	0.102	0.101	0.082
Ti(s)spot	0.155	0.017	0.086	0.026	0.146	0.224	0.000	0.000	0.027	0.035	0.083	0.121	0.331	0.073	0.118	0.058
Fe	2.187	1.012	2.561	1.187	0.760	8.580	2.783	20.806	1.311	3.047	1.207	0.726	2.522	2.053	1.887	1.657
Fe(2σ)inst	0.031	0.024	0.033	0.027	0.024	0.066	0.029	0.422	0.026	0.035	0.026	0.023	0.033	0.029	0.029	0.027
Fe(s)spot	1.083	0.158	0.304	0.210	0.169	5.749	0.373	15.386	0.483	0.597	0.116	0.324	1.217	0.811	1.088	1.582
Cu	0.027	0.442	0.212	1.658	0.445	0.019	0.273	0.414	0.064	0.186	0.309	0.257	0.063	0.664	0.006	0.674
Cu(2σ)inst	0.002	0.006	0.005	0.016	0.007	0.002	0.005	0.014	0.003	0.005	0.005	0.005	0.003	0.008	0.002	0.007
Cu(s)spot	0.034	0.632	0.032	0.513	0.116	0.010	0.069	0.262	0.019	0.045	0.061	0.115	0.040	0.439	0.006	1.114
Mo	0.000	0.000	0.000	0.005	0.000	0.000	0.000	0.002	0.000	0.000	0.000	0.000	0.000	0.002	0.005	0.043
Mo(2σ)inst	0.002	0.002	0.002	0.001	0.002	0.002	0.002	0.002	0.002	0.002	0.002	0.002	0.002	0.002	0.002	0.001
Mo(s)spot	0.000	0.000	0.000	0.002	0.000	0.000	0.000	0.003	0.000	0.000	0.000	0.000	0.000	0.003	0.008	0.042
balance	50.397	45.743	53.974	43.805	45.910	34.650	39.374	15.422	45.370	42.181	45.960	49.860	47.192	40.553	43.746	46.146

Table 4. Cont.

Sample	IP-3	IP-10	IP-12	L-1	Mo-2	Mo-4	Mo-6	Mo-8	OH-1	S-1	S-2	S-4	SM-3	SM-4	Y-3
Mg	3.466	0.000	0.000	0.061	0.269	0.000	0.377	0.303	1.051	1.292	0.202	0.595	0.497	2.504	0.646
Mg(2σ)$_{inst}$	0.349	0.460	1.146	0.201	0.261	0.243	0.273	0.307	0.357	0.317	0.440	0.261	0.287	0.401	0.208
Mg(s)$_{spot}$	2.952	0.000	0.000	0.106	0.233	0.000	0.327	0.270	0.493	0.421	0.350	0.166	0.038	0.410	0.315
Al	3.792	0.630	2.089	0.516	7.597	0.552	5.698	6.713	5.670	7.591	4.372	6.134	7.896	8.039	4.013
Al(2σ)$_{inst}$	0.132	0.075	0.153	0.058	0.165	0.066	0.148	0.167	0.167	0.187	0.150	0.156	0.182	0.212	0.114
Al(s)$_{spot}$	2.270	0.657	3.105	0.802	0.232	0.072	3.146	0.808	0.968	0.344	0.426	0.395	0.254	0.424	0.729
Si	33.855	42.872	27.146	46.957	39.581	53.778	39.729	36.751	29.638	32.075	36.292	36.938	34.163	28.977	38.218
Si(2σ)$_{inst}$	0.226	0.176	0.167	0.210	0.227	0.228	0.205	0.204	0.182	0.216	0.203	0.202	0.187	0.243	0.212
Si(s)$_{spot}$	10.773	6.181	4.108	4.859	0.925	2.258	7.880	3.732	4.311	1.404	1.954	2.825	0.809	1.918	0.839
P	0.417	0.045	0.047	0.085	0.137	0.000	0.030	0.056	0.014	0.108	0.239	0.071	0.167	0.153	0.057
P(2σ)$_{inst}$	0.025	0.038	0.043	0.035	0.024	0.042	0.031	0.024	0.051	0.024	0.027	0.024	0.026	0.023	0.028
P(s)$_{spot}$	0.436	0.042	0.081	0.023	0.044	0.000	0.052	0.016	0.024	0.004	0.161	0.028	0.044	0.067	0.059
S	1.424	11.951	21.206	0.825	2.322	2.969	0.866	3.018	11.947	3.284	5.937	4.012	7.376	1.453	0.323
S(2σ)$_{inst}$	0.023	0.080	0.150	0.014	0.031	0.033	0.018	0.037	0.091	0.041	0.057	0.041	0.064	0.027	0.011
S(s)$_{spot}$	1.185	8.437	6.641	0.642	0.295	3.238	0.515	1.362	8.311	0.953	1.410	3.240	0.765	0.552	0.197
K	5.778	0.301	0.087	0.361	4.530	0.296	5.385	5.082	1.676	3.363	3.247	5.932	4.304	5.508	3.039
K(2σ)$_{inst}$	0.301	0.015	0.026	0.010	0.057	0.014	0.057	0.058	0.043	0.056	0.050	0.065	0.056	0.093	0.043
K(s)$_{spot}$	0.075	0.358	0.077	0.573	0.073	0.042	2.497	1.210	0.058	2.083	0.498	1.774	0.105	0.334	0.905
Ca	2.875	0.007	0.032	0.011	0.257	0.030	2.443	2.069	2.309	3.203	2.203	0.888	1.359	0.498	0.729
Ca(2σ)$_{inst}$	0.558	0.009	0.012	0.004	0.022	0.005	0.044	0.043	0.054	0.062	0.046	0.034	0.039	0.040	0.026
Ca(s)$_{spot}$	0.033	0.012	0.014	0.001	0.082	0.003	0.892	1.472	0.387	0.719	0.473	0.402	0.696	0.085	0.215
Ti	0.624	0.000	0.000	0.000	0.358	0.126	0.263	0.241	0.137	0.404	0.465	0.168	0.307	1.228	0.232
Ti(2σ)$_{inst}$	0.698	0.066	0.150	0.031	0.102	0.045	0.084	0.091	0.131	0.135	0.110	0.106	0.102	0.235	0.088
Ti(s)$_{spot}$	0.148	0.000	0.000	0.000	0.032	0.016	0.171	0.030	0.120	0.018	0.325	0.185	0.036	0.027	0.118
Fe	3.156	4.299	11.918	0.498	2.275	0.856	1.167	2.134	6.721	4.254	2.922	2.792	3.619	8.000	3.543
Fe(2σ)$_{inst}$	0.039	0.037	0.089	0.018	0.034	0.023	0.026	0.031	0.055	0.045	0.036	0.035	0.037	0.066	0.037
Fe(s)$_{spot}$	2.385	3.306	1.197	0.247	0.097	0.986	0.576	0.828	2.205	0.253	1.337	1.392	0.222	0.907	1.221
Cu	0.022	0.003	14.803	0.199	0.000	0.003	0.409	0.757	0.024	0.033	0.930	0.032	0.607	0.400	0.116
Cu(2σ)$_{inst}$	0.002		0.160	0.004	0.003	0.055	0.006	0.009	0.002	0.002	0.010	0.002	0.008	0.007	0.004
Cu(s)$_{spot}$	0.006	0.139	3.822	0.166	0.000	0.000	0.235	0.545	0.008	0.017	0.640	0.038	0.097	0.218	0.085
Mo	0.273	0.065	0.000	0.009	0.000	0.000	0.007	0.008	0.009	0.002	0.000	0.034	0.002	0.001	0.000
Mo(2σ)$_{inst}$	0.003	0.001	0.002	0.002	0.002	0.002	0.001	0.001	0.001	0.002	0.000	0.001	0.001	0.002	0.002
Mo(s)$_{spot}$	0.372	0.069	0.000	0.016	0.000	0.000	0.004	0.006	0.006	0.000	0.000	0.005	0.002	0.002	0.000
balance	46.561	39.708	22.672	50.479	42.674	41.333	43.625	42.867	40.805	44.392	43.190	42.405	39.705	43.238	49.085

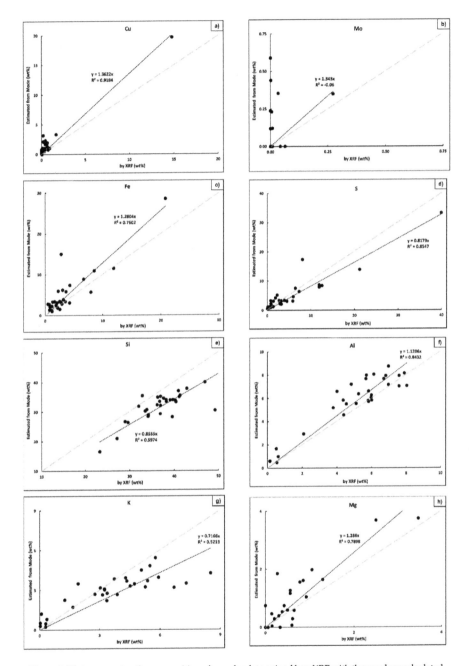

Figure 5. Plots comparing the composition of samples determined by pXRF with those values calculated from mineral modes: (**a**) Cu, (**b**) Mo, (**c**) Fe, (**d**) S, (**e**) Si, (**f**) Al, (**g**) K, and (**h**) Mg. The grey, coarsely dashed lines show 1:1 equivalency for both methods.

The degree of correlation between the two methods depends on the particular element. The lightest and the heaviest elements are systematically overestimated by optical petrography compared to pXRF: Cu, Mo, Fe, Al, Mg, and Ti (Figure 5a–c,f,h). In contrast, the mid-atomic number elements are underestimated by optical petrography compared to pXRF: S, Si, K, Ca, and P (Figure 5d,e,g). The element that shows the closest 1:1 correlation between the two methods is Al (Figure 5f), followed by Si (Figure 5e). The poorest correlations are shown for Cu and Mo (Figure 5a,b). The correlation for Cu is far worse when the single high-Cu sample (IP-12) is omitted from the fit (Figure 6). The best trendline fits are for Cu, S, and Al (Figure 5a,d,f). However, the fit is very poor in terms of Cu if sample IP-12 is omitted (Figure 6). The worst fit is shown by Mo (R^2 = −0.06, Figure 5b), followed by Ca (R^2 = 0.443), Ti (R^2 = 0.4662), and P (R^2 = 0.4721). The balances (remaining wt% not attributed to a single element) are slightly overestimated by mode compared to pXRF by a factor of 1.0623, and with an adequate fit (R^2 = 0.7361). Significant outliers from these trends are Bu-06: higher Fe and S, but lower Si by mode than by pXRF, Bi-02: lower K by mode than by pXRF, F-02, higher K by mode than by pXRF, L-01: higher Cu by mode than by pXRF, and Mo-02: higher Mg by mode than by pXRF.

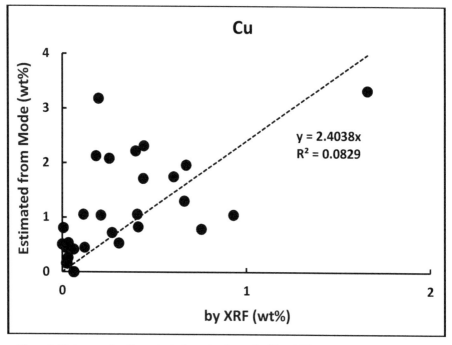

Figure 6. Plot comparing Cu content of samples determined by pXRF with those values calculated from mineral modes but omitting high-Cu sample IP-12.

4. Discussion

There are multiple factors affecting each method that may bias results. These affect the accuracy and precision of the mode-based or pXRF-based methods, and the correlation of the two.

4.1. Causes of Poor Correlation

In terms of optical petrography, melanocratic (i.e., mafic minerals, sulfides, and oxides) minerals are typically overestimated compared to leucocratic minerals (i.e., felsic minerals, carbonates, and phosphates; c.f., [27]). Sulfide minerals are exclusive hosts for Cu and Mo in the samples. Similarly, biotite-chlorite and rutile are the major hosts for Mg and Ti, respectively. Biotite, chlorite, hematite, and sulfides are also the hosts for Fe. These minerals are all dark-colored or opaque in TPPL and readily noted. Colorless or light-colored minerals in TPPL such as quartz, muscovite (as sericite), orthoclase, plagioclase, calcite, and apatite collectively host the bulk of Si, K, Ca, and P. The residency of Al is less clear as it is overestimated by optical petrography (Figure 5f) and is an important constituent (along with K and Si) of biotite-chlorite but is a more significant component of the feldpars and sericite. Human error as systematic visual overestimation of biotite and chlorite during optical petrography would give a positive bias to the Mg, Fe, and Al calculated by mode; particularly if it was at the expense of quartz content. The other element with unusual behavior is S in that it is underestimated by optical petrography. As the only S-bearing phases noted are sulfides, it would be expected that optical mineralogy would overestimate this element as with Fe, Cu, and Mo. These elements track with S (Figure 4), but it is possible another S-bearing phase was overlooked, such as barite, anhydrite, or gypsum. Barite is not likely as a negligible Ba content was noted by pXRF in all samples. Hypogene anhydrite occurs in porphyry-type deposits (e.g., [28]), and can be fairly common [5,29]. Another cause for S underestimation from the mode is misidentifying sulfide minerals that are S-rich (chalcopyrite and covellite) for S-poor phases (bornite, digenite, and chalcocite). However, the most abundant S-bearing mineral is pyrite, and a similar underestimation is not present with Fe; nor are those two groups of Cu-sulfides easily confused in RPPL. It may be that the pXRF systematically overestimates S content of the samples and further calibration should have been applied. Systematic overestimation of S (along with P, Cu, and Mg) by pXRF has been observed with field analysis of soils [30]. However, Roman Alday et al. [16] found that for samples from the Elatsite porphyry-type deposit, pXRF overestimates Fe and K, and underestimates Mg compared to more intensive laboratory methods. Matrix effects on pXRF vary with mineral assemblages and development of calibration standards is reliant on rock type [23,31]. The poorest correlation of the chalcophile elements is for Mo. Economic grades in porphyry-type deposits range from approximately 0.1 to 0.3 wt% Mo, although lower grades in the presence of another commodity such as Cu are common [32]. This is a lower grade than Cu due to the higher valuation of Mo, typically 2× to 3× the per-pound metal price. This grade range approaches the limit of detection for Mo by most pXRF units in a typical silicate–sulfide matrix. The result is that although molybdenite is fairly distinct and easy to identify in reflected light, it may be in trace concentrations that contribute grades too low to be detected by pXRF. Thus, mode-based estimates of grade will likely be higher than those detected by pXRF. The selection of the pXRF analytical spots may also play a role (see Section 4.4).

4.2. Outliers from Trends

Five samples, in particular, plotted as major outliers from the trends established by the rest of the data for one or more elements. These are most likely the result of errors in visual estimation of modes as discussed above. Bu-06 has 2.8 wt% Fe by pXRF and 15 wt% Fe by mode. This is accompanied by a similar overestimation of S by mode and underestimation of Si by mode. This is most likely due to the overestimation of pyrite at the expense of quartz. Bi-02 has 8.5 wt% K by pXRF and 4.2 wt% K by mode. This is likely from an underestimation of muscovite (as sericite). F-02 has 1.9 wt% K by pXRF and 3.5 wt% by mode. The overestimation of potassic phases is likely. Orthoclase is most probable as it has a higher proportion of K to Al than muscovite and there is not a significant Al outlier for this sample. There is no accompanying Fe and/or Mg outlier, so biotite-chlorite is not likely. L-01 has 0.2 wt% Cu by pXRF and 3.2 wt% Cu by mode. This is most likely due to an overestimate of the modal abundance of chalcopyrite. Mo-02 has 0.3 wt% Mg by pXRF and 1.9 wt% Mg by mode, most likely due to an overestimation of a mafic mineral, such as chlorite. It should be noted, that the element with the highest instrumental error is Mg (Table 4).

4.3. Assumptions of Mineral Composition

Other than errors in visually estimating the mode of a sample or in misidentifying phases, another source of error in calculating the bulk rock composition from the mode is the assumed composition of the mineral phases. Chlorite, biotite, and plagioclase all have assumed compositions in terms of solid solution components for the purposes of translating the modes into compositional data. These assumptions are not supported by any mineral analysis such as electron microprobe analysis. Minerals such as biotite can vary in composition between different deposits and within single deposits. Magmatic biotite of the host is typically more Fe-rich than the later hydrothermal biotite [6], largely depending on what alteration zone (i.e., potassic vs. phyllic) they are from (e.g., [33]). There was no distinction between these parageneses of the same minerals during the determination of the modes. As chlorite is largely a product of biotite alteration in these deposits, a similar degree of compositional variation is expected. Likewise, plagioclase in porphyry deposits varies in composition. For example, plagioclase from Sierrita ranges from Ab_{25} to Ab_{45} [34]. The simplified estimation of a molar 1:1 ratio of Na:Ca in plagioclase in this study is likely why bulk compositions based on modes overestimates Ca content. Muscovite and chlorite from the Highland Valley district (that includes the Lornex and Bethlehem deposits) range in composition 2.2–2.6 Al apfu and 0.5–3.5 Mg apfu, respectively [35].

4.4. Sampling Biases

A further source of disparity between the results of the two methods lies in sampling. Though the thin section is cut from the billet, there may be fairly distinct differences in mode between a 0.03 mm thick section and a ~10 mm thick billet. The thin section is essentially a 2D representation of a 3D rock, whereas the pXRF analysis is an average of three different spots on the billet, each 8 mm in diameter (the default aperture size on the Niton XL3t), with an effective sampling depth (d) expressed by Equation (1):

$$d = m/\rho \qquad (1)$$

where d is the sampling depth from which 99% of the signal is sourced, m is the mass per unit area sampled and ρ is sample density [23]. For the samples, the least dense abundant mineral is plagioclase (ρ_{albite} = 2.61 g/cm^3) and the densest is pyrite (ρ_{pyrite} = 5.01 g/cm^3). Denser samples (i.e., sulfide-rich) will approach more surficial analyses, whereas sampling depth will be deeper for samples dominated by felsic minerals. If the billet has compositional variations in the vertical (depth) sense, then the spot analyses become pseudo-bulk analyses of a cylinder 8 mm in diameter a a depth that is a function of sample density. Similarly, in the horizontal sense, three spots at 50.27 mm^2 totals ~150.8 mm^2, whereas the thin section is up to 1196 mm^2 in total area. Accounting for 3 mm margins around the billet, this reduces to 800 mm^2 of sample area. Thus, this method in the best case, sampled only ~19% of the sample area and assumes it is representative of the other 21%. Sampling precision for finer-grained rocks by pXRF has been shown to be ≤5% relative standard deviation, but much higher for coarser-grained samples [36]. The thin section is a ~60 μm slice off of the top of the billet that is subsequently ground down to 30 μm. Sample heterogeneity is typically on a scale an order of magnitude larger. Differences in thin section and sample billet surface are negligible in comparison to the other discussed sources of uncertainty. Although most of the rocks in this sample suite are hypabyssal to volcanic, the presence of phenocrysts and veins/veinlets increases the number of measurements required to achieve the same precision that would be possible compared to an un-veined, aphyric sample. The large sample standard deviations across the spot analyses ((s)$_{spot}$) for many samples illustrates this (Table 4). Samples with a high degree of heterogeneity in terms of multiple elements are all the Ithaca Peak samples, Be-4, Bu-2, Bu-7, F-3. Mo-6, OH-1, and S-4.

4.5. Applicability of pXRF to Mineral Exploration

Compared to the classic standard of optical petrography, pXRF is faster, more accessible, portable, and more easily conducted with little training. It is not a replacement for laboratory methods as it cannot reliably reach the levels of precision and accuracy required to meet standards for mineral resource reporting in terms of NI 43-101 (National Instrument 43-101 is the Canadian code for the Standards of Disclosure for Mineral Projects. It sets standards for reporting mineral resource information such as grade, tonnage, etc., and what methods are suitable for measuring them to an adequate degree of certainty) or JORC (Joint Ore Reserves Committee Code is essentially the Australian equivalent to NI 43-101) guidelines, but with calibration, it can achieve a low but useful level of quantitative certainty (e.g., [16,20,21]). However, for quick decisions regarding unknown samples in the field or core shed, semi-quantitative analysis is satisfactory in order to make rough interpretations on whether to collect a sample for more detailed assay/examination. Similarly, pXRF can be used to guide petrographic examinations in terms of possible (and impossible) phases present as with this study.

Although this study attempts to simulate "field" use of pXRF, the analytical conditions for this study were more optimal and controlled than those typical of field or even core shed settings. Analyzed surfaces of the billets were fresh, clean, and flat. Samples from outcrop, float, or trenches may have soil, lichen, or other debris covering the rock. Thick weathering rinds may also be present. Surfaces are typically irregular. Although conditions for drill core analysis are more controlled, sample geometry is not flat, but curved. This is especially relevant for smaller diameter drill core such as AQ, BQ, or NQ. pXRF analysis of reverse circulation (RC) chips suffers from especially variable conditions in terms of sample geometry and open space between the chips. These sources of uncertainty can be mitigated by cleaning surfaces, preparing flat surfaces, and even pulping and packing them into homogenized pucks, but all at the cost of significant time.

4.6. Directions for Future Research

The certainty of pXRF results would be much improved from non-factory calibration specific to porphyry-type samples. On the petrographic side, determination of the variations in mineral composition, at least of biotite, chlorite, plagioclase, and muscovite is needed in order to provide more accuracy in determining a chemical composition from the mode. The use of a scanning electron microscope with energy-dispersive X-ray spectra would provide sufficient accuracy. More detailed petrographic work, along with cathodoluminescence would help distinguish between different parageneses of minerals and better characterize the range of compositions present (c.f., [29,37]).

Author Contributions: Conceptualization, A.D.V.; methodology, A.D.V.; formal analysis, C.A.G.; investigation, C.A.G.; resources, A.D.V.; data curation, C.A.G.; writing—original draft preparation, C.A.G.; writing—review and editing, A.D.V. and C.A.G.; supervision, A.D.V.; project administration, A.D.V.; funding acquisition, C.A.G. and A.D.V. All authors have read and agreed to the published version of the manuscript.

Funding: Funding for Connor Gray was provided by a Bloomsburg University Undergraduate Research, Scholarly, and Creative Activity (URSCA) award.

Acknowledgments: Samples were collected by Peter H. Kirwin and John Ray. Funding for the Leica microscope was provided by the Bloomsburg University College of Science and Technology Dean's office via Troy Prutzman and Robert Aronstam. I, Connor Gray, want to personally thank my undergraduate research advisor Adrian Van Rythoven for the opportunity to work on this project. Two anonymous reviewers and the editorial staff at *Minerals* provided comments that improved the quality of the manuscript.

Conflicts of Interest: The authors declare no conflict of interest.

References

1. Titley, S.R.; Hicks, C.L. *Geology of the Porphyry Copper Deposits, Southwestern North America*; University of Arizona Press: Tuscon, AZ, USA, 1966.
2. Lowell, J.D.; Guilbert, J.M. Lateral and vertical alteration-mineralization zoning in porphyry ore deposits. *Econ. Geol.* **1970**, *65*, 373–408. [CrossRef]
3. Sillitoe, R.H. A Plate Tectonic Model for the Origin of Porphyry Copper Deposits. *Econ. Geol.* **1972**, *67*, 184–197. [CrossRef]
4. Sillitoe, R.H.; Bonham, H.F. Volcanic landforms and ore deposits. *Econ. Geol.* **1984**, *79*, 1286–1298. [CrossRef]
5. Berger, B.R.; Ayuso, R.A.; Wynn, J.C.; Seal, R.R. *Preliminary Model of Porphyry Copper Deposits. Open-File Report 2008–1321*; USGS: Reston, VA, USA, 2008.

6. John, D.A.; Ayuso, R.A.; Barton, M.D.; Blakely, R.J.; Bodnar, R.J.; Dilles, J.H.; Gray, F.; Graybeal, F.T.; Mars, J.C.; McPhee, D.K.; et al. Porphyry Copper Deposit Model: Scientific Investigations Report 2010–5070–B. In *Mineral Deposit Models for Resource Assessment*; USGS: Reston, VA, USA, 2010; p. 169.
7. Sillitoe, R.H. Porphyry Copper Systems. *Econ. Geol.* **2010**, *105*, 3–41. [CrossRef]
8. Richards, J.P. Tectono-Magmatic Precursors for Porphyry Cu-(Mo-Au) Deposit Formation. *Econ. Geol.* **2003**, *98*, 1515–1533. [CrossRef]
9. Eliopoulos, D.G.; Economou-Eliopoulos, M. Platinum-group element and gold contents in the Skouries porphyry copper deposit, Chalkidiki Peninsula, northern Greece. *Econ. Geol.* **1991**, *86*, 740–749. [CrossRef]
10. Tarkian, M.; Stribrny, B. Platinum-group elements in porphyry copper deposits: A reconnaissance study. *Mineral. Petrol.* **1999**, *65*, 161–183. [CrossRef]
11. Babcock, R.C., Jr.; Ballantyne, G.H.; Phillips, H. Summary of the geology of the Bingham District, Utah. *Arizona Geol. Soc. Dig.* **1995**, *20*, 316–325.
12. Hou, Z.; Cook, N.J. Metallogenesis of the Tibetan collisional orogen: A review and introduction to the special issue. *Ore Geol. Rev.* **2009**, *36*, 2–24. [CrossRef]
13. Copper Alliance. *The Impacts of Copper Mining in Chile: Economic and Social Implications for the Country*; International Copper Association: Santiago, Chile, 2017.
14. Potts, P.J. Introduction, Analytical Instrumentation and Application Overview. In *Portable X-ray Fluorescence Spectrometry*; Potts, P.J., West, M., Eds.; Royal Society of Chemistry: Cambridge, UK, 2008; pp. 1–12, ISBN 978-0-85404-552-5.
15. Liangquan, G. Geochemical Prospecting. In *Portable X-ray Fluorescence Spectrometry*; Potts, P.J., West, M., Eds.; Royal Society of Chemistry: Cambridge, UK, 2008; pp. 141–173, ISBN 978-0-85404-552-5.
16. Roman Alday, M.C.; Kouzmanov, K.; Harlaux, M.; Stefanova, E. Comparative study of XRF and portable XRF analysis and application in hydrothermal alteration geochemistry: The Elatsite porphyry Cu-Au-PGE deposit, Bulgaria. In Proceedings of the 16th Swiss Geoscience Meeting, Bern, Switzerland, 30 November–1 December 2018; p. 2.
17. Lemière, B. A review of pXRF (field portable X-ray fluorescence) applications for applied geochemistry. *J. Geochemical Explor.* **2018**, *188*, 350–363. [CrossRef]
18. Kalnicky, D.J.; Singhvi, R. Field portable XRF analysis of environmental samples. *J. Hazard. Mater.* **2001**, *83*, 93–122. [CrossRef]
19. Uvarova, Y.; Cleverley, J.; Baensch, A.; Verrall, M. Coupled XRF and XRD analyses for rapid and low-cost characterization of geological materials in the mineral exploration and mining industry. *Explor. Newsl. Assoc. Appl. Geochemists* **2014**, *162*, 1–14.
20. Ahmed, A.; Crawford, A.J.; Leslie, C.; Phillips, J.; Wells, T.; Garay, A.; Hood, S.B.; Cooke, D.R. Assessing copper fertility of intrusive rocks using field portable X-ray fluorescence (pXRF) data. *Geochem. Explor. Environ. Anal.* **2019**, *20*, 81–97. [CrossRef]
21. Andrew, B.S.; Barker, S.L.L. Determination of carbonate vein chemistry using portable X-ray fluorescence and its application to mineral exploration. *Geochem. Explor. Environ. Anal.* **2018**, *18*, 85–93. [CrossRef]
22. Gallhofer, D.; Lottermoser, B.G. The Influence of Spectral Interferences on Critical Element Determination with Portable X-Ray Fluorescence (pXRF). *Minerals* **2018**, *8*, 320. [CrossRef]
23. Markowicz, A.A. Quantification and Correction Procedures. In *Portable X-ray Fluorescence Spectrometry*; Potts, P.J., West, M., Eds.; Royal Society of Chemistry: Cambridge, UK, 2008; pp. 13–38, ISBN 978-0-85404-552-5.
24. Taylor, R. *Ore Textures*; Springer: Berlin/Heidelberg, Germany, 2009; ISBN 978-3-642-01783-4.
25. Whitney, D.L.; Evans, B.W. Abbreviations for names of rock-forming minerals. *Am. Mineral.* **2010**, *95*, 185–187. [CrossRef]
26. Le Maitre, R.W.; Streckeisen, A.; Zanettin, B.; Le Bas, M.J.; Bonin, B.; Bateman, P.; Bellieni, G.; Dudek, A.; Efremova, S.; Keller, J.; et al. *Igneous Rocks: A Classification and Glossary of Terms: Recommendations of the International Union of Geological Sciences Subcommission on the Systematics of Igneous Rocks*, 2nd ed.; Le Maitre, R.W., Streckeisen, A., Zanettin, B., Le Bas, M.J., Bonin, B., Bateman, P., Eds.; Cambridge University Press: Cambridge, UK, 2002; ISBN 9780511535581.
27. Wright, T.L.; Peck, D.L. *Crystallization and Differentiation of the Alae Magma, Alae Lava Lake, Hawaii. Professional Paper 935-C*; USGS: Reston, VA, USA, 1976.

28. Arnott, A.M.; Zentilli, M. Significance of calcium sulphates in the Chuquicamata porphyry copper system, Chile. In Proceedings of the Atlantic Geoscience Society Annual Meeting Abstracts - Atlantic Geology, Amherst, NS, Canada, 5–6 February 1999; Volume 35, p. 86.

29. Hutchinson, M.C.; Dilles, J.H. Evidence for Magmatic Anhydrite in Porphyry Copper Intrusions. *Econ. Geol.* **2019**, *114*, 143–152. [CrossRef]

30. Declercq, Y.; Delbecque, N.; De Grave, J.; De Smedt, P.; Finke, P.; Mouazen, A.M.; Nawar, S.; Vandenberghe, D.; Van Meirvenne, M.; Verdoodt, A. A Comprehensive Study of Three Different Portable XRF Scanners to Assess the Soil Geochemistry of an Extensive Sample Dataset. *Remote Sens.* **2019**, *11*, 2490. [CrossRef]

31. Pinto, A.H. Portable X-Ray Fluorescence Spectrometry: Principles and Applications for Analysis of Mineralogical and Environmental Materials. *Asp. Min. Miner. Sci.* **2018**, *1*, 42–47. [CrossRef]

32. Ludington, S.; Plumlee, G.S. *Climax-Type Porphyry Molybdenum Deposits. Open-File Report 2009-1215*; USGS: Reston, VA, USA, 2009.

33. Ayati, F.; Yavuz, F.; Noghreyan, M.; Haroni, H.A.; Yavuz, R. Chemical characteristics and composition of hydrothermal biotite from the Dalli porphyry copper prospect, Arak, central province of Iran. *Mineral. Petrol.* **2008**, *94*, 107–122. [CrossRef]

34. Preece, R.K., III. *Paragenesis, Geochemistry, and Temperatures of Formation of Alteration Assemblages of the Sierrita Depost, Pima County, Arizona*; University of Arizona: Tucson, AZ, USA, 1979.

35. Alva Jimenez, T.R. Variation in hydrothermal muscovite and chlorite composition in the Highland Valley Porphyry Cu-Mo District, British Columbia, Canada. Ph.D. Thesis, University of British Columbia, Vancouver, BC, Canada, 2011.

36. Potts, P.J.; Williams-Thorpe, O.; Webb, P.C. The Bulk Analysis of Silicate Rocks by Portable X-Ray Fluorescence: Effect of Sample Mineralogy in Relation to the Size of the Excited Volume. *Geostand. Geoanalytical Res.* **1997**, *21*, 29–41. [CrossRef]

37. Götze, J. Application of Cathodoluminescence Microscopy and Spectroscopy in Geosciences. *Microsc. Microanal.* **2012**, *18*, 1270–1284. [CrossRef] [PubMed]

MDPI

St. Alban-Anlage 66

4052 Basel

Switzerland

Tel. +41 61 683 77 34

Fax +41 61 302 89 18

www.mdpi.com

Minerals Editorial Office

E-mail: minerals@mdpi.com

www.mdpi.com/journal/minerals

Lightning Source UK Ltd.
Milton Keynes UK
UKHW050802120820
368059UK00006B/132